Low Power Semiconductor Devices and Processes for Emerging Applications in Communications, Computing, and Sensing

Devices, Circuits, and Systems

Series Editor Krzysztof Iniewski

Wireless Technologies
Circuits, Systems, and Devices
Krzysztof Iniewski

Circuits at the Nanoscale
Communications, Imaging, and Sensing
Krzysztof Iniewski

Internet Networks
Wired, Wireless, and Optical Technologies
Krzysztof Iniewski

Semiconductor Radiation Detection Systems
Krzysztof Iniewski

Electronics for Radiation Detection
Krzysztof Iniewski

Radiation Effects in Semiconductors
Krzysztof Iniewski

Electrical Solitons
Theory, Design, and Applications
David Ricketts and Donhee Ham

Semiconductors
Integrated Circuit Design for Manufacturability
Artur Balasinski

Integrated Microsystems
Electronics, Photonics, and Biotechnology
Krzysztof Iniewski

Nano-Semiconductors
Devices and Technology
Krzysztof Iniewski

Atomic Nanoscale Technology in the Nuclear Industry
Taeho Woo

Telecommunication Networks
Eugenio Iannone

For more information about this series, please visit: https://www.crcpress.com/
Devices-Circuits-and-Systems/book-series/CRCDEVCIRSYS

Low Power Semiconductor Devices and Processes for Emerging Applications in Communications, Computing, and Sensing

Edited by
Sumeet Walia

Managing Editor
Krzysztof Iniewski

CRC Press
Taylor & Francis Group
Boca Raton London New York

CRC Press is an imprint of the
Taylor & Francis Group, an **informa** business

MATLAB® is a trademark of The MathWorks, Inc. and is used with permission. The MathWorks does not warrant the accuracy of the text or exercises in this book. This book's use or discussion of MATLAB® software or related products does not constitute endorsement or sponsorship by The MathWorks of a particular pedagogical approach or particular use of the MATLAB® software.

CRC Press
Taylor & Francis Group
6000 Broken Sound Parkway NW, Suite 300
Boca Raton, FL 33487-2742

First issued in paperback 2020

© 2019 by Taylor & Francis Group, LLC
CRC Press is an imprint of Taylor & Francis Group, an Informa business

No claim to original U.S. Government works

ISBN-13: 978-1-138-58798-4 (hbk)
ISBN-13: 978-0-367-73384-1 (pbk)

Library of Congress Cataloging-in-Publication Data

Names: Walia, Sumeet, editor.
Title: Low power semiconductor devices and processes for emerging
applications in communications, computing, and sensing / [edited by]
Sumeet Walia.
Description: Boca Raton : Taylor & Francis, a CRC title, part of the Taylor &
Francis imprint, a member of the Taylor & Francis Group, the academic
division of T&F Informa, plc, 2018. | Series: Devices, circuits, and
systems | Includes bibliographical references and index.
Identifiers: LCCN 2018024678| ISBN 9781138587984 (hardback : alk. paper) |
ISBN 9780429503634 (ebook : alk. paper)
Subjects: LCSH: Low voltage integrated circuits.
Classification: LCC TK7874.66 .L6735 2018 | DDC 621.3815--dc23
LC record available at https://lccn.loc.gov/2018024678

Visit the Taylor & Francis Web site at
http://www.taylorandfrancis.com

and the CRC Press Web site at
http://www.crcpress.com

Contents

Preface

Making processing information more energy-efficient would save money, reduce energy use and permit batteries that provide power to mobile devices to run longer or be smaller in size. New approaches to the lower energy requirement in computing, communication and sensing need to be investigated. This book addresses this need in multiple application areas and will serve as a guide in emerging circuit technologies.

Revolutionary device concepts, sensors and associated circuits and architectures that will greatly extend the practical engineering limits of energy-efficient computation are being investigated. Disruptive new device architectures, semiconductor processes and emerging new materials aimed at achieving the highest level of computational energy efficiency for general purpose computing systems need to be developed. This book will provide chapters dedicated to such efforts from process to device.

Series Editor

Krzysztof (Kris) Iniewski is managing R&D at Redlen Technologies Inc., a start-up company in Vancouver, Canada. Redlen's revolutionary production process for advanced semiconductor materials enables a new generation of more accurate, all-digital, radiation-based imaging solutions. Kris is also a president of CMOS Emerging Technologies (www.cmoset.com), an organization of high-tech events covering communications, microsystems, optoelectronics, and sensors. In his career, Dr. Iniewski has held numerous faculty and management positions at the University of Toronto, the University of Alberta, Simon Fraser University, and PMC-Sierra Inc. He has published over 100 research papers in international journals and conferences. He holds 18 international patents granted in the United States, Canada, France, Germany, and Japan. He is a frequent invited speaker and has consulted for multiple organizations internationally. He has written and edited several books for IEEE Press, Wiley, CRC Press, McGraw-Hill, Artech House, and Springer. His personal goal is to contribute to healthy living and sustainability through innovative engineering solutions. In his leisure time, Kris can be found hiking, sailing, skiing, or biking in beautiful British Columbia. He can be reached at kris.iniewski@gmail.com.

Editor

Sumeet Walia is a vice chancellor's fellow at the Royal Melbourne Institute of Technology in Australia. Dr. Walia earned his PhD in the multidisciplinary field of functional materials and devices. His research focuses on low-dimensional nanoelectronics, including micro/nanoscale energy sources, electronic memories, sensors, and transistors. He holds three patents and has been recognized as one of the top 10 innovators under-35 in Asia by *MIT Technology Review*. He has published several high-impact research articles and is a reviewer for a number of international peer-reviewed journals and government grant bodies. He can be reached at waliasumeet@gmail.com and sumeet.walia@rmit.edu.au.

Contributors

Mohammad Karbalaei Akbari
University of Ghent Global Campus
Incheon, South Korea

Mark B.H. Breese
Singapore Synchrotron Light Source (SSLS)
and
Department of Physics
and
NUSNNI-NanoCore
National University of Singapore
Singapore

Daniel R. Brennan
School of Engineering
Newcastle University
Newcastle upon Tyne, UK

Wayne Burleson
Department of Electrical and Computer
    Engineering
University of Massachusetts Amherst
Amherst, Massachusetts

Taw Kuei Chan
Department of Physics
National University of Singapore
Singapore

Hua-Khee Chan
School of Engineering
Newcastle University
Newcastle upon Tyne, UK

Hong Chen
Institute of Microelectronics
Tsinghua University
Beijing, China

Lawrence T. Clark
School of Electrical, Computer and Energy
    Engineering
Arizona State University
Tempe, Arizona

Rémi Comyn
Université Côte d'Azur, CNRS, CRHEA
Valbonne, France

Yvon Cordier
Université Côte d'Azur, CNRS, CRHEA
Valbonne, France

Roy Dagher
Université Côte d'Azur, CNRS, CRHEA
Valbonne, France

Bhupendra Nath Dev
Department of Physics and School of Nano
    Science and Technology
Indian Institute of Technology Kharagpur
Kharagpur, India

C.Y. Fong
Department of Physics
University of California, Davis
Davis, California

Eric Frayssinet
Université Côte d'Azur, CNRS, CRHEA
Valbonne, France

Filippo Giannazzo
Consiglio Nazionale delle Ricerche –
    Istituto per la Microelettronica e
    Microsistemi (CNR-IMM)
Catania, Italy

Jonathan P. Goss
School of Engineering
Newcastle University
Newcastle upon Tyne, UK

Giuseppe Greco
Consiglio Nazionale delle Ricerche –
    Istituto per la Microelettronica e
    Microsistemi (CNR-IMM)
Catania, Italy

Philipp Gutruf
Center for Bio-Integrated Electronics
Simpson Querrey Institute for
  BioNanotechnology
and
Department of Materials Science and
  Engineering
Northwestern University
Evanston, Illinois

Noel Healy
School of Engineering
Newcastle University
Newcastle upon Tyne, UK

Matthew D. Higgins
Warwick Manufacturing Group
University of Warwick
Coventry, UK

Ben R. Horrocks
School of Chemistry
Newcastle University
Newcastle upon Tyne, UK

Alton B. Horsfall
Department of Engineering
University of Durham
Durham, UK

Hanjun Jiang
Institute of Microelectronics
Tsinghua University
Beijing, China

Raghavan Kumar
Intel Lab
Portland, Oregon

Mark S. Leeson
School of Engineering
University of Warwick
Coventry, UK

Shuo Li
Department of Electrical and Computer
  Engineering
University of Massachusetts Amherst
Amherst, Massachusetts

Yi Lu
Warwick Manufacturing Group
University of Warwick
Coventry, UK

Sinu Mathew
NUSNNI-NanoCore
National University of Singapore
Singapore
and
St. Berchmans College
Changanassery, India

Adrien Michon
Université Côte d'Azur, CNRS, CRHEA
Valbonne, France

Marzaini Rashid
School of Physics
Universiti Sains Malaysia
Penang, Malaysia

Fabrizio Roccaforte
Consiglio Nazionale delle Ricerche –
  Istituto per la Microelettronica e
  Microsistemi (CNR-IMM)
Catania, Italy

Shaojie Su
Institute of Microelectronics
Tsinghua University
Beijing, China

John T.L. Thong
Department of Electrical and Computer
  Engineering
National University of Singapore
Singapore

Vinay Vashishtha
School of Electrical, Computer and Energy
  Engineering
Arizona State University
Tempe, Arizona

T. Venkatesan
NUSNNI-NanoCore
and
Department of Physics
and
Department of Electrical and Computer
  Engineering
National University of Singapore
Singapore

Zhihua Wang
Institute of Microelectronics
Tsinghua University
Beijing, China

Nicholas G. Wright
School of Engineering
Newcastle University
Newcastle upon Tyne, UK

Xiaolin Xu
Department of Electrical and Computer
  Engineering
University of Florida
Gainesville, Florida

Lin H. Yang
Physics Division, Lawrence Livermore
  National Laboratory
Livermore, California

Y.J. Zeng
School of Physics
Sun Yat-sen University
Guangzhou, China

R.L. Zhang
School of Physics
Nanjing University
Nanjing, China

Peng Zhou
Fudan University
Shanghai, China

Serge Zhuiykov
University of Ghent Global Campus
Incheon, South Korea

1

ASAP7: A finFET-Based Framework for Academic VLSI Design at the 7 nm Node

Vinay Vashishtha and Lawrence T. Clark

CONTENTS

1.1 Introduction

Recent years have seen fin field-effect transistors (finFETs) dominate highly scaled, e.g., sub-20 nm, complementary metal-oxide-semiconductor (CMOS) processes (Wu et al., 2013; Lin et al., 2014) due to their ability to alleviate short channel effects, provide lower leakage, and enable some continued V_{DD} scaling. However, the availability of a realistic finFET-based predictive process design kit (PDK) for academic use that supports investigation into both circuit as well as physical design, encompassing all aspects of digital design, has been lacking. While the finFET-based FreePDK15 was supplemented with a standard cell library, it lacked full physical verification, layout vs. schematic check (LVS) and parasitic extraction (Bhanushali et al., 2015; Martins et al., 2015). Consequently, the only available sub-45 nm educational PDKs are the planar CMOS-based Synopsys 32/28 nm and FreePDK45 (45 nm PDK) (Goldman et al., 2013; Stine et al., 2007). The cell libraries available for those processes are not very realistic since they use very large cell heights, in contrast to recent industry trends. Additionally, the static random access memory (SRAM) rules and cells provided by these PDKs are not realistic. Because finFETs have a three-dimensional (3-D) structure and there have been significant density impacts in their adoption, using planar libraries scaled to sub-22 nm dimensions for research is likely to give poor accuracy.

    Commercial libraries and PDKs, especially for advanced nodes, are often difficult to obtain for academic use, and access to the actual physical layouts is even more restricted. Furthermore, the necessary non-disclosure agreements (NDAs) are unmanageable for large university classes and the plethora of design rules (DRs) can distract from the key points. NDAs also make it difficult for the publication of physical design as these may disclose proprietary DRs and structures.

    This chapter focuses on the development of a realistic PDK for academic use that overcomes these limitations. The PDK, developed for the N7 node before 7 nm processes were available even in industry, is thus *predictive*. The predictions have been based on publications of the continually improving lithography, as well as our estimates of what would be available at N7. The original assumptions are described in Clark et al. (2016). For the most part, these assumptions have been accurate, except for the expectation that extreme ultraviolet lithography (EUVL) would be widely available, which has turned out to be optimistic. The background and impact on design technology co-optimization (DTCO) for standard cells and SRAM comprise this chapter. The treatment here includes learning from using the cells originally derived in Clark et al. (2016) in realistic designs of SRAM arrays and large digital designs using automated place and route tools.

1.1.1 Chapter Outline

The chapter first outlines the important lithography considerations in Section 1.3. Metrics for overlay, mask errors and other effects that limit are described first. Then, modern liquid immersion optical lithography and its use in multiple patterning (MP) techniques that extend it beyond the standard 80 nm feature limit are discussed. This sets the stage for a discussion of EUV lithography, which can expose features down to about 16 nm in a single exposure (SE), but at a high capital and throughput cost. This section ends with a brief overview of DTCO. DTCO has been required on recent processes to ensure that the very limited possible structures that can be practically fabricated are usable to build real designs. Thus, a key part of a process development is not just to determine transistor and interconnect structures that are lithographically possible, but also to ensure that successful designs can be built with those structures. This discussion is carried out by separating the front end of line (FEOL), middle of line (MOL), and back end of line (BEOL) portions of the process, which fabricate the transistors, contacts and local interconnect, and global interconnect metallization, respectively. The cell library architecture and automated placement and routing (APR) aspects comprise the next section, which with the SRAM results comprise most of the discussion. The penultimate section describes the SRAM DTCO and array development and performance in the ASAP7 predictive PDK. The final section summarizes.

1.2 ASAP7 Electrical Performance

The PDK uses BSIM-CMG SPICE models and the values used are derived from publicly available sources with appropriate assumptions (Paydavosi et al., 2013). A drive current increase from 14 to 7 nm node is assumed to be 15%, which corresponds to the diminished I_{dsat} improvement over time. In accordance with modern devices, the saturation current is assumed to be 4.5× larger than that in the linear region (Clark et al., 2016). A relaxed 54 nm contacted poly pitch (CPP) allows a longer channel length and helps with the assumption of a near ideal subthreshold slope (SS) of 60 mV/decade at room temperature, along with a drain-induced barrier lowering (DIBL) of approximately 30 mV/V. P-type metal-oxide semiconductor (PMOS) strain seems to be easier to obtain according to the 16 and 14 nm foundry data and larger I_{dsat} values for PMOS than those for a n-type metal-oxide semiconductor (NMOS) have been reported (Wu et al., 2013; Lin et al., 2014). Following this trend, we assume a PMOS-to-NMOS drive ratio of 0.9:1. This value provides good slew rates at a fan-out of six (FO6), instead of the traditional four.

Despite the same drawn gate length, the PDK and library timing abstract views support four threshold voltage flavors, viz. super low voltage threshold (SLVT), low voltage threshold (LVT), regular threshold voltage (RVT), and SRAM, to allow investigation into both high-performance and low-power designs. The threshold voltage is assumed to be

TABLE 1.1

NMOS Typical Corner Parameters (Per Fin) at 25°C

Parameter	SRAM	RVT	LVT	SLVT
I_{dsat} (µA)	28.57	37.85	45.19	50.79
I_{eff} (µA)	13.07	18.13	23.56	28.67
I_{off} (nA)	0.001	0.019	0.242	2.444
V_{tsat} (V)	0.25	0.17	0.10	0.04
V_{tlin} (V)	0.27	0.19	0.12	0.06
SS (mV/decade)	62.44	63.03	62.90	63.33
DIBL (mV/V)	19.23	21.31	22.32	22.55

TABLE 1.2

PMOS Typical Corner Parameters (Per Fin) at 25°C

Parameter	SRAM	RVT	LVT	SLVT
I_{dsat} (µA)	26.90	32.88	39.88	45.60
I_{eff} (µA)	11.37	14.08	18.18	22.64
I_{off} (nA)	0.004	0.023	0.230	2.410
V_{tsat} (V)	−0.20	−0.16	−0.10	−0.04
V_{tlin} (V)	−0.22	−0.19	−0.13	−0.07
SS (mV/decade)	64.34	64.48	64.44	64.94
DIBL (mV/V)	24.10	30.36	31.06	31.76

changed through work function engineering. For SRAM devices, the very low leakage uses both a work function change and lightly doped drain (LDD) implant removal. The latter results in an effective channel length (Leff) increase, gate-induced drain leakage (GIDL) reduction, and an overlap capacitance reduction. The drive strength reduces from SLVT to SRAM. The SRAM V_{th} transistors are a convenient option for use in retention latches and designs that prioritize low-standby power. In addition to typical-typical (TT) models, fast-fast (FF) and slow-slow (SS) models are also provided for multi-corner APR optimization. Tables 1.1 and 1.2 show the electrical parameters for single fin NMOS and PMOS, respectively, for the TT corner at 25°C (Clark et al., 2016). The nominal operating voltage is $V_{DD} = 700$ mV.

1.3 Lithography Considerations

Photolithography, hereinafter referred to simply as lithography, in a semiconductor industry context, refers to a process whereby a desired pattern is transferred to a target layer on the wafer through use of light. Interconnect metal, via, source-drain regions, and gate layers in a CMOS process stack are a few examples of the patterns defined, or "printed", using lithography.

A simplified pattern transfer flow is as follows. From among the pattern information that is stored in an electronic database file (GDSII) corresponding to all the layers of a given integrated circuit (IC) design, the enlarged pattern, or its photographic negative, corresponding to a single layer is inscribed onto a photomask or reticle. The shapes on the photomask, hereinafter referred to as mask, define the regions that are either opaque or transparent to light. Light from a suitable source is shone on the mask through an illuminator, which modifies the effective manner of illumination, and passes through the transparent mask regions. Thereafter, light passes through a projection lens, which shrinks the enlarged pattern geometries on the mask to their intended size, and exposes the photoresist that has been coated on the wafer atop the layer to be patterned. The photoresist is developed to either discard or retain its exposed regions corresponding to the pattern. This is followed by an etch that removes portions of the target layer not covered by the photoresist, which is then removed, leaving behind the intended pattern on the layer. Both lines and spaces can be patterned through this approach with some variations in the process steps.

Lithography plays a leading role in the scaling process, which is the industry's primary growth driver, as it determines the extent to which feature geometries can be shrunk in successive technology nodes. Lithography is one of the most expensive and complex procedures in semiconductor manufacturing, with mask manufacturing being the most expensive processing steps within lithography (Ma et al., 2010). Both the complexity and the number of masks used for manufacturing at a node affect the cost, and an increase in either of these can increase the cost to the point of becoming the limiting factor in the overall cost of the product.

As in any other manufacturing process, the various lithography steps also suffer from variability. The lithographic resolution determines the minimum feature dimension, called the critical dimension (CD), for a given layer and is based on the lithography technique employed at a particular technology node. DRs constitute design guidelines to minimize the effects from mask manufacturability issues, the impact of variability and layer misalignment, and ensure printed pattern fidelity to guarantee circuit operation at good yield. Ascertaining these DRs thus requires consideration of the following lithography-related metrics that can cause final printed pattern on a layer to deviate from the intent and/or result in reliability issues.

1.3.1 Lithography Metrics and Other Considerations for Design Rule Determination

1.3.1.1 Critical Dimension Uniformity (CDU)

CDU relates to the consistency in the dimensions of a feature printed in resist. CD variations arise due to a number of factors—wafer temperature and photoresist thickness, to name a few. It is defined by

$$\text{CDU} = \frac{\sqrt{\text{CDU}_E^2 + \text{CDU}_F^2 + \text{CDU}_M^2}}{2}, \tag{1.1}$$

where CDU_E, CDU_F, and CDU_M are the CD variation due to dose, focus, and mask variations (Chiou et al., 2013). The required CDU is typically calculated as 7% of the target CD requirement, but modern scanner systems continue to push the envelope beyond that requirement. The 3σ CDU for 40 nm isolated and dense lines can be as small as 0.58 and 0.55 nm (DeGraff et al., 2016), respectively, for ASML's TWINSCAN NXT:1980Di optical immersion lithography scanner released in 2016. For the ASAP7 PDK, we assumed a CDU of 2 nm for optical immersion lithography, which is in line with Vandeweyer et al. (2010). For EUVL patterned layers, we assumed the CDU to be 1 nm, which is close to the 1.2 nm CDU estimated by Van Setten et al. (2014) and the later CDU specification of 1.1 nm for ASML's TWINSCAN NXE:3400B EUV scanner (ASML, 2017).

1.3.1.2 Overlay

Overlay refers to the positional inaccuracy resulting from the misalignment between two subsequent mask steps and denotes the worst-case spacing between two non-self-aligned mask layers (Servin et al., 2009). Single machine overlay (SMO) refers to the overlay arising from both layers being printed on the same machine (scanner), which results in better alignment accuracy and thus smaller overlay. However, using the same machine for two layers or masks is slower from a processing perspective, and consequently, more expensive. Matched-machine overlay (MMO) refers to that arising from two successive layers being printed on different machines, resulting in a larger value than SMO. As two separate scanners are employed, the overall processing rate in an assembly line setting is faster.

Lin et al. (2015) predicted 3σ SMO and MMO values at N7 to be 1.5 and 2 nm, respectively. ASML's TWINSCAN NXT:1980Di optical immersion lithography scanner released in 2016 has a 3σ SMO and MMO of 1.6 and 2.5 nm, respectively, while its Twinscan NXE:3400B EUV scanner released in 2017 has 3σ SMO and MMO of 1.4 and 2 nm, respectively (ASML, 2017). For the PDK, we assumed a 3σ MMO of 3.5 nm for optical immersion lithography, based on ASML (2015). We assumed a 3σ MMO of 1.7 nm for the EUVL, based on the estimates by Van Setten et al. (2014).

1.3.1.3 Mask Error Enhancement Factor (MEEF) and Edge Placement Error (EPE)

MEEF refers to the ratio of wafer or resist CD error to the mask CD error and is given as

$$\text{MEEF} = \frac{\Delta \text{CD}_{\text{wafer}}}{\Delta \text{CD}_{\text{mask}}}. \tag{1.2}$$

Thus, it denotes the amount by which errors on the mask are magnified when they are transferred to the wafer and it depends on the mask, optics, and the process. Its effects are more pronounced near the resolution limit for a specific patterning technique (Yeh and Loong, 2006). Features such as the metal line-ends or tips are typically more adversely affected in optical immersion lithography (193i) systems. Van Setten et al. (2014) found the MEEF for 193i patterned layers to range from five to seven, but found it to be nearly one for EUVL patterned layers.

The EPE gives the deviation in edge placement of one layer relative to another, while accounting for both CDU and overlay contributions. For two layers, each patterned through SE steps, the EPE is given as

$$\text{EPE} = \sqrt{\left(\frac{3\sigma \text{CDU}_{\text{layer 1}}}{2}\right)^2 + \left(\frac{3\sigma \text{CDU}_{\text{layer 2}}}{2}\right)^2 + \left(3\sigma \text{Overlay}_{\text{layer 1-2}}\right)^2}. \tag{1.3}$$

1.3.1.4 Time-Dependent Dielectric Breakdown (TDDB)

The primary DR limiter for metal layers is TDDB. At very small fabrication dimensions, very high electric fields are generated not just in the gate dielectric, but also in all isolating dielectrics between metals. A key issue in the DTCO process is determining the worst-case spacing between any two metal structures with misalignment, so that the resulting process is reliable against TDDB. TDDB occurs due to the presence of a large (although not as high as in gate dielectrics) electric field between two conductors over a long duration. Its severity is more readily pronounced in conductor layers with large overlay issues, for instance between a via and metal at disparate voltages, or in the MOL layers (Standiford and Bürgel, 2013). Obviously, sharp edges exacerbate the fields and are thus also an important issue.

Although layer self-alignment can alleviate the TDDB to some extent, it does not guarantee complete mitigation and necessitates other measures. One such case is the self-aligned raised source-drain contact to gate separation, that must be increased through the addition of extra spacer thickness and gate cap (Demuynck et al., 2014). This is partially in anticipation of some erosion of the self-aligning spacer material. The final separation between two layers must not, therefore, be based on just overlay and CDU, i.e., EPE, but also on the TDDB requirement, i.e., the expected potential differences between the structures. For the PDK, we assumed a 9 nm spacing requirement for TDDB prevention, a value similar to that assumed by Standiford and Bürgel (2013). Given that operating voltages are well below 1 V V_{DD}, this is conservative, which hopefully covers for any other small errors in the analysis.

1.3.2 Single Exposure Optical Immersion Lithography

The conventional lithography resolution limit, which determines the CD, is given by the Rayleigh equation as follows (Ito and Okazaki, 2000):

$$CD = k_1 \frac{\lambda}{NA} \tag{1.4}$$

where λ is the illumination source wavelength and NA is the projection lens numerical aperture. NA is given as

$$NA = n_1 \sin \theta, \tag{1.5}$$

where θ is the maximum angle of the light diffracted from transparent mask regions, which can be captured by the lens; and n_1 is the refractive index of the material between the projection lens and wafer. The value of processing factor k_1 in Equation 1.4 depends on the illumination method and the resist process.

The term optical lithography has become nearly synonymous with the use of ArF light sources in the industry, employed since the 90 nm technology node (Liebmann et al., 2014a). The use of water to boost the NA leads to the technique being termed as optical immersion lithography. NA for the present 193i toolsets is 1.35 and the k_1 value is 0.28. The present set of values for these terms are a result of enhancements over the years, arising from resist process improvements and resolution enhancement techniques such as optical proximity correction (OPC), off-axis illumination (OAI), and source mask optimization (SMO) among others, each enabling a smaller CD at successive technology nodes. Ultimate limits for NA and k_1 are 1.35 and 0.25, respectively, but operating at these limits is challenging (Lin, 2015). Thus, as it stands, CD or half pitch for the layers patterned using an SE in 193i is about 40 nm. Although, the technical specifications for ASML's TWINSCAN NXT:1980Di optical immersion lithography scanner suggests that it can attain a resolution of about 38 nm (ASML, 2016).

1.3.3 Multi-Patterning Approaches

The use of 193i SE to achieve CD targets for all the patterned layers ended at the 22 nm technology node. This marked a severe restriction to continuing with the scaling trends for the subsequent nodes. Overcoming this limitation requires the use of multi-patterning (MP) techniques.

1.3.3.1 Litho-Etch^x (LE^x)

One of the most straightforward approaches to MP involves using multiple independent lithography and etch steps, where one litho-etch (LE) step refers to patterning shapes on a given layer through SE and etch step. The technique is termed LE^x, where x represents the number of LE steps. It applies to any light source and is not specific to just 193i. LE^2, or LELE, technique is of more immediate use since it is used to pattern the target shapes just below the SE patterning limit. To prepare a design for the LELE process, the design layout (Figure 1.1a)—containing target shapes at a pitch that is smaller than the SE limit—is decomposed into two separate layers, e.g., A and B, with different "colors" (Figure 1.1b). This decomposition or "coloring" must produce shapes assigned to a specific color layer at a pitch that can be patterned through an SE, thereby "splitting" the pitch. Consequently, the two color layers, resulting from the LELE decomposition step, correspond to two masks that are used in consecutive LE steps to pattern all the shapes on a single layer. This approach results in the LELE steps defining the line CD. Yet another approach involves specifying the space CD through LELE steps, instead of the line CD. While the process steps are simpler for the latter, the layout decomposition is more complex compared to the former approach.

The same basic principles used for LELE can be extended to the LE^x process with an x value that is larger than two, where x denotes the number of distinct colors and corresponding masks. As multiple masks are used to pattern a single layer, the cost associated with LE^x is higher than the SE lithography and increases with the number of masks. Unsurprisingly, the greater the number of LE steps to pattern a single layer, the higher the complexity and overlay concerns. Moreover, misalignment between steps must be considered, and jagged edges can result. LE^x is prone to odd-cycle conflicts, whereby decomposition may result in a coloring conflict when more than two shapes geometries exist so as to preclude topology patterning through SE. Such odd-cycle conflicts will be discussed in later sections. Stitching can alleviate these odd-cycle conflicts to some extent by patterning a contiguous shape through different exposures. It requires that the disparate fragments of the same shape have a reasonable overlap to counter EPE. Stitching does not completely

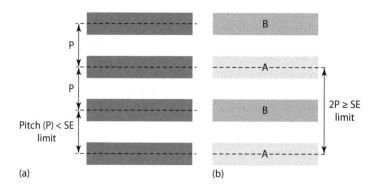

FIGURE 1.1
Litho-etch-litho-etch (LELE) multi-patterning (MP) approach. (a) Target layout shapes with inter-shape pitch below the single exposure (SE) patterning limit. (b) Target layout after decomposition of shapes into separate colors. Same-color shapes are at a pitch above the 193i SE patterning limit.

mitigate odd-cycle conflicts in all topologies. One-dimensional (1-D) patterns are specially challenging from this perspective.

1.3.3.2 Self-Aligned Multiple Patterning

Self-aligned multiple patterning (SAMP) represents another MP approach that seeks to limit the mask-defined line or space CD, thereby reducing the overlay error as compared to LE^x. The technique derives its name from spacers that are deposited along the sidewalls of a mask-defined 1-D or bidirectional (2-D) line, and are thus self-aligned to it. These spacers subsequently define the layer as the actual mask. Self-aligned double patterning (SADP) and self-aligned quadruple patterning (SAQP) are two of the more common forms of SAMP technique and denote whether the pitch is split by a factor of two or four, respectively. In the latter case, a first spacer is used to produce two second spacers, i.e., pitch splitting, that is used to pattern the actual lines.

SAMP can be broadly categorized into "spacer positive tone" and "spacer negative tone" process flows (Ma et al., 2010). In the former, the spacers define the dielectric isolation or space between the lines; therefore, the process is also called "spacer-is-dielectric" (SID). It allows for multiple line and space CDs. In the latter process flow, also called "spacer-is-metal" (SIM), the spacers define the line. However, the latter only allows for two line widths and allows more variability in the intra-layer line spacing, which is a reliability risk for TDDB.

Figure 1.2 shows a generic SADP SID flow. Similar to LE^x, decomposition is also necessary for the SADP process (Figure 1.2b). In decomposition, one of these two masks (e.g., mask A) is selected as a candidate for a derivative photomask, using which the shape is patterned on the resist using an SE (Figure 1.2c). The resist is trimmed in the event the target line CD is under the SE resolution limit, followed by an etch and resist strip for mandrel formation (Figure 1.2d) (Oyama et al., 2015). The mandrel is a sacrificial feature, around which the spacers are deposited as shown in Figure 1.2e. Note that the result includes loops. A second photomask, called the block mask, is then used in conjunction with the spacers to "block" the regions where the feature should not be present. In the case of metal lines, the remaining regions define the trenches in a damascene process and subsequently define the line widths. The SAQP process involves two spacer deposition steps, where the first set of spacers serve as mandrels for the second set of spacers. This is evident in the following fin examples.

Note that the decomposition criterion for SAMP is different from LE^x, since the mandrel is continuous and any discontinuities in it can be marked using the block mask, unlike LE^x, where the lines with such discontinuities must be patterned through a separate exposure, resulting in a potential odd-cycle conflict for certain topologies. It must also be noted that for SAMP, shapes with different colors do not correspond to separate lithography masks, as the deposition of other colored shapes is through spacer deposition and the number of masks employed is lower than the number of decomposition colors.

1.3.3.3 Multiple Patterning Approach Comparison

Before selecting a particular MP technique for layer patterning, consideration must be given to the variability, complexity, and cost associated with it, and the way in which it differs from another MP technique based on these metrics. Being the simplest and most commonly used MP techniques, LELE and SADP are contrasted here to determine their favorability as the preferred MP technique. In terms of cost, the SADP process is more

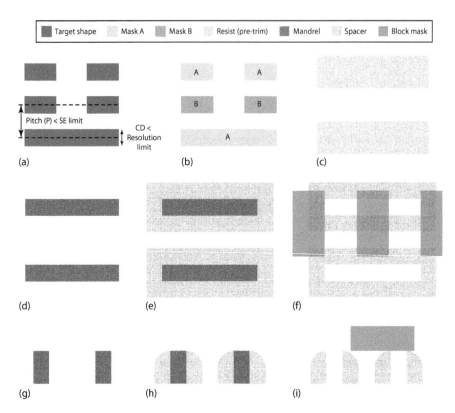

FIGURE 1.2

A "spacer-is-dielectric" (SID) or "spacer positive tone" self-aligned double patterning (SADP) process. (a) Target layout shapes with inter-shape pitch and line CD below the single exposure (SE) patterning limit. (b) Target layout after decomposition of shapes into separate colors. (c) Resist patterning (pre-trim) with CD obtainable through single exposure resolution limit. (d) Mandrel formation (post resist trim and etch). (e) Sidewall spacer deposition. (f) Spacer and block mask–defined trench formation for line patterning. (g) Mandrel cross section. (h) Mandrel and spacer cross section after spacer deposition. (i) Spacer cross section after mandrel strip. Block mask is overlaid on top of the spacer to denote the regions where metal will not be deposited, but is not indicative of the process.

expensive than the LELE due to its sequential etch and deposition steps (Vandeweyer et al., 2010). Liebmann et al. (2015) put the 193i LELE and 193i SADP normalized wafer costs at 2.5× and 3×, respectively, of the 193i SE cost. However, the lower variability and resulting smaller values for similar DRs that determine design density weigh in favor of SADP.

The LELE steps define the line CD and as the two line populations are distinct, their CDU are generally uncorrelated (Arnold 2008). Any overlay between the two masks affects the space CD. When LELE steps define the space CD instead, the CDU between the two space populations is uncorrelated and overlay error affects the line CD. Thus, notwithstanding its use to define either the line or space CD, the CDU is entangled with overlay in the LELE process. The EPE in LELE can be calculated as

$$\mathrm{EPE_{LELE}} = \sqrt{\left(\frac{3\sigma\mathrm{CDU_{line\,1}}}{2}\right)^2 + \left(\frac{3\sigma\mathrm{CDU_{line\,2}}}{2}\right)^2 + \left(3\sigma\mathrm{Overlay_{line\,1\text{-}2}}\right)^2} \qquad (1.6)$$

for metal lines. On the other hand, in an SADP process, as the line or space CD is mostly defined through a single mask and spacer, the overlay or misalignment does not play a significant role in CD determination unless the target shape edges that determine CD are defined by the block mask. Consequently, block mask edge definition should be avoided. The EPE in SADP, for spacer-defined features, can thus be given as

$$\text{EPE}_{\text{SADP}} = \sqrt{\left(3\sigma\text{CDU}_{\text{litho}}\right)^2 + \left(3\sigma\text{spacer}_{\text{left\_edge}} + 3\sigma\text{spacer}_{\text{right\_edge}}\right)^2}. \tag{1.7}$$

The spacer edge–related terms in Equation 1.7 are associated with the spacer 3σ CDU, which can be as small as 1 nm (Ma et al., 2010). The absence of overlay due to the absence of block mask–defined edges, together with the small spacer CDU, gives SADP an advantage over LELE in terms of EPE control for spacer-defined edges, assuming greater than 3 nm LELE overlay and a well-controlled SADP spacer CDU of about 1 nm. However, if LELE overlay can be made small enough (~2.5 nm), then the EPE for the LELE overlay-influenced edges approach that for SADP spacer-defined edges, assuming that the remaining EPE constituent terms remain unchanged (Jung et al., 2007). As mentioned earlier, ASML's TWINSCAN NXT:1980Di 193i scanner has a 3σ MMO of 2.5 nm, but this is accompanied by a change in the CDU value, so that the advantage still lies with SADP. Finally, the spacers do not suffer from much line edge roughness (LER) (Oyama et al., 2012), which is another advantage that SADP offers over LELE. From an electrical perspective, the larger lithography-related variations for LELE patterned layers translate into larger RC variations, which can be nearly twice as much as that for SADP patterned layers employed for critical net routing, primarily due to capacitance variations (Ma et al., 2012). Thus, a choice of LELE creates a potential plethora of metal corners for designers to deal with.

The impact of topology imposed DR constraints must also be considered when selecting an MP technique, so as to limit any density penalty. Double patterning can typically overcome the large tip-to-tip (T2T) or tip-to-side (T2S) spacing requirements inherent to SE lithography by avoiding these features from being assigned to the same color. The SE T2T or T2S spacing values are larger than the minimum SE-defined width to spatially accommodate the hammerheads used for OPC applied to ensure pattern fidelity (Wong et al., 2008). Large spacing also prevents tips from shorting due to bridging by ensuring sufficient contrast, as the regions between tips have low image contrast. However, even with double patterning, certain feature topologies, such as gridded metal routes with discontinuities, can result in T2T or T2S features on the same mask (Ma et al., 2010). This occurs because shapes on adjacent routing tracks must be colored alternatingly, which forces the shapes along the same track to be the same color. For an LELE patterned layer, as illustrated in Figure 1.3a, this color assignment necessitates the use of DR values related to SE T2T spacing, x, that are even larger than the minimum SE-defined width and nullify the advantage of double exposure. By comparison, the T2T and T2S features in SADP are block/cut mask defined (see Figure 1.3b). Although also SE-defined, the block/cut mask width is similar to the SE-defined line width, y, instead of the larger T2T or T2S SE spacing requirement. As identified by Ma et al. (2012), this enables SADP patterned metal routes on the same track to connect to pins on a lower metal interdigitating another lower metal route as shown in Figure 1.4a, while different tracks must be used as in Figure 1.4b to realize the same connections for LELE patterned metal routes. Thus, SADP can lower the density penalty in certain design scenarios.

Furthermore, as shown in Figure 1.5a, LELE can produce odd-cycle conflicts during mask decomposition into two colors. Such a conflict requires increasing the spacing between the features to the SE T2T value (x) in order to resolve the conflict. In contrast, SADP is largely free of odd-cycle conflicts for 1-D topologies with equal metal width along routing tracks,

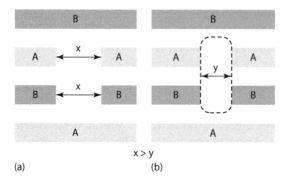

FIGURE 1.3
Comparison of tip-to-tip (T2T) spacing cases for different double patterning approaches. (a) Scenario where T2T spacing, x, is single exposure limited for LELE patterned layers is not too uncommon for designs with 1-D, gridded routing. (b) T2T spacing, y, for SADP patterned layer is defined by the block or cut mask (dotted polygon). Being a line-like feature, the block/cut mask width is similar to the single exposure–defined line width, which is smaller than the spacing required for single exposure–defined tips.

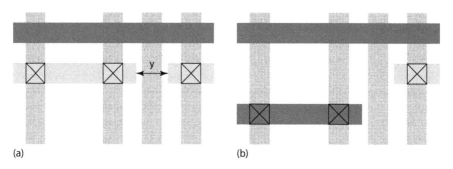

FIGURE 1.4
Pin connection scenarios for double patterned layers. (a) Connections to next neighboring pins can be made through metals on the same track by leveraging the smaller block mask–defined T2T spacing for an SADP patterned metal layer. (b) The same becomes impossible with an LELE patterned metal layer due to larger T2T spacing requirements for same-color shapes and different tracks must be utilized, which results in density penalty.

as evident in Figure 1.5b. However, odd-cycle conflicts may arise during mask decomposition into two colors for SADP, when unequal metal widths or 2-D features are used as shown in Figure 1.5c. Thus, the simpler SADP is adopted for BEOL in the ASAP7 PDK. Some simple DRs, such as limiting line widths to specific values and pitches, make automatic decomposition possible. This is handled by the Calibre design rule checking (DRC) flows automatically, to simplify usage.

1.3.4 Extreme Ultraviolet Lithography (EUVL)

1.3.4.1 EUVL Necessity

Keeping in line with the area scaling trends requires manufacturing capability enhancement through technological advancement, with pitch scaling through photolithographic improvements constituting the major effort. This increases the wafer cost initially by nearly 20%–25% at a new technology node (Mallik et al., 2014). As a particular technology node matures, the subsequent process and yield optimization bring down the wafer cost. These improvements, together with the increased transistor density, eventually result in

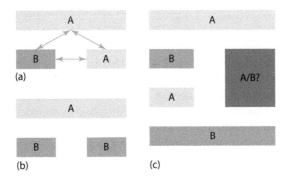

FIGURE 1.5
(a) Even the most common topologies suffer from LELE odd-cycle conflicts. (b) SADP is largely free of such conflicts, given correct coloring and considering 1-D equal width metals. (c) Conflicts with SADP may arise for wide metals used in power or clock routing.

the overall cost reduction per transistor. 193i MP adoption to pattern increasingly small FEOL and BEOL pitches has led to a larger than conventional wafer cost increment. This is due to an increased number of photomasks and process steps required to pattern a given layer with MP—hitherto patterned through a single LE step. MOL layer and finFET introduction have led to further cost escalation as they constitute critical layers that require MP. The transistor cost-reduction trend has slowed down over the previous few nodes, to which MP has contributed to some extent. This contribution will become even larger with an increase in layers patterned through MP techniques with each new node, resulting in a further slow-down in the per transistor cost-reduction trend. Thus, MP can potentially undermine the cost-effectiveness associated with transitioning to a new node.

In addition to cost, MP is also challenging from a variability perspective, which leads to stringent CDU and overlay considerations, as discussed in previous sections. Accommodating these requires guard-bending the projected pitch/spacing targets for relaxed process margins. Patterning a critical BEOL layer through MP is one such case. Patterning a 2-D shape using LE^x necessitates stitching, which complicates overlay requirements and requires that the pitch target be relaxed to ensure sufficient shape overlap. An alternative is using SADP patterning, which is not very amenable to 2-D shapes, and a 1-D patterning approach using SADP becomes the other choice (Ryckaert et al., 2014). This detrimentally affects the cell circuit density, but is easier to manufacture (Vaidyanathan et al., 2014). Mallik et al. (2015) estimate a 16% and 5%–15% lower area penalty for 2-D, instead of 1-D, metal layers for SRAM cell and standard cells, respectively. A pure 1-D approach makes the layout of even relatively simple logic gates more difficult and results in poor input pin accessibility.

EUVL can mitigate some of the MP-related issues. It uses a 13.5 nm light wavelength, instead of 193 nm used for ArF immersion lithography. This enables patterning the features at a much smaller pitch and resolution, so that a single EUV exposure suffices for patterning at the target pitch only attainable through multiple exposure with 193i, thus greatly simplifying the DRs. Since academic use was a primary goal of the PDK, opting for simpler EUV rules was a key consideration, even at the risk of optimism in EUV availability. However, we are presently working on additions and libraries to support pure 1-D metallization.

1.3.4.2 EUVL Description and Challenges

ASML's EUVL scanners have a 0.33 NA and can operate with a processing factor k_1 of nearly 0.4 (Van Setten et al., 2015; Kerkhof et al., 2017). Using these values, together with a 13.5 nm wavelength, in Equation 1.4 gives a CD of approximately 16 nm. Improvements have led the CD to be reduced to 13 nm (ASML, 2017; Kerkhof et al., 2017). An NA improvement to 0.5 can lead to a further CD reduction to 8 nm (VanSchoot et al., 2015).

EUVL differs from 193i in a number of ways that make it a more challenging patterning approach and have contributed to the delay in EUVL being production-ready. Light at EUV wavelength is generated as follows. A droplet generator releases tin droplets, which are irradiated by a laser to create plasma containing highly ionized tin that emits light at 13.5 nm wavelength that is gathered by a collector for further transmission (Tallents et al., 2010). Light in the EUV spectrum is absorbed in the air, which necessitates manufacturing under vacuum condition. It is also absorbed by nearly all materials, which precludes optical lens use to prevent high energy loss. Instead, reflective optics, i.e., mirrors, are used. These mirrors have a reflectivity of around 70% and an EUV system can contain over 10 such mirrors, resulting in only around 2% of the optical transmission to reach the wafer (Tallents et al., 2010). Inefficiency in the power source reduces the transmitted optical power even further, thus creating a demand for a high-power light source. It is also desirable for the photoresist to have a high sensitivity to EUV light.

Production throughput for a photolithography system, given in wafers per hour (WPH), is closely associated with cost-effectiveness. To a considerable extent, it relies on the optical power transmitted to the wafer, and thus on the EUV light source power. It also depends on the amount of time the system is available for production, i.e., system availability. System downtime adversely affects the cost. For EUVL patterning to be cost-effective, a source power exceeding 250 W is desired for over 100 WPH throughput at a 15 mJ/cm^2 photoresist sensitivity (Mallik et al., 2014). Currently, the source power is around 205 W and droplet generator, hitherto a major factor in EUVL system unavailability, has become a smaller concern in more modern EUVL systems (Kim et al., 2017). These and other improvements have brought EUVL systems close to the HVM production goals by increasing the throughput to 125 WPH (Kerkhof et al., 2017). Collector lifetime is the biggest contributor to the system unavailability at the moment and a number of other issues must be surmounted for further cost-effectiveness (Kim et al., 2017). Overall, EUVL systems continue to improve and are slated to be deployed by some foundries for production at N7 (Ha et al., 2017; Xie et al., 2016).

1.3.4.3 EUVL Advantages

The decision to choose SE EUV over 193i MP approaches comes down to both cost and complexity concerns. The EUV mask cost alone is approximately 1.5× that of a 193i mask (Mallik et al., 2015). Other operational expenses bring up the EUVL cost to nearly 3× of 193i SE (Liebmann et al., 2015). The number of masks used in each technology node has increased almost linearly up until N10, but the continuation of MP use will result in an abrupt departure from this trend at N7 (Dicker et al., 2015). The issue is compounded by an increase in the associated process steps, which further adds to the cost. Liebmann et al. (2015) estimate the normalized LE^2, LE^3, SADP, and SAQP cost to be 2.5×, 3.5×, 3×, and 4.5× that of 193i SE, respectively. Dicker et al. (2015) estimate a 50% patterning cost reduction with EUV as opposed to SAQP. They also estimate a faster time to yield and time to market with EUVL due to cycle time reduction as compared to 193i MP approaches that suffer from a large learning cycle time—as large as 30% compared to 2-D EUV, thus improving EUVL cost-effectiveness.

EUVL SE also reduces the process complexity by virtue of reduced overlay. The small EUV wavelength allows the processing factor k_1 to be relatively large—in the range 0.4–0.5.

This enables EUVL to have a high contrast, given by normalized aerial image slope (NILS), than 193i and allows features to be printed with higher fidelity (Kerkhof et al., 2017; Ha et al., 2017). The high feature fidelity, better corner rounding, and an SE use with EUVL cause fewer line and space CD variations. Consequently, metals and vias patterned through EUVL have more uniform sheet resistance (Ha et al., 2017). They also have lower capacitance as compared to SADP patterned shapes, as EUV SE obviates dummy fills and metals cuts. These improvements contribute to improved scalability and better performance (Kim et al., 2017).

Mallik et al. (2014) estimate the normalized wafer cost to increase by 32% at N10 and by a further 14% at N7 without EUV insertion at these nodes. They also estimate a 27% cost reduction, as compared to the latter case, due to EUVL use at N7 for critical BEOL layer patterning with a 150 WPH throughput as a best-case scenario. Ha et al. (2017) put the number of mask reduction at N7 due to EUV use at 25%. Dicker et al. estimate over 40% cost per function reduction in moving from N10 to N7, and further to N5 as a consequence of EUV insertion for critical BEOL layers. Thus, EUVL deployment at these nodes will likely help ensure the economic viability of process node transition.

1.3.5 Patterning Cliffs

Patterning cliffs mark the pitch limits for a given lithography technique or MP approach. Table 1.3 summarizes these pitch limits for the metal layers (Sherazi et al., 2016; VanSchoot et al., 2015).

It must be noted that EUV scanners are being continuously refined and their capabilities may vary, resulting in different final patterning cliffs. However, Table 1.3 gives good rule of thumb values.

1.3.6 DTCO

When a process is still in development, designers are faced with "what if?" scenarios, where the process developers naturally wish to limit the process complexity, but excessive limitation may make design overly difficult or lacking the needed density to make a new process node worthwhile. Those who must make decisions regarding cell architecture for future processes face significant challenges, as the target process is not fully defined. Specifically, as bends in diffusions and gates have become increasingly untenable, the

TABLE 1.3

Patterning Cliffs, i.e., Minimum Feature Pitch for a Particular Patterning Technique

Patterning Technique	Minimum Pitch (nm)
193i	80
193i LELE	64
193i LELELE	45
193i SADP	40
EUV SE (2-D, NA = 0.33)	36
EUV SE (1-D, NA = 0.33)	26
EUV SE (2-D, NA = 0.55)	22
193i SAQP	20
EUV SE (1-D, NA = 0.55)	16

MOL layers have been introduced to connect source-drain regions and replace or augment some structures such as poly crossovers. Decisions that were once purely up to the technology developers increasingly affect design possibilities. Consequently, DTCO is used to feed the impact of such process structure support decisions on the actual designs back into the technology decision-making process (Aitken et al., 2014; Chava et al., 2015; Liebmann et al., 2014b, 2015). It is increasingly important as finFET width discretization and MP constrain the possible layouts. Consequently, determining DRs progressed in this predictive PDK development by setting rules based on the equipment capabilities, designing cell layouts to use them, and iterating the rules based on the outcomes. This includes the APR and SRAM array aspects as described later.

1.4 FEOL and MOL Layers

Transistors are assumed to be fabricated using a standard finFET-type process: a high-K metal gate replaces an initial polysilicon gate, allowing different work functions for NMOS and PMOS, as well as different threshold voltages (V_{th}) (Vandeweyer et al., 2010; Seo et al., 2014; Lin et al., 2014; Schuegraf et al., 2013). Fins are assumed to be patterned at a 27 nm pitch and have a 7 nm drawn (6.5 nm actual) thickness. The layer active is drawn so as to be analogous to the diffusion in a conventional process and encloses the fins—over which raised source-drain is grown—by 10 nm on either side along the direction perpendicular to the fin run length (Clark et al., 2016). The drawn active layer differs from the actual active layer, which is derived by extending it halfway underneath the gates—perpendicular to the fins. The actual active layer, therefore, corresponds to the fin "keep" mask, with its horizontal extent marking the place where fins are cut and its vertical extent denoting the raised source-drain regions.

Gates are uniformly spaced on a grid with a relatively conservative 54 nm CPP. Gates are 20 nm wide (21 nm actual). Spacer formation follows poly gate deposition (Hody et al., 2015). Cutting gate polysilicon with the gate cut mask, in a manner that keeps the spacers intact with a dielectric deposition following, ensures that fin cuts are buried under gates or the gate cut fill dielectric, so source/drain growth is on full fins. A double diffusion break (DDB) is assumed to be required to keep fin cuts under the gate. Recently announced processes have removed the DDB requirement, improving standard cell density. Adding this into the PDK as an option is under consideration for a future release. A 20 nm gate cap layer thickness is assumed. This thickness provides adequate distance to avoid TDDB after self-aligned contact etch sidewall spacer erosion, accounting for gate metal thickness non-uniformity (Demuynck et al., 2014). The dual spacer width is 9 nm.

The resulting FEOL and MOL process cross section comprises Figure 1.6. Figure 1.6a shows the (trapezoidal) source/drains grown on the fins. The MOL layers can be used for functions typically reserved for the first interconnect metal (M1) layer (Lin et al., 2014) and serve to lower M1 routing congestion, thereby improving standard cell pin accessibility (Ye et al., 2015). The connection to the MOL local interconnect source-drain (LISD) layer is through the source-drain trench (SDT) contact layer. The minimum SDT vertical width of 17 nm is required in a SRAM cell, which necessitates patterning using EUVL. It has a 24 nm drawn horizontal width, with the actual width being 25 nm. This width is larger than the 15 nm gap between the spacers, so as to ensure complete gap coverage and contact with the raised source-drain (RSD), despite the 5 nm 3σ EPE for EUV.

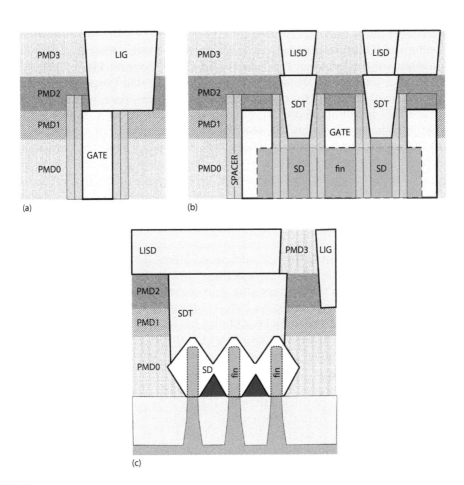

FIGURE 1.6
FEOL and MOL cross sections. (a) LIG connection to the gate. (b) LISD to SDT and SDT to source-drain (SD) connection. LISD location in the stack allows it to cross over gates to be used for routing. (c) Fin and SD cross section. LIG is shown here to illustrate its necessary separation from LISD and SDT. Sub-fin and shallow trench isolation (STI) are evident underneath PMD0.

LISD provides the means of connecting RSD regions to power rails and other equipotential RSDs within a standard cell. It is drawn at the same horizontal width as SDT for lower resistance when connecting to RSD through SDT. LISD may also be used for routing purposes within a standard cell at 18 nm width and 36 nm pitch, as it can pass over gates—further lowering M1 usage. The layer thicknesses are defined by different dielectric layers, so that appropriate etch stops can be used. This implies 2-D LISD routing, which, when combined with width and pitch assumptions, means that EUVL must be used for patterning. LISD connects to M1 through via 0 (V0), which is another MOL layer.

The local interconnect gate (LIG) layer is used for connecting gates to M1 through V0 and for power delivery to standard cells by connection to LISD upon contact. The minimum LIG width of 16 nm is dictated by the LIG power rail spacing from the gate. The width value implies that LIG is also patterned using EUVL. Recent advances in EUVL have demonstrated resolutions lower than 18 nm (Neumann et al., 2015; VanSchoot et al., 2015) even for 2-D patterns, but the PDK restricts this value as the minimum line width for 2-D layers, in accordance with a more conservative 36 nm pitch for 2-D EUVL (Mallik et al., 2015).

However, smaller line width is permitted for LIG as its usage is limited to unidirectional, i.e., 1-D, patterns. LIG connects to the gate through the cap layer in Figure 1.6b, which cuts through a standard cell between the NMOS and PMOS devices. Figure 1.6c shows the same view, but at the fins, so the source/drains are illustrated perpendicular to that in Figure 1.6a. As mentioned, the fin cuts occur under a dummy gate, which comprises 1/2 of a DDB. The SDT is self-aligned to the gate spacer as shown. MOL rules turn out to be limiting for both standard cells and SRAM. The details are described in Sections 1.6.3 and 1.8.1. More details and the transistor electrical behavior are presented in Clark et al. (2016).

1.5 BEOL Layers

The ASAP7 PDK assumes nine interconnect metal layers (M1-M9) for routing purposes and corresponding vias (V1-V8) to connect these metals. Figure 1.7 shows a representative BEOL stack cross section, comprising a lower metal layer (Mx), via (Vx), and an upper metal layer (Mx+1). Following the industry trend (Lin et al., 2014), all BEOL layers assume copper (Cu) interconnects. Metal and vias have a 2:1 aspect ratio, in line with the ITRS roadmap (ITRS, 2015). Table 1.4 enumerates the thickness of the interconnect layers. Figure 1.7 also shows the barrier layers that increasingly consume the damascene trench, increasing resistance.

In the ASAP7 PDK, metals have the same thickness as the corresponding inter-layer dielectric (ILD), but the vias are thicker than the ILD by 10%—an amount corresponding to the assumed HM thickness. Actual processes at N7 include as many or more than 14 metals, with more metal layers at each thickness and pitch value, culminating in two layers that are much thicker and wider at the top than in our PDK. The layers here are representative of all but the very thick top layers, which are primarily for power distribution. We had initially not foreseen PDK use for large die power analysis, but are considering adding layers to make the PDK more amenable to full die analysis.

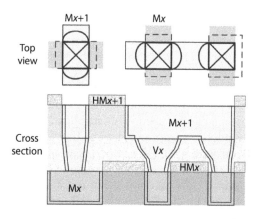

FIGURE 1.7

Representative BEOL cross section. The dotted lines represent the actual self-aligned via (SAV) masks top view. Vias with (left two) and without end-cap (rightmost) are shown. Arcs along Mx+1 length denote via widening at the non-hard mask edges, evident in the cross section for the two vias at the right.

TABLE 1.4

BEOL Layer Thickness and Metal Pitches

Metal/Via Layer	Thickness (nm)
M1–M3	36
V1–V3	39.6
M4–M5	48
V4–V5	52.8
M6–M7	64
V6–V7	70.4
M8–M9	80
V8	88

1.5.1 SAV and Barrier Layer

In a typical via-first flow, the ILD corresponding to a via sustains damage and erosion—caused by dry etch and cleaning steps, respectively, during both via and metal patterning. The metal width itself is also affected, since via formation goes through the metal damascene trench. This results in via widening that can cause shorts to adjacent, non-equipotential metal lines (Baklanov et al., 2012; Brain et al., 2009). This issue is exacerbated by small metal pitches at lower technology nodes. Consequently, in the ASAP7 DRs, a self-aligned via (SAV) formation flow, as described by Brain et al. (2009), is assumed, whereby vias are patterned after the upper interconnect metal layer is patterned on a hard mask (HM) that is relatively unaffected by the via etch. The HM greatly limits via widening perpendicular to its edges, although some widening occurs along the metal, as evident in Figure 1.7. The resulting via edges are delineated by the upper metal HM, i.e., via self-alignment. Via mask edges tend to extend outward where they are defined by the via rather than a HM. Nonetheless, the via is perfectly aligned perpendicular to the upper metal direction despite via and upper metal overlay errors. The dashed lines in Figure 1.7 show that the actual via mask overlaps so that the HM defines the width even with misalignment. However, for simplicity the vias are drawn conventionally in the PDK. They are sized as part of the DRC flows.

Barrier materials, such as tantalum nitride (TaN), are required at the Cu and ILD interface to prevent Cu diffusion. The thickness of the barrier—composed of more resistive TaN—does not scale commensurately with the interconnect scaling. This, together with diffuse electron scattering at interfaces, causes a greater increase in line resistance than is expected as a consequence of scaling (Im et al., 2005). Additionally, the presence of TaN at the via interface has the undesirable effect of increased resistance (Schuegraf et al., 2013). For this reason, the PDK assumes the use of manganese-based self-forming barriers (SFBs) that assuage the shortcoming of TaN by virtue of their conformity, surface smoothness, smaller thickness in the total interconnect fraction, and high diffusivity in Cu (Schuegraf et al., 2013; Au et al., 2010).

The metal resistivity, as specified for extraction purposes, is calculated based on (Pyzyna et al., 2015)

$$\rho = \rho_0 \frac{3}{8} C (1-p) \left(\frac{1}{h} + \frac{h}{A} \right) \lambda + \rho_0 \left[1 - \frac{3}{2}\alpha + 3\alpha^2 - 3\alpha^3 \ln\left(1 + \frac{1}{\alpha}\right) \right]^{-1} \quad (1.8)$$

where ρ_0 denotes the bulk resistivity; C is a geometry-based constant; p is the electron collision specularity with surfaces; λ is the bulk electron mean free path; h is the line height; A is the cross-section area; α is given as $\lambda R/[G(1 - R)]$, where R is the electron reflection coefficient at the grain boundaries and G is the average grain size. The first and second terms are the resistivity due to surface electron scattering and grain boundary scattering, respectively. The latter dominates for the 7 nm node (Pyzyna et al., 2015).

1.5.2 EUV Lithography Assumptions and Design Rules

In addition to some of the FEOL, i.e., fin cut, and MOL layers in the PDK, EUVL is also assumed for patterning M1-M3 and vias corresponding to these metals, i.e., V1 through V3. The choice of an EUVL assumption and the accompanying M1-M3 pitch of 36 nm is based on the premise that this pitch may be attained using single EUV exposure (Mallik et al., 2014). Meeting the same target using optical immersion lithography requires the use of MP techniques, such as SAQP with LE or LELE block mask, which also pushes toward all 1-D topologies. While EUVL is costlier than optical immersion lithography when considering an SE, the use of multiple masks in SAQP with block means that the MP approach becomes nearly as expensive as EUVL due to expensive mask tooling and associated processing steps (Liebmann et al., 2015). It is noteworthy that several ITRS target pitch values at N7 are at or over the EUV SE cliff. As a result, beyond N7, MP will be required even for EUV lithography-defined layers. With N7 at the cutoff, we felt EUVL would be appropriate.

As EUVL permits the single mask use, employing it simplifies the design process by circumventing the issues related to MP, such as complicated cell pin optimization issues due to SADP and odd-cycle conflicts as a consequence of LELE use (Xu et al., 2015). As mentioned earlier, EUVL use at a slightly relaxed pitch of 36 nm also permits 2-D routing, which has the effect of further simplifying both standard cell and SRAM cell design. Thus, given similar cost and reduced design complexity, the PDK assumed EUVL over 193i MP schemes for a number of layers that complicate standard cell design, viz. SDT, LISD, LIG, V0, and M1. However, subsequent iterations of the PDK will revise this choice, so as to use SAQP for 1-D M1-M3, as a consequence of further delays in EUV readiness for high volume manufacturing. While the aggressive pitch value of 36 nm that EUV affords is not required for M2 and M3, we considered these layers to be EUVL patterned due to the ease that this assumption lends when routing to standard cells during APR. Having the vertical M3 match the M2 also allows better routing density vs. a choice of say, the gate pitch for vertical M3. This also allows relative flexibility in metal directions. We foresee vertical M2 as a better choice for 1-D cells.

A 3σ EPE of 5 nm for two EUV layers, when determining inter-layer DRs, was calculated assuming a 3σ mixed machine overlay (MMO) of 1.7 nm for the EUV scanner, a 3σ error in the placement of 2 nm due to process variations and through additional guardbanding (Van Setten et al., 2014). Patterned M1-M3 lines have a 36 nm 2-D pitch and a minimum line width of 18 nm is enforced by the DRs. T2T spacing for narrow lines is 31 nm following Van Setten et al., while wider lines can have a smaller T2T spacing at 27 nm (Van Setten et al., 2015; Mulkens et al., 2015). As per the PDK DRs, lines narrower than 24 nm are considered thin lines and those wider than this value are considered wide lines. This threshold value was determined based on the minimum LISD width, since LISD routes near the power rails become the limiting cases for T2T spacing. A moderate T2S spacing of 25 nm follows the results demonstrated by Van Setten et al. (2014). A corner-to-corner EUV metal spacing of 20 nm enables via placement to metals on parallel tracks at the minimum possible via spacing of 26 with a 5 nm EUV upper metal end-cap to allow full enclosure.

1.5.3 MP Optical Lithography Assumptions and Design Rules

Metal interconnect layers above the intermediate metal layers, i.e., M3, are assumed to be patterned using 193 nm optical immersion lithography and MP. The metal pitches are the same as the thickness values described in Table 1.3 (the 2:1 aspect ratio). The pitch values for M4-M5 and M6-M7 correspond to the targets defined by Liebmann et al. (2015) for 1.5× and 2× metal, respectively. The same metal pitch ratios do not apply to our PDK since the 1× metal pitch is 36 nm to ensure 2-D routing instead of the roadmap value of 32 nm. Moreover, recent foundry releases seem to be tending to the more conservative 36 nm pitch as well. Incrementing the pitch in multiples other than 0.5× does not have any design implication as proven by our DTCO APR experiments. The metal and via aspect ratios are in line with ITRS roadmap projections.

1.5.3.1 Patterning Choice

The M4-M5 pitch target of 48 nm can be attained using either SADP or LE³, and that of 64 nm for M6-M7 using either SADP or LELE. LE³ was dismissed outright for M4-M5 patterning, since it is costlier than SADP (Liebmann et al., 2015), and even though the lower LELE cost is appealing for M6-M7 patterning, SADP has lower EPE, LER, and hence RC variations as mentioned in Section 1.3.3.3. Therefore, we chose SADP over LELE for patterning the layers M4-M6. Furthermore, Ma et al. (2010) also found that LELE fails the TDDB reliability test at 64 nm pitch with 6 nm overlay while SADP appears reliable. Although we used a 3.5 nm overlay for 193i patterned layers that would allay the TDDB severity, we still assumed SADP patterning for M6-M7 so as to be more cautious, given the TDDB concern in addition to the other aforementioned SADP advantages. We chose SID over SIM as it permits multiple metal widths and spaces, which is beneficial for clock and most importantly, power routing. Patterning 2-D shapes is possible in SADP but presents challenges, as certain topologies, such as the odd-pitch U and Z constructs, may either not be patterned altogether or contain block mask–defined metal edges that result in overlay issues and adversely affect metal CD (Ma et al., 2012). Consequently, we chose to restrict M4 through M7 to 1-D (straight line) routing in the DRs.

The ASAP7 PDK supports DRCs based on actual mask decomposition into two different masks (colors) for the purpose of SADP. The mask decomposition is performed as part of the rule decks using the Mentor Graphics Calibre MP solution. An automated decomposition methodology (Pikus, 2016) is employed and does not require coloring by the designer, which greatly simplifies the design effort. As mentioned, we consider this key in an academic environment. To the best of our knowledge, this is the first educational PDK that offers DRs based on such a decomposition flow for SADP. In contrast, the FreePDK15 also used multi-colored DRs (Bhanushali et al., 2015), but required decomposition by a designer by employing different-colored metals for the same layer.

1.5.3.2 SADP Design Rules and Derivations

DRs must ensure that shapes patterned using the two photomasks, viz. the block and the mandrel, can be resolved. This entails writing DRCs in terms of the derived photomasks and perhaps even showing these masks to the designer. However, fixing DRCs by looking at these masks, and the spacer, is both non-intuitive and confusing. In the flows here, the colors are generated automatically, and the flows can produce the block and mandrel masks created by metal layer decomposition. We formulated restrictive design rules (RDRs) that ensure correct-by-construction metal topologies and guarantee resolvable shapes created using mandrel and block masks referring only to the metals as drawn, so that the rules are color agnostic. Nonetheless, to validate the DRCs, the mandrel, block mask, and spacer shapes are completely derived in our separate validation rule decks (Vashishtha et al., 2017a).

The decomposition criteria for assigning shapes to separate colors for further SADP masks and spacer derivation is as follows. In addition to assigning shapes at a pitch below the 193i pitch limit of 80 nm to separate colors, a single CD routing grid–based decomposition criterion is also used. The latter prevents isolated shapes, which do not share a common run length with another shape, from causing coloring conflict with a continuous mandrel through the design extent. Any off-track single CD shapes are marked invalid to enforce APR compatibility. Metals wider than a single CD are checked against the single CD metal grid to prevent them from causing incorrect decomposition. Such cases are essentially odd-cycle conflicts between wide metals and a single CD metal grid.

The ASAP7 PDK supports metals wider than a single CD, but the DRs stipulate that the width be such that the metal spans an odd number of routing tracks. This prevents odd-cycle coloring conflicts between wide metals and single CD metal grids with a pre-assigned color. Wide metals are particularly useful for power distribution purpose. The PDK also supports rules to ensure correct block mask patterning. The minimum block mask width is considered as 40 nm, which corresponds to the minimum 193i resolution. The same T2T spacing value between two SADP patterned metals on the same routing tracks, as well as on the adjacent routing tracks, prevents minimum block mask width violation. A 44 nm minimum parallel run length between metals on adjacent tracks is also enforced to ensure sufficient block mask T2T spacing. These rules are described in greater detail in Vashishtha et al. (2017a).

1.6 Cell Library Architecture

A standard cell library has become the usual way to construct general digital circuits, by synthesizing a hardware description language (HDL; typically Verilog) behavioral description and then performing APR of the gates and interconnections. This section describes the cell library. We also explain the cells provided and discuss changes to the DRs based on APR results.

1.6.1 Gear Ratio and Cell Height

As in planar CMOS with gridded gates, which was introduced by Intel at the 45 nm node and other foundries at 28 nm, the selection of standard cell height in a finFET process is also application specific. Low power application-specific integrated circuits (ASICs) tend to use short cells for density, and high-performance systems, e.g., microprocessors, have historically used taller cells. The cell height constrains the number of usable fins per transistor and thus the drive strength as well as the number of M1 tracks for cell internal routing. Assuming horizontal M2, an adequate number of M2 pitches, with sufficient M2 track access to cell pins, are also required. Thus, the horizontal metal (here M2) to fin pitch ratio, known as the gear ratio, becomes an important factor.

The 3/4 M2 to fin pitch ratio in the ASAP7 predictive PDK used for our libraries allows designing standard cells at 6, 7.5, 9, or 12 M2 tracks and with 8, 10, 12, or 16 fins fitting in the cell vertical height, respectively. Multiple foundries have mentioned "fin depopulation" on finFET processes, i.e., using fewer fins per gate over time. This appears primarily driven by the diminishing need for higher speed, increasingly power-constrained designs, high drive per fin, MP lithography considerations, and the near 1:1 PMOS:NMOS drive ratio, as well as power density considerations. Moreover, for academic use and teaching cell design issues, a large number of tracks make tall cells trivial and their excessive drive capability makes them unsuitable for low power applications. The two fin wide transistors in 6-track library cells have limited M1 routes and pin accessibility, but the best density. 1-D layout, with two metals layers for intra-cell routing, is well suited for 6-track cells, but is not very amenable for classroom use. Therefore, we chose a 7.5-track cell library, which is in line with other publications targeted at N7 (Liebmann et al., 2015). This choice also allows wider M2 power follow-rails at the APR stage, which are preferred for robust power delivery. However, wide M2, while supportable (and indeed required for SAQP M2), requires non-SAV V1 on the power follow-rails. This feature is currently not supported.

Utilizing all of the 10 fins in a 7.5-track library is not possible as sufficient separation is required between the layers SDT and LISD that connect to the transistor source-drain. Also, transistors must have sufficient enclosure by the select layers used for doping. The minimum transistor separation thus dictates that the middle two fins in a standard cell may not be used, as apparent in Figure 1.8, which shows a NAND2 and inverter cell adjacent to each other. Figure 1.8 also illustrates the cell boundaries producing a DDB between the cells. The fin closest to the power rails cannot be used either. Thus, each transistor has three fins and this choice is aided by the nearly equal PMOS-to-NMOS ratio.

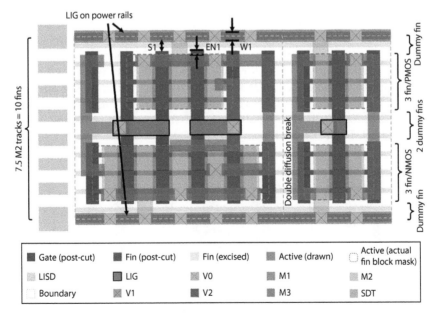

FIGURE 1.8
A 7.5 M2 track standard cell template, with three fins per device type, shows the FEOL, MOL, and M1 in adjacent NAND2 and inverter. A double diffusion break is required between the cells. Fins are tucked underneath the gate when breaking the diffusion. LIG, V0, and M1 are not shown here. S1 = minimum LIG to GATE spacing, EN1 = minimum GATE end-cap, W1 = minimum LIG width.

Notwithstanding the number of fins, the transistors have sufficient drive capability due to the large per-fin drive current value (Clark et al., 2016). ASAP7 standard cells are classified based on their drive strengths, so that those with three-fin transistors are said to have a 1× drive. Those with less drive strength have fraction values, e.g., a one-fin transistor corresponds to a p33× (0.33×) drive strength.

1.6.2 Fin Cut Implications

Again referring to Figure 1.8, the vertical extent of the drawn active layer in the PDK denotes the raised source-drain region. As evident in Figures 1.6, 1.8, and 1.9, the actual active fin "keep" mask extends horizontally halfway underneath the non-channel forming (dummy) gates, where fins are actually cut. Implementing the fin cuts in such a manner necessitates a DDB requirement for the cases where diffusions at disparate voltages must be separated. Therefore, all standard cells end in a single diffusion break (SDB), resulting in a DDB upon abutment with other cells to ensure fin separation. Apart from the cell boundaries, an SDB may exist elsewhere in a cell if equipotential diffusion regions are present on either side of a dummy gate and is equivalent to shorted fins underneath such a gate. Nonetheless, this is a useful structure, particularly in the latch and flip-flop CMOS pass-gates.

As shown in Figure 1.9, block mask rounding effects, when implementing fin cuts, create sharp fin edges that have a high charge density and, consequently, a high electric field (Vashishtha et al., 2017d). This creates a possible TDDB issue between the fin at the cut edge and the dummy gate, potentially causing leakage through the gate oxide. To mitigate this, in the libraries provided with the PDK, the dummy gates are cut at the middle in the standard cells to preclude an electrical path between the PMOS and NMOS transistors through the dummy gate. The gate cut also allows longer LIG for nearby routing without attaching to the cut dummy gate, as evident in Figure 1.9c.

SAQP requires a fixed width and constrains the fin spacing of fins (Figure 1.10). Assuming an EUVL fin cut/keep mask enables the 7.5T library as designed (Figure 1.10a). If MP is assumed, LELE cut/keep is complicated as in Figure 1.10b and c, where the middle fins are excised. An even number of fins for PMOS and NMOS transistors allows the SAQP mandrel to completely define the fin locations, greatly easing the subsequent fin masking as illustrated in Figure 1.10d. A 6T (or 6.5T) or 9T library, which has an even number of fins per device at two or four, is more amenable to SAQP fin patterning than the 7.5T library we primarily used for DTCO.

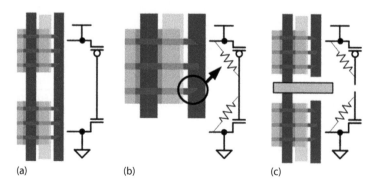

(a)   (b)   (c)

FIGURE 1.9
Dummy gate at DDB (a). Sharp fin edges arise due to mask rounding when cutting fins. This creates a TDDB scenario between the fins and the dummy gate (b). This is avoided where possible by cutting the dummy gates (c).

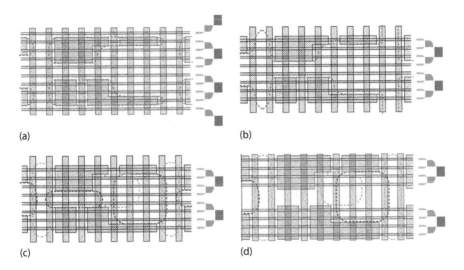

(a)   (b)

(c)   (d)

FIGURE 1.10
The original EUV fin keep mask assumptions and SAQP spacer/mandrels (a). Another fin keep option that allows larger patterns (b). LELE fin cut (c) may be problematic, but an even number of fins eases this (d).

1.6.3 Standard Cell MOL Usage

LISD is primarily used to connect diffusion regions to power rails and other equipotential diffusions within the cell through M1, somewhat relieving M1 routing congestion. However, connections to the M1 power rails cannot be completed using LISD alone, since that would require extending LISD past the cell boundary to allow sufficient V0 landing on LISD, or a 2-D layout that would have issues with the T2S spacing. Larger LISD overlap (past the cell boundary) would violate the 27 nm LISD T2T spacing by interaction with the abutting cell LISD that is not connected to the power rails. Therefore, an LIG power rail is used within the cells to connect LISD to M1, as LISD and LIG short upon contact, but being different layers, do not present 2-D MEEF issues. The PDK DRs require a 4 nm vertical gate end-cap past active and a 14 nm LIG to gate spacing (S1). These rules restrict the LIG power rail to its 16 nm minimum width (W1). This width does not provide fully landed V0, but the LIG power rail is fully populated with V0 to ensure robustness and low resistance.

LISD can also serve the secondary role of a routing layer within a cell, but is used sparingly, as local interconnects favor tungsten over copper, and therefore suffer from higher resistivity (Sherazi et al., 2016). However, LISD routing is helpful in complex cells, especially sequential cells, due to M1 routing congestion and allows us to limit M2 track usage. Figure 1.11 shows a D-latch, in which LISD use to connect diffusions across an SDB of a constituent tristate inverter becomes necessary, since M1 cannot be used due to intervening M1 routes or their large T2T and T2S spacing requirements. The SDBs arise due to the inability to accommodate two gate contacts along a single gate track, which also necessitates an M1 crossover. SDBs are used in the cell CMOS pass-gate (crossover) composed of a CLKN-CLKB-CLKN gate combination evident in the schematic at the bottom of Figure 1.11. DDBs are used in the cell, as all diffusions cannot be shared.

1.6.4 Standard Cell Pin and Signal Routing

The cell library architecture emphasizes maximizing pin accessibility, since it has a direct impact on the block density after APR. Where possible, standard cell pins are extended to

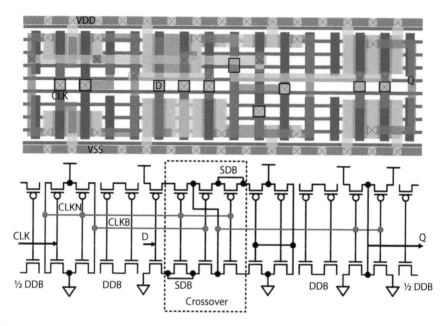

FIGURE 1.11
Cell layout of a transparent high D-latch. Double and single diffusion breaks are shown. The latter require LISD crossovers. Fins shown are prior to the cut.

span at least two and preferably three M2 routing tracks. This ensures that even standard cells with a large number of pins, such as the AOI333 in Figure 1.12, have all of their M1 pins accessible to multiple M2 tracks. The figure demonstrates the use of staggered M2 routes to allow access to the cell pins by M3. Good pin access is forced by our use of the M1 template for cell design (Vashishtha et al., 2017d). The M1 template was employed for rapid cell library development, as the pre-delineated metal constructs take the M1 DRs into consideration, which aids in DRC error-free M1 placement. Most, but not all, of the M1 routing in the standard cells is template based.

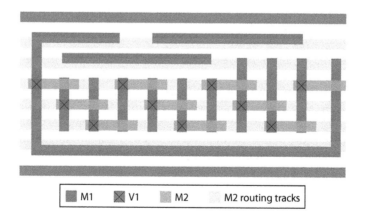

FIGURE 1.12
An AOI333 cell with all of its input M1 pins connected to M2. The staggered M2 allows further connections to M3. Note that this version does not use the bends on M1, following the post-APR DTCO changes. These vias may be slightly unlanded.

Referring to Figure 1.8, the non-power M1 lines closest to the cell boundary are horizontal, which alleviates the T2S DR spacing requirement for such routes as they constitute a side rather than a tip feature. This approach was used for the initial cell version in the library, as it allowed better V1 landing from M2. However, this feature was removed after extensive APR validation, as mentioned later. The 7.5-track tall standard cell results in equidistant M2 routing tracks that are all 18 nm wide, except for the wider spacing of 36 nm around the M2 power follow-rails. The cells accommodate seven horizontal M1 routes that are arranged as two groups of three equidistant tracks between the M1 power rails, with a larger spacing at the center of the cell where vertical M1 is required for pins. Accounting for the extra M1 spacing near the center, instead of the power rails, helps to maximize pin access through M1 pin extension past the M2 tracks adjacent to the power rails. The vertical M1 template locations, corresponding to the pins, match the CPP. Originally, they always ended at a spacing of 25 nm from the nearest horizontal M1 constructs that they do not overlap, so as to honor the M1 T2S spacing. However, L-shaped pins formed using solely vertical M1 constructs on either one of their ends, do not provide the stipulated 5 nm V1 enclosure necessary when connecting to an M2 track. Consequently, the final topology in Figure 1.12, which does not allow full V1 landing to input pins, is adopted for releases after the first, as described in Section 1.6.6.

Connecting to M1 near the power rails is enabled by via merging. While the example presented here is for V0, all layers support via merging. The presence of V0 on the power rails may engender a case similar to that in Figure 1.13a, where the unmerged via mask spacing "*d*" may be smaller than the lithographically allowed minimum metal side-to-side spacing "*y*". Merging these vias (Figure 1.13b) in such cases permits the patterning of V0 containing M1 tracks at minimum spacing. It is enabled by keeping the vias on grid as evident in Figure 1.13c.

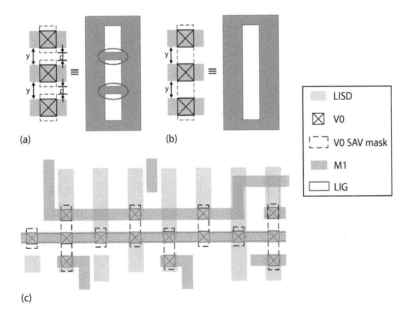

(a) (b) (c)

LISD
V0
V0 SAV mask
M1
LIG

FIGURE 1.13
(a) Sub-lithographic features (ellipses) on an SAV mask. (b) Merging SAV masks precludes such features. (c) Merging enables M1 signal routes to be at the minimum spacing from the power rails, which would otherwise have not been possible due to unmerged SAV mask shapes.

1.6.5 Library Collaterals

A cell library requires liberty timing characterization files (compiled into .db for Synopsys tools) as well as layout exchange format (LEF) of the cell pins and blockages (FRAM for Synopsys). We have focused primarily on Cadence collateral, since we are most familiar with those tools. The technology LEF provides basic DR and via construction information to the APR tool. This is a key enabling file for high quality, i.e., low DRC count, APR results. Additional files are the .cdl (spice) cell netlists. We provide Calibre PEX extracted, and LVS versions. The former allow full gate-level circuit simulation without needing to re-extract the cells. Data sheets of the library cells are also provided. These are generated automatically by the Liberate cell characterization tool. Mentor Graphics Calibre is used for DRC, LVS, and parasitic RC extraction. The parasitic extraction decks allow accurate circuit performance evaluation.

1.6.6 DTCO-Driven DR Changes Based on APR Results

The original M1 standard cell template shown in Figure 1.14a was meant to provide fully landed V1 on M1 in all cases. However, this meant that where an L shape could not be provided, the track over that M1 portion of the pin was lost. APR tools generally complete the routing task, resulting in DRC violations when the layout cannot be completed without producing one. The APR exercises illuminated some significant improvements that could be made, as well as pointing out some substandard cells, which have been revised. A key change to the library templates is illustrated here.

Based on reviewing the APR results described in Section 1.7.3, the V1 landing DR was changed to 2 nm, which does not provide full landing with worst-case misalignment. However, most standard cell pins are inputs and thus drive gates. We decided that gaining the useful pin locations was desirable for multiple reasons. Firstly, a single gate load does not present much capacitance, so the extra RC delay from a misaligned (partially landed) via will not significantly affect delay. Secondly, some gates, notably those with a large number of inputs such as the AOI22 cell, have considerably better pin access, as shown in Figure 1.14b. Initially, two of the pins only had one usable M2-M1 intersection at the middle. These also conflicted for the same M2 track. The improved layout is simpler and has lower capacitance, as well as moving the pins to a regular horizontal grid identical

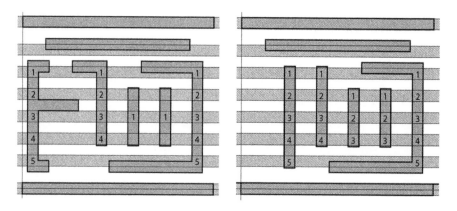

FIGURE 1.14
The original AOI22x1 M1 layout and M2 track overlay (a) and the new layout (b) showing the improved access by relaxing the V1 to M1 overlap rule from 5 nm (full landing) to 2 nm.

to the 54 nm gate grid. The cell output nodes still require 2-D routing to reach the correct diffusions. This maintains the original full 5 nm landing for the output node that drives a much larger load and is subject to self-heating.

1.7 Automated Place and Route with ASAP7

1.7.1 Power Routing and Self-Aligned Via (SAV)

Wide power rails are important in their ability to provide a low resistance power grid. Low resistance alleviates IR drop, particularly the dynamic droop in each clock cycle. The ASAP7 PDK comprehends this by providing wide metal DRs that are compatible with the restrictive metal DRs. While the M2 and M3 are assumed EUV, they can easily be converted to SAQP assumptions and the place and route collateral keeps them on a grid that facilitates this change (Vashishtha et al., 2017b). Consequently, all M2 and M3 routes are, like M4 through M7, on grid with no bends in the provided APR flows. On the SADP M4 through M7, wide metals must be 5×, 9×, 13× of the minimum metal widths, so that the double patterning coloring is maintained. These rules are not enforced by DRCs on M2 and M3 as they are on M4-M7, but the APR flows follow the convention via the techLEF and appropriate power gridding commands in the APR tool.

A portion of the V_{SS} and V_{DD} power grid of the fully routed digital design is shown in Figure 1.15. In Figure 1.15a, the top down view shows the minimum width horizontal M2 power rails over cell top and bottom boundaries. A vertically oriented M3 is attached to each power rail. The V_{SS} on the left and the V_{DD} on the right are connected by a single wide self-aligned V2 to each respective M2 rail. A horizontal M4 is connected to M3 with five SAVs, which can be placed close together for minimum resistance due to the M3 EUV lithography assumption. The five vias fit perfectly in the 9× width M3 power routes. The similar V4 connections from M4 to M5 at the lower right of Figure 1.15a have only three vias, since these are assumed to be patterned by LELE. The technology LEF file provides via definitions at this spacing.

All vias are consistently SAV type, with the same minimum width as other vias on the layer, and are laid out on the routing grid. Ideally, wide M2 could be used for the follow-rails to minimize the resistance. This, however, would require non-SAV V1 on the follow-rails,

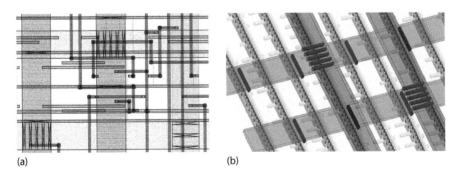

(a)　　　　　　　　　　　　　(b)

FIGURE 1.15
Layout portion of an APR block (a). The wide vias for power distribution are evident connecting M2, M3, and M4. A 3-D extraction of a section of a fully routed design showing M4 down to the MOL layers (b). The way that LISD connects to LIG beneath the M1 power rails is evident.

since SAVs would be too close to the underlying horizontal M1 tracks. The reader will also note that in an SAQP M3 and M2 scenario, all spaces are identical, so a wider M2 would be required. Figure 1.15b shows a 3-D extracted view of the V_{SS} and V_{DD} power rails. The perspective clearly shows the M2-V1-M1-V0-LIG follow-rail stack, with the LISD protruding into the cell gate areas to provide current paths to the NMOS and PMOS sources from V_{SS} and V_{DD}, respectively. The figure emphasizes the M2 over M1 follow-rails, but the M3 and M4 power distribution grid is evident. There is substantial redundancy in the power scheme—if a break forms in M2, M1, or LIG due to electromigration or a defect, the other layers act as shunts.

1.7.2 Scaled LEF and QRC TechFile

We use Cadence Innovus and Genus for the library APR and synthesis validation, respectively. Standard academic licensing did not allow features below 20 nm in 2016–2017, moving to support 14 nm in 2017. Neither is adequate for the 7 nm PDK. Synopsys academic licensing has the same issue as of this writing. To work around this, APR collateral is scaled 4× for APR, and the output .gds is scaled when streaming into Virtuoso to run Calibre LVS and DRC. Consequently, the technology LEF and macro LEF are scaled up by a factor of four, as is the technology LEF file. Since APR is performed at 4× size but Calibre parasitic extraction is at actual size, a scaled QRC technology file is used in APR. We ran APR on the EDAC design and constrained the metal layers to provide high wire density. The QRC-based SPEF was then compared to the SPEF obtained from Calibre PEX after importing the design into Virtuoso. The QRC technology file was then iteratively "dialed in" to match Calibre. We had expected this to be a straightforward (linear) scaling, but it was not, presumably due to capacitance non-linearity from fringing fields. Figure 1.16 shows that good resistance and capacitance correlation was obtained, with nearly 98% total net capacitance correlation and better than 99% total net resistance correlation.

The technology LEF has fixed via definitions and via generation rule definitions for all layers. As mentioned, they are assumed SAV, allowing upper metal width vias. High-quality (low resistance) multi-cut via generation for wide power stripes and rings, essential to low resistance power delivery, are also provided for one set of metal widths. User modification for different sizes should be straightforward, following the examples in the provided technology LEF file.

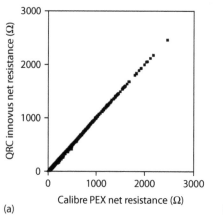
(a)

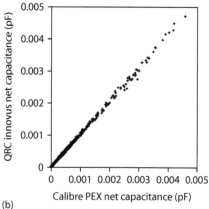
(b)

FIGURE 1.16
Correlation of Calibre and scaled QRC techFile resistance (a) and capacitance (b).

1.7.3 Design Experiments and Results

We performed numerous APR experiments in iterative DTCO on the library, to develop the technology LEF to drive routing and via generation in the Innovus APR tool, and to evaluate the library richness and performance. Three basic designs were used. The first is a small L2 cache error detection and correction (EDAC) block, which generates single error correct, double error correct (SEC-DEC) Hamming codes in the input and output pipelines. In the output direction, the syndrome is generated and in the event of incorrect data, the output is corrected by decoding the syndrome. This design has also been used in class laboratory exercises to use the APR (Clark et al., 2017). It has about 2k gates. The second design is a triple modular redundant (TMR) fully pipelined advanced encryption standard (AES), intended as a soft-error mitigation test vehicle (Chellappa et al., 2015; Ramamurthy et al., 2015). This is a large design, with about 350k gates in most iterations. Finally, we used an MIPS M14k, with the Verilog adapted to SRAMs designed on the ASAP7 PDK (Vashishtha et al., 2017c) to test the integration of SRAMs and their collateral (.lib and LEF). This design requires about 50k gates.

Versions of the designs are shown post-APR in Figure 1.17. In general, there are less than about 20 DRCs after importing these designs back into Virtuoso and running Calibre. Most, if not all of the DRCs are found by Innovus. Because they were the first arrays fully designed, 8kB SRAM arrays are used for the M14k instruction cache (left) and data cache

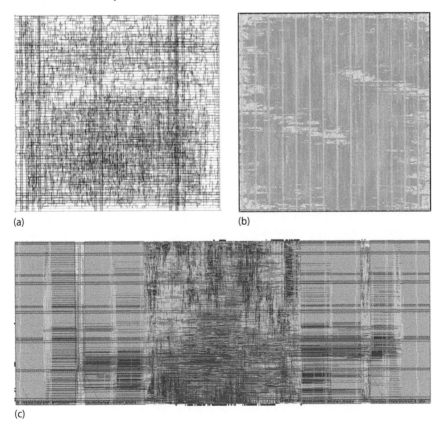

(a)   (b)

(c)

FIGURE 1.17
APR layouts of finished APR blocks. (a) L2 cache EDAC, which is approximately $22 \times 20\,\mu m$. (b) The TMR AES engine, which is $215 \times 215\,\mu m$. (c) The MIPS processor, which is $208 \times 80\,\mu m$ in this version.

(right) tag and data arrays. Ideally, these would be smaller, but following field-programmable gate array (FPGA) convention, some address inputs and many storage locations are unused in this example. The EDAC (Figure 1.17a) allows fast debug of APR problems, running in minutes. This design can route in three metal layers (Clark et al., 2017) but the figure includes seven. This design can reach 6 GHz at the TT process corner (25°C) with extensive use of the SLVT cells.

The AES (Figure 1.17b) has 14 pipeline stages with full loop unrolling for both key and data encryption/decryption and requires at least four routing layers, i.e., M2-M5. The design shown uses seven. It has 1596 input and output pins and contains three independent clock domains to support the TMR. TMR storage increases the number of flip-flops to over 15k. This design thus has large clock trees to ensure adequate library support of clock tree synthesis.

The MIPS M14k processor comprises Figure 1.17c and includes large SRAM arrays. As mentioned, the code was adapted from that provided for FPGA implementation by changing the cache arrays from block random access memory (BRAM) to SRAM arrays designed for the ASAP7 PDK. The translation lookaside buffers (TLBs) and register file are synthesized, occupying about one-quarter of the standard cell area. The 8kB SRAM arrays are apparent in the figure. They are too large for the design, which only needs 1kB tag arrays and 2kB data memory arrays for each cache, but confirmed the liberty and LEF files, as well as the APR routing over the arrays to the pins located in the sense/IO circuits at the center. The control logic, between the left and right storage arrays and the decoders, is laid out using Innovus APR, as are the pre-decoders.

1.8 SRAM Design

SRAMs are essential circuit components in modern digital ICs. Due to their ubiquity, foundries provide special array rules that allow smaller geometries than in random logic for SRAM cells to minimize their area. SRAM addressability makes them ideal vehicles for defect analysis to improve yield in early production. Moreover, running early production validates the issues arising from the tighter DRs. One focus of this section is how SRAMs affected the DTCO analysis for the ASAP7 DRs. They turn out to be more limiting, and thus more important than the previously discussed cell library constructs.

Due to the use of the smallest geometries possible, SRAMs are especially prone to random microscopic variations that affect SRAM cell leakage, static noise margin (SNM), read current (speed), and write margin. Consequently, they also provide a place to discuss the PDK use in statistical analysis, as well as our assumptions at N7. Historically, the 6-T SRAM transistor drive ratios required to ensure cell write-ability and read stability have been provided by very careful sizing of the constituent transistors, i.e., the pull-down is largest, providing a favorable ratio with the access transistor to provide read stability, while the pull-up PMOS transistor is smallest, so that the access transistor can overpower it when writing. Improving PMOS vs. NMOS strain has led to near identical, or in some literature greater PMOS drive strengths, which in combination with discrete finFET sizing, requires read- or write-assist techniques for a robust design. The yield limiting cases occur for cells that are far out on the tail of the statistical distribution, due to the large number of SRAM cells used in a modern device. Consequently, statistical analysis is required.

1.8.1 FinFET Implications and Fin Patterning

Besides the constraint on transistor width to discrete fin count, MP techniques, e.g., SADP or SAQP, further complicate the allowed cell geometries. On finFET processes, SRAM cells are divided into classes based on the ratios of the pull-up, pass-gate (access), and pull-down ratios, represented by PU, PG, and PD fin counts. The different-sized cells that we used for DTCO in the PDK comprise Figure 1.18. For instance, the smallest cell is 111 (Figure 1.18a) and has nominally equal drive strength for each device. The cell that most easily meets the read stability and write-ability requirements previously outlined is thus the 123 cell whose layout is illustrated in Figure 1.18d. As in the standard cells, there must be adequate spacing between the NMOS and PMOS devices for well boundaries, as well as active region separation. Thus, at least one fin spacing is lost between adjacent NMOS transistors in separate cells and between the NMOS and PMOS devices. The 112 and 122 cells comprise Figure 18b and c, respectively

The SRAM DR active mask spacing, which is optimistically set at a single fin, is evident in Figure 1.19a and c. The single fin spacing allows the SAQP fin patterning to be uniform across the ASAP7 die. However, the fin patterning can be changed (and often is on foundry processes) by adjusting the mandrels. The EUV assumption drives some of the metal patterning. Referring to Figure 1.19b, note that the M1 is not 1-D, in that the cell V_{DD} connection zigzags through the array on M1. This M1 is redundant with the straight M2 route. This also provides full M1 landing for the via 1 (V1) connection in each cell. The zigzag is forced by the MEEF constraints described earlier (the M1 lines vertical in the figure that connect the V_{SS} and BL to the MOL). Since the MOL provides a great deal of the connections, M1 BL designs are also possible and given the slow actual roll out of EUV, are probably dominant at the 10 nm and possibly the 7 nm nodes. However, they are not compatible with the horizontal M2 direction previously outlined for ASAP7. Consequently, we focus here on M2 BL designs, which maintain the same metal directions that are used in the standard cell areas across the SRAM arrays. The cell gate layers through metallization are illustrated in Figure 1.19d, which emphasizes the high aspect ratios of modern metallization. The fins are omitted from the figure.

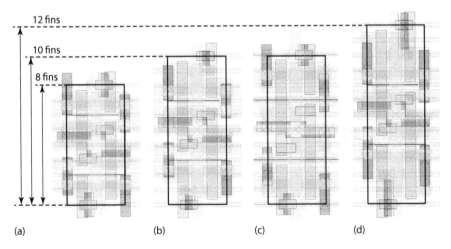

FIGURE 1.18
Layouts for 111 (a), 112 (b), 122 (c), and 123 (d) ASAP7 SRAM cells.

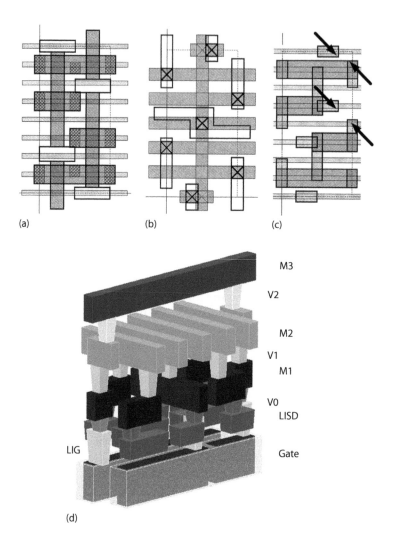

FIGURE 1.19
ASAP7 111 SRAM cell layout showing fin, active, SDT, gate, and gate cut layers (a); M1 and M2 as well as via 1 (b); and active and MOL layers (c). A 3-D cell view with the correct metal aspect ratios (d). The gate spacer is shown as transparent, but fins are not shown.

1.8.2 Statistical Analysis

AVT has long been used as a measure of local mismatch. AVT is a transistor channel area normalized $\sigma\Delta V_T$, where $\sigma\Delta V_T$ is the V_T difference variance as measured between nearby, identical transistors (Pelgrom et al., 1989; Kuhn et al., 2011). Thus,

$$\text{AVT} = \sigma\Delta V_T \times \sqrt{A} \qquad (1.9)$$

$$\sigma\Delta V_T = \sigma\left(\left|V_{T1} - V_{T2}\right|\right). \qquad (1.10)$$

In non-fully depleted devices, the dopant atoms are proportional to the area but vary statistically as random dopant fluctuations that in aggregate affect the transistor V_{th}, and dominate the mismatch. Other parameters affect the matching, but can often be driven out by improvements in the manufacturing processes and design. SRAM memories use the smallest devices that can be fabricated (in this case a single fin). Consequently, SRAMs are strongly affected and statistical yield analysis is critical to the overall design (Liu et al., 2009). As a result, SRAM yield falls off rapidly as V_{DD} is lowered toward V_{DDmin}, the minimum yielding SRAM V_{DD}. This is due to increasing variability in the drive ratios as the transistor gate overdrive $V_{GS} - V_{th}$ diminishes, due to V_{th} variations. In metal gate and finFET devices, the variability is primarily due to fin roughness and metal gate grain size variations (Liu et al., 2009; Matsukawa et al., 2009). Nonetheless, due to its ease of measurement and historical significance, AVT is still used to characterize mismatch (Kuhn et al., 2011). FinFET devices, while having different sources of variability since they are fully depleted, continue to use the AVT rubric for analysis. In general, AVT tracks the inversion layer thickness, including the effective oxide thickness. It improved by nearly a half with the advent of high-K metal gate processes, which eliminated the poly depletion effect, and again with finFETs, as they are fully depleted (Kuhn et al., 2011). The primary sources of variability in finFETs turn out to be metal gate grain size and distribution, followed by fin roughness and oxide thickness and work function variations. We use AVT = 1.1 mV.µm, which we chose as a compromise between the NMOS and PMOS values published for a finFET 14-nm process (Giles et al., 2015). The literature has reported pretty steady values across key device transistor fabrication technologies, i.e., poly gate, metal gate, and finFET. Of course, the overall variability increases in our 7 nm SRAM transistors vs. the 14 nm devices, since the channel area is less.

For sense amplifier analysis, we use a conventional Monte Carlo (MC) approach, applying different differentials to a fixed variability case to find the input-referred offset, then applying the next fixed variability, to reach the desired statistics. To determine the offset mean and sigma, 1k MC points are used. For SRAM cell analysis, we use a combination of MC and stratified sampling (Clark et al., 2013). In general, unless huge numbers of simulations are used, MC is not adequate for estimations at the tails of the distributions. It is, however, adequate to estimate the standard deviation σ and mean µ of the distributions. We confirm the needed σ/µ using the stratified sampling approach, which applies all of the statistical variability at each circuit (sigma) strata, to all the possible combinations of the transistors within that strata, using a full factorial. Thus, 728 simulations are performed, one for each possible 1, 0, or –1 impact on each combination of devices at each strata, attempting to find failures in that circuit strata (at that circuit sigma). As an example, if the circuit sigma is distributed evenly across six devices, then there is minimal variation in the circuit. However, if all of the circuit-level variability is applied to just the access and pull-down transistors, and in opposite fashion, then the read stability may be jeopardized. In this manner, the technique evaluates the tail of the circuit distribution directly and efficiently. At each strata, the transistor variation applied changes with the number of non-zero variations, following the circuit σ_{DEVICE} as

$$\sigma_{DEVICE} = \frac{\sigma_{CELL}}{\sqrt{\sum a_i^2}}. \tag{1.11}$$

The technique has been successful at predicting the SRAM yield for foundry processes (Clark et al., 2013).

1.8.3 SRAM Cell Design and DTCO Considerations

1.8.3.1 MOL Patterning

Referring to Figure 1.19c, the key DTCO limitation is the corner of LISD and SDT to LIG spacing, as indicated by the arrows at the upper right. EUV single patterning would result in some rounding, which helps by reducing the peak electric field and increasing the spacing slightly. As before, we use conservative assumptions. Another result of maintaining adequate spacing for TDDB with worst-case misalignment of the LISD/SDT and LIG layers is that the SDT does not fully cover the active areas, evident at the top and bottom of the NMOS stacks (Figure 1.19a). This led us to separate the LISD- and LIG-drawn layers. We believe that given the late introduction of EUV, it will most likely be used for MOL layer patterning, as well as vias and cuts. The former is due to projections that MOL layers may require as many as five block masks for SRAM at N7, which makes the expense of EUV more favorable (Sakhare et al., 2015). Foundry presentations have also shown MOL experimental EUV results.

1.8.3.2 1-D Cell Metallization

The 111 cell (Figure 1.19) has 1-D M2 and M3, potentially making it very amenable with a 1-D cell library, and as mentioned, M1 BL designs are compatible with horizontal M1 1-D cell library architectures. The 2-D cell layouts of the 122 cell are shown in Figure 1.20a–d. The similarity to the overall architecture of the 111 cell is apparent. The same MOL limitations exist, but the two fin NMOS devices make the overall source/drain connections to

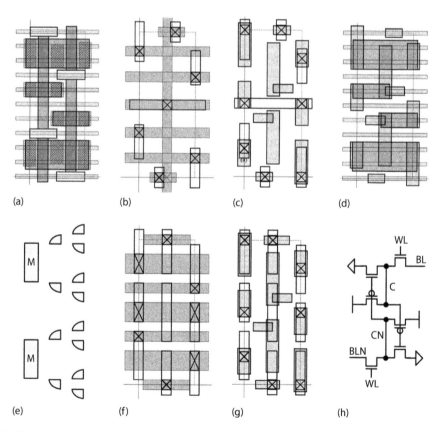

(a)   (b)   (c)   (d)

(e)   (f)   (g)   (h)

FIGURE 1.20
ASAP7 122 SRAM cell layout showing fin, active, SDT, gate, gate cut, active, and SDT layers (a); M3, V2, M2, V1, and M1 layers (b); M1, V0, and MOL (c); and fin, active, and MOL (d). The SAQP mandrel and spacers produced for SID M2 (e) and the resulting M2, V1, and M1 layers (f). M1 through MOL (g) and the schematic (h).

the NMOS stacks considerably better, as they have more coverage and lower resistance at worst-case layer misalignments. The EUV M2 assumption allows narrow metals with wide spacing. This in turn reduces line-to-line coupling and overall capacitance. Other widths are possible.

Figure 1.20f and g show 1-D metal impact on the cell design at the M2 through MOL layers. Here, M1 and M2 are produced by 1-D stripes using SADP or SAQP, respectively, and then cut to produce separate metal segments. The cuts are assumed the same as SAV, i.e., 16nm. Cuts are aligned to each other where possible. Wider M2 is required given the SID assumptions. The SAQP mandrel and the first- and second-level spacers to produce this layout are shown in Figure 1.20e. The spacers walk across the cells, but the wide, thin, wide, thin repetitive pattern is easily produced. The lines/cuts 1-D metal approach requires a dummy M1 in the middle of the cell, evident in Figure 1.20f and g. The SAV V1 are wider, but still aligned by the M2 HM, following the convention used for the APR power routes above. Note that these vias are un-landed on M1. Figure 1.20h is provided as an aid to the transistor layout pattern for readers who are not readily familiar with these modern, standard SRAM layouts.

1.8.3.3 Stability and Yield Analysis

Read mode SNM analysis following Seevinck et al. (1978) at the typical process corner is remarkably similar for the four cells (Vashishtha et al., 2017c). The 112 cell has the best SNM due to the 1:2 PG to PD ratio, followed by the 123 cell with its 2:3 ratio. At TT, the 122 cell has the lowest value, but it is very close to the 111 cell. Referring to Table 1.5, the read SNM σ is inversely proportional to the number of fins as expected. The read SNM overall quality can be ranked by the standard μ/σ of each, since it gives an indication of the sigma level at which the margin vanishes. Using this metric, the largest (123) cell is the best, as expected. The 111 cell is the worst, also as expected, entirely due to greater variability. The 112 and 122 cells are nearly even, with the latter making up for a lower baseline SNM with lower variability. The lithography for the 122 cell is easier, and as shown in Table 1.5, it has better write margins, so it becomes the preferred cell. With the possible exception of the 111 cell, all cells have good read margins.

TABLE 1.5

Mean, Variation, and Mean/Sigma of Read SNM, Hold SNM, and Write Margin Simulations for 111, 112, 122, and 123 SRAM Cells at Nominal $V_{DD} = 0.5\,V$ (Hold at $V_{DD} = 350\,mV$) and 25°C

Margin Type	Quantity	Cell Type			
		111	112	122	123
Read SNM	μ	105.2	116	100.4	106.6
	σ	11.2	9.48	8.3	7.5
	μ/σ	9.4	12.2	12.1	14.2
Hold SNM	μ	119.0	126.1	126.1	128.8
	σ	8.2	6.2	6.2	5.4
	μ/σ	14.4	20.3	20.3	23.9
Write Margin	μ	68.1	69.3	103.6	98.4
	σ	20.9	23.8	18.3	15.8
	μ/σ	3.3	2.9	5.7	6.2

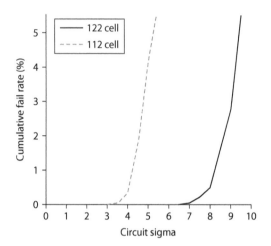

FIGURE 1.21
Comparing the failure rates of the 112 and 122 SRAM cell write margins using the stratified sampling approach.

In the arrays presented here, a CMOS Y-multiplexer allows a single tristate write circuit per column and eliminates the need for separate write enables (Vashishtha et al., 2017c). Write margin is essentially the BL voltage at which the cell writes under the worst-case voltage and variability conditions. For this analysis, we use an approach that includes the write driver and Y-multiplexer variability, since the write signal passes through it. The analysis thus includes not only the write-ability of the cell, but also the ability of the column circuits to drive the BL low. MC-generated V_{th} offsets are applied to the cell, Y-mux, and write driver transistors. The write driver input swings from V_{SS} to V_{DD}, sweeping the driver output voltage low. We tie the gate of the opposing PMOS transistor in the SRAM cell low, so that the circuit ratio varies only with the write driver voltage. The margin is the BL difference between the point at which the far SRAM node (CN) rises past $V_{DD}/2$ and the lowest BL voltage that can be driven. Table 1.5 shows that the 122 cell has good margin. The high sigma margins are confirmed by the stratified sampling approach as shown in Figure 1.21. The first fails occur at the sigma predicted in Table 1.5, after 2.9 μ/σ for the 112 cell, and after 5.7 (with some added margin) for the 122 cell. While five sigma margins have been suggested in academic papers (Qazi et al., 2011) and this value has been used in commercial designs at the column group level, we consider it inadequate at the cell level since a column group can have as many as 8 × 256 cells. The conclusion is that the preferred 122 cell, which has good density and read stability, would benefit from write assist.

The hold margin controls V_{DDmin}, as well as limiting write assist using reduced V_{DD} or raised V_{SS}. Our target V_{DDmin} is 0.5 V, so we use a V_{DDcol} of 350 mV as the hold SNM evaluation point to provide guard band. The results are also shown in Table 1.5. A low SRAM transistor I_{off} allows a good I_{on} to I_{off} ratio at low voltage, providing large hold SNM at 350 mV V_{DD} μ/σ.

1.8.4 Array Organization and Column Design

Further analysis comprehends more than one SRAM cell, so we proceed by describing the column group, which is the unit that contains the write, sense, and column multiplexing circuits, as well as multiple columns of SRAM above and below it. Typical choices are four or eight columns of SRAM per sense amplifier, so we (somewhat arbitrarily) chose four.

The basic circuit is shown in Figure 1.22, including the "DEC" style sense amplifier. This sense circuit is chosen since it is naturally isolated from the BLs during writes and one of the author's prior experience has shown it to have similar mismatch, but simpler timing than the sense amplifier using just cross-coupled inverters. Note that changing the array size merely requires changing the height of the SRAM columns and the number of column groups included in the array. As shown in Figure 1.22, the differential sense amplifier drives a simple set-reset (SR) latch to provide a pseudo-static output from the array. The SR latch has a differential input, so it will be well behaved, and does not need a delayed clock that a conventional D-latch would in this function. There is thus no race condition on the sense precharge.

The sense amplifier input referred offset voltage was determined by SPICE MC simulation. As noted previously, the simulations apply between −100 and 100 mV to V(SA) − V(SAN) in

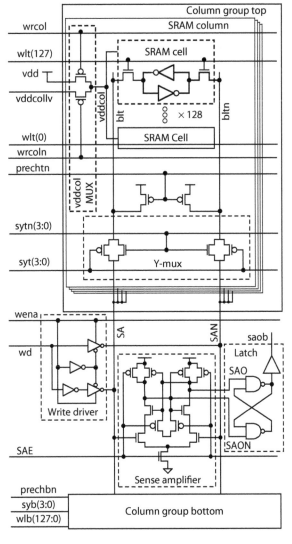

FIGURE 1.22
Column group circuits. Four SRAM columns are attached at the top and bottom of the sense and I/O circuits, sharing a common CMOS pass-gate Y-mux for reads and writes.

1 mV increments for each random (MC chosen) amplifier transistor mismatch selection. The point at which the amplifier output changes direction is the input-referred offset for that mismatch selection. The results produce a mean offset less than 1 mV, which is near the expectation of zero, also indicating no serious systematic offsets due to the layout. The distribution is Gaussian. The input-referred offset standard deviation is 16.5 mV. Using the aforementioned five sigma offset for the sense, 82.6 mV of voltage difference is required at the sense nodes for correct operation. We guard band this up to 100 mV since extra signal is required to provide adequate speed—the sense operation with no residual offset starts the circuit in the metastable state and can produce significant output delays.

1.8.5 Write Assist

The limitations of transistor width and length possible given the MP and other limitations essentially require the need for read- or write-assist techniques in the ASAP7 SRAM memory designs. Thus, circuit design techniques provide the needed statistical yield margins instead of the cell geometries (Chandra et al., 2010). While many assist approaches have been published, the impact on the overall SRAM size, as well as the variability of the assist techniques themselves, drove the choice for our ASAP7 SRAM arrays. We evaluated two approaches. First, lowering the V_{DD} of the column being written. Previous designs have used timed pull-downs to V_{SS}. In the SRAMs here, we use charge sharing V_{DDcol} voltage generation as in Chandra et al. (2010). The resulting column V_{DD} voltage is capacitively matched to the SRAM columns and thus tracks corners well. However, this scheme adds eight poly tracks to the top and bottom of the column sense/write circuits.

This area is recovered by using a negative BL write assist that provides the write driver with a low supply of less than V_{SS}. The same charge sharing circuit is used, with minor polarity changes, to drive a negative BL. We see excellent margin improvement with −150 mV on the low-driven BL. The CMOS BL multiplexer is unchanged and passes the negative voltage well. Leakage increases at lower voltages. For either scheme, sufficient capacitance is provided by eight SRAM width columns, which are integrated into the dummy cells at the left and right of each array. The circuit occupies the same columns in the read/write/IO circuit height. The reader is referred to Vashishtha et al. (2017c) for details.

1.9 Chapter Summary

This chapter described the ASAP7 PDK and its development. It discussed the electrical performance characteristics of the ASAP7 device. The chapter also provided a basic overview of lithography variability concerns and double patterning approaches, and how they compare with each other. Furthermore, it covered the EUV lithographic basics and challenges associated with it. PDK details, such as patterning choices for the various layers, cell library architecture, and DTCO considerations for developing the architecture were also discussed. The chapter concluded with APR experiments and SRAM designs based on the PDK, which demonstrated the PDKs suitability for research into various VLSI circuit and system design–related aspects.

The ASAP7 PDK has been deployed in graduate-level VLSI courses at the Arizona State University since 2015. The PDK is available for free to universities. It has received attention from research groups and faculty at a number of other universities. This indicates a high likelihood of its adoption in the classroom. Thus, we hope that the ASAP7 PDK will fulfill its development intent of enabling sub-10 nm CMOS research in academia at a much larger scale and beyond a few university research groups with access to advanced foundry PDKs. We intend to continue our work for other process nodes and ASAP5 PDK development for the 5 nm technology node is currently underway.

References

Aitken, R., et al., Physical design and FinFETs, *Proc. ISPD*, pp. 65–68, 2014.

Arnold, W., Toward 3 nm overlay and critical dimension uniformity: An integrated error budget for double patterning lithography, *Proc. SPIE*, vol. 6924, pp. 692404-1-9, 2008.

ASML, TWINSCAN NXT:1970Ci, https://www.asml.com/asml/show.do?lang=EN&ctx=46772& dfp_product_id=8036, 2015.

ASML, TWINSCAN NXT:1980Di, https://www.asml.com/products/systems/twinscan-nxt/en/ s46772?dfp_product_id=10567, 2016.

ASML, TWINSCAN NXE:3400B, https://www.asml.com/products/systems/twinscan-nxe/twin-scan-nxe3400b/en/s46772?dfp_product_id=10850, 2017.

Au, Y., et al., Selective chemical vapor deposition of manganese self-aligned capping layer for Cu interconnections in microelectronics, *J. Electrochem. Soc.*, vol. 157, no. 6, p. D341–345, 2010.

Baklanov, M., P. S. Ho, and E. Zschech, eds, *Advanced Interconnects for ULSI Technology*, Wiley, 2012.

Bhanushali, K., et al., FreePDK15: An open-source predictive process design kit for 15nm FinFET technology, *Proc. ISPD*, pp. 165–170, 2015.

Brain, R., et al., Low-k interconnect stack with a novel self-aligned via patterning process for 32 nm high volume manufacturing, *Proc. IITC*, pp. 249–251, 2009.

Chandra, V., et al., On the efficacy of write-assist techniques in low voltage nanoscale SRAMs, *Proc. DATE*, vol. 1, pp. 345–350, 2010.

Chava, B., et al., Standard cell design in N7: EUV vs. immersion, *Proc. SPIE*, vol. 9427, 94270E-1-9, 2015.

Chellappa, S., et al., Advanced encryption system with dynamic pipeline reconfiguration for minimum energy operation, *Proc. ISQED*, pp. 201–206, 2015.

Chiou, T., et al., Lithographic challenges and their solutions for critical layers in sub-14 nm node logic devices, *Proc. SPIE*, vol. 8683, pp. 86830R-1-15, 2013.

Clark, L. T., et al., SRAM cell optimization for low AVT transistors, *Proc. ISLPED*, pp. 57–63, 2013.

Clark, L. T., et al., ASAP7: A 7-nm finFET predictive process design kit, *Microelectron. J.*, vol. 53, pp. 105–115, 2016.

Clark, L. T., et al., Design flows and collateral for the ASAP7 7 nm FinFET predictive process design kit, *Proc. MSE*, pp. 1–4, 2017.

DeGraff, R., et al., NXT:1980Di immersion scanner for 7 nm and 5 nm production nodes, *Proc. SPIE*, vol. 9780, pp. 978011-1-9, 2016.

Demuynck, S., et al., Contact module at dense gate pitch technology challenges, *Proc. IITC*, pp. 307–310, 2014.

Dicker, G., et al., Getting ready for EUV in HVM, *Proc. SPIE*, vol. 9661, pp. 96610F-1-7, 2015.

Giles, M. D., et al., High sigma measurement of random threshold voltage variation in 14 nm logic FinFET technology, *IEEE Symp. VLSI Tech.*, pp. 150–151, 2015.

Goldman, R., et al., 32/28 nm educational design kit: Capabilities, deployment and future, *Proc. PrimeAsia*, pp. 284–288, 2013.

Ha, D., et al., Highly manufacturable 7 nm FinFET technology featuring EUV lithography for low power and high performance applications, *Proc. VLSIT*, pp. T68–T69, 2017.

Hody, H., et al., Gate double patterning strategies for 10-nm node FinFET devices, *Proc. SPIE*, vol. 9054, pp. 905407-1-7, 2015.

Im, S., et al., Scaling analysis of multilevel interconnect temperatures for high-performance ICs, *IEEE Trans. Electron. Devices*, vol. 52, no. 12, pp. 2710–2719, 2005.

Ito, T., and S. Okazaki, Pushing the limits of lithography, *Nature*, vol. 406, no. 6799, pp. 1027–1031, 2000.

ITRS. http://www.itrs2.net/, 2015.

Jung, W., et al., Patterning with amorphous carbon spacer for expanding the resolution limit of current lithography tool, *Proc. SPIE*, vol. 6520, no. 2007, pp. 1–9, 2007.

Kerkhof, M., et al., Enabling sub-10 nm node lithography: Presenting the NXE:3400B EUV scanner with improved overlay, imaging, and throughput, *Proc. SPIE*, vol. 10143, pp. 101430D-1-14, 2017.

Kim, S.-S., et al., Progress in EUV lithography toward manufacturing, *Proc. SPIE*, vol. 10143, pp. 1014306-1-10, 2017.

Kuhn, J., et al., Process technology variation, *IEEE Trans. Elec. Devices*, vol. 58, no. 8, pp. 2197–2208, 2011.

Liebmann, L., et al., Demonstrating production quality multiple exposure patterning aware routing for the 10 nm node, *Proc. SPIE*, vol. 9053, pp. 905309–1-10, 2014a.

Liebmann, L., et al., Design and technology co-optimization near single-digit nodes, *Proc. ICCAD*, pp. 582–585, 2014b.

Liebmann, L., et al., The daunting complexity of scaling to 7NM without EUV: Pushing DTCO to the extreme, *Proc. SPIE*, vol. 9427, pp. 942701-1–12, 2015.

Lin, C., et al., High performance 14 nm SOI FinFET CMOS technology with 0.0174 μm2 embedded DRAM and 15 levels of Cu metallization, *IEDM*, pp. 74–76, 2014.

Lin, B. J., Optical lithography with and without NGL for single-digit nanometer nodes, *Proc. SPIE*, vol. 9426, pp. 942602-1-10, 2015.

Liu, Y., K. Endo, and O. Shinichi, On the gate-stack origin of threshold voltage variability in scaled FinFETs and multi-FinFETs, *IEEE Symp. VLSI Tech.*, pp. 101–102, 2009.

Ma, Y., et al., Decomposition strategies for self-aligned double patterning, *Proc. SPIE*, vol. 7641, pp. 76410T–1-13, 2010.

Ma, Y., et al., Self-aligned double patterning (SADP) compliant design flow, *Proc. SPIE*, vol. 8327, pp. 832706-1-13, 2012.

Mallik, A., et al., The economic impact of EUV lithography on critical process modules, *Proc. SPIE*, vol. 9048, pp. 90481R-1-12, 2014.

Mallik, A., et al., Maintaining Moore's law: Enabling cost-friendly dimensional scaling, *Proc. SPIE*, vol. 9422, pp. 94221N–94221N-12, 2015.

Martins, M., et al., Open cell library in 15 nm FreePDK technology, *Proc. ISPD*, pp. 171–178, 2015.

Matsukawa, T., et al., Comprehensive analysis of variability sources of FinFET characteristics, *IEEE Symp. VLSI Tech.*, pp. 159–160, 2009.

Mulkens, J., et al., Overlay and edge placement control strategies for the 7 nm node using EUV and ArF lithography, *Proc. SPIE*, vol. 9422, pp. 94221Q-1-13, 2015.

Neumann, J. T., et al., Imaging performance of EUV lithography optics configuration for sub-9 nm resolution, *Proc. SPIE*, vol. 9422, pp. 94221H–1-9, 2015.

Oyama, K., et al., CD error budget analysis for self-aligned multiple patterning, *Proc. SPIE*, vol. 8325, pp. 832517-1-8, 2012.

Oyama, K., et al., Sustainability and applicability of spacer-related patterning towards 7nm node, *Proc. SPIE*, vol. 9425, 942514-1-10, 2015.

Paydavosi, N., et al., BSIM—SPICE models enable FinFET and UTB IC designs, *IEEE Access*, vol. 1, pp. 201–215, 2013.

Pelgrom, M. J. M., et al., Matching properties of MOS transistors, *IEEE J. Solid-state Circ.*, vol. 24, no. 5, pp. 1433–1439, Oct. 1989.

Pikus, F. G., Decomposition technologies for advanced nodes, *Proc. ISQED*, pp. 284–288, 2016.

Pyzyna, A., R. Bruce, M. Lofaro, H. Tsai, C. Witt, L. Gignac, M. Brink, and M. Guillorn, Resistivity of copper interconnects beyond the 7 nm node, *Symp. VLSI Circuits*, vol. 1, no. 1, pp. 120–121, 2015.

Qazi, M., K. Stawiasz, L. Chang and A. P. Chandrakasan, A 512kb 8T SRAM macro operating down to 0.57 V with an AC-coupled sense amplifier and embedded data-retention-voltage sensor in 45 nm SOI CMOS, *IEEE J. Solid-State Circuits*, vol. 46, no. 1, pp. 85–96, 2011.

Ramamurthy, C., et al., High performance low power pulse-clocked TMR circuits for soft-error hardness, *IEEE Trans. Nucl. Sci.*, vol. 6, pp. 3040–3048, 2015.

Ryckaert, J., et al., Design technology co-optimization for N10, *CICC*, pp. 1–8, 2014.

Sakhare, S., et al., Layout optimization and trade-off between 193i and EUV-based patterning for SRAM cells to improve performance and process variability at 7 nm technology node, *Proc. SPIE*, vol. 9427, p. 94270O-1-10, 2015.

Schuegraf, K., et al., Semiconductor logic technology innovation to achieve sub-10 nm manufacturing, *IEEE J. Electron. Dev. Soc.*, vol. 1, no. 3, pp. 66–75, 2013.

Seevinck, E., et al., Static noise margin analysis of MOS SRAM cells, *IEEE J. Solid-State Circuits*, vol. SC-22, no. 5, pp. 748–754, 1978.

Seo, S., et al., A 10 nm platform technology for low power and high performance application featuring FINFET devices with multi workfunction gate stack on bulk and SOI, *Proc. VLSIT*, pp. 1–2, 2014.

Servin, I., et al., Mask contribution on CD and OVL errors budgets for double patterning lithography, *Proc. SPIE*, vol. 7470, pp. 747009-1-13, 2009.

Sherazi, S. M. Y., et al., Architectural strategies in standard-cell design for the 7 nm and beyond technology node, *Proc. SPIE*, vol. 15, no. 1, pp. 13507-1-11, 2016.

Standiford, K., and C. Bürgel, A new mask linearity specification for EUV masks based on time-dependent dielectric breakdown requirements, *Proc. SPIE*, vol. 8880, pp. 88801M-1-7, 2013.

Stine, J. E., et al., FreePDK: An open-source variation aware design kit, *Proc. MSE*, pp. 173–174, 2007.

Tallents, G., E. Wagenaars and G. Pert, Optical lithography: Lithography at EUV wavelengths, *Nat. Photonics*, vol. 4, no. 12, pp. 809–811, 2010.

Vaidyanathan, K., et al., Design implications of extremely restricted patterning, *J. Micro/Nanolith. MEMS MOEMS*, vol. 13, pp. 031309-1-13, 2014.

Vandeweyer, T., et al., Immersion lithography and double patterning in advanced microelectronics, *Proc. SPIE*, vol. 7521, pp. 752102-1-11, 2010.

Van Schoot, J., et al., EUV lithography scanner for sub-8 nm resolution, *Proc. SPIE*, vol. 9422, pp. 94221F-1-12, 2015.

Van Setten, E., et al., Imaging performance and challenges of 10 nm and 7 nm logic nodes with 0.33 NA EUV, *Proc. SPIE*, vol. 9231, pp. 923108-1-14, 2014.

Van Setten, E., et al., Patterning options for N7 logic: Prospects and challenges for EUV, *Proc. SPIE*, vol. 9661, pp. 96610G-1-13, 2015.

Vashishtha, V., et al., Design technology co-optimization of back end of line design rules for a 7 nm predictive process design kit, *Proc. ISQED*, pp. 149–154, 2017a.

Vashishtha, V., et al., Systematic analysis of the timing and power impact of pure lines and cuts routing for multiple patterning, *Proc. SPIE*, vol. 10148, pp. 101480P-1-8, 2017b.

Vashishtha, V., et al., Robust 7-nm SRAM design on a predictive PDK, *Proc. ISCAS*, pp. 360–363, 2017c.

Vashishtha, V., et al., ASAP7 predictive design kit development and cell design technology co-optimization, *Proc. ICCAD*, pp. 992–998, 2017d.

Wong, B. P., et al., *Nano-CMOS Design for Manufacturability: Robust Circuit and Physical Design for Sub-65nm Technology Nodes*, Wiley, 2008.

Wu, S. Y. et al., A 16 nm FinFET CMOS technology for mobile SoC and computing applications, *Proc. IEDM*, pp. 224–227, 2013.

Xie, R., et al., A 7 nm FinFET technology featuring EUV patterning and dual strained high mobility channels, *Proc. IEDM*, vol. 12, pp. 2.7.1–2.7.4, 2016.

Xu, X., et al., Self-aligned double patterning aware pin access and standard cell layout co-optimization, *IEEE Trans. Comput. Des. Integr. Circuits Syst.*, vol. 34, no. 5, pp. 699–712, 2015.

Ye, W., et al., Standard cell layout regularity and pin access optimization considering middle-of-line, *Proc. GLSVLSI*, pp. 289–294, 2015.

Yeh, K., and W. Loong, Simulations of mask error enhancement factor in 193 nm immersion lithography, *Jpn. J. Appl. Phys.*, vol. 45, pp. 2481–2496, 2006.

2

When the Physical Disorder of CMOS Meets Machine Learning

Xiaolin Xu, Shuo Li, Raghavan Kumar, and Wayne Burleson

CONTENTS

While the development of semiconductor technology is advancing into the nanometer regime, one significant characteristic of today's complementary metal-oxide-semiconductor (CMOS) fabrication is the random nature of process variability, i.e., the physical disorder of CMOS transistors. It is observed that with the continuous development of semiconductor technology, the physical disorder has become an important factor in CMOS design during the last decade, and will likely continue in the forthcoming years. The low cost of modern semiconductor design and fabrication techniques benefits the ubiquitous applications, but also poses strict constraints on the area and energy of these systems. In this context, many traditional circuitry design rules should be reconsidered to tolerate the possible negative effects caused by the physical disorder of CMOS devices. As the physical deviation from nominal specifications becomes a big concern for many electronic systems [1], the performance of electronic blocks that have high requirements on the symmetric design and fabrication process is greatly impacted. For example, the resolution of time-to-digital converters (TDC) will be decreased by the physical process variations, which deviates the fabricated delay elements from the designed (nominal) delay length. On the one hand, such process variations introduce uncertainty into the standard CMOS design [2], while on the other hand, it also becomes a promising way to leverage such properties for constructive purposes. One major production of this philosophy is the development of physically unclonable functions (PUFs) in the literature, which extract secret keys from uncontrollable manufacturing variabilities on integrated circuits (ICs).

As electronic designs like PUF and TDC are impacted by the process variations of fabrication, many issues including variability, modeling attacks and noise sensitivity also need to be reconsidered and addressed. Due to the microscopic characteristic of such physical disorder, it is either infeasible or very expensive to measure and mitigate it with tractional electronic techniques. For example, any physical probing into a PUF instance would change the original physical disorder and therefore the measured results are not necessarily correct. In this context, employing the *non-invasive* way to characterize the internal physical disorder becomes a promising solution. This chapter presents some recent work on advancing this physical disorder modeling with the help of machine learning techniques. More specifically, it shows that through modeling the physical disorder, machine learning techniques can benefit the performance (i.e., reliability improvement) of PUFs by filtering out unreliable challenges and responses (challenge–response pairs [CRPs]). As for a fabricated TDC circuitry with a given internal physical disorder, it is demonstrated that a backpropagation-based machine learning framework can be utilized to mitigate the process variations and optimize the resolution.

This chapter is structured as follows. Section 2.1 briefly describes the sources of process variations, i.e., the physical disorder in modern CMOS circuits. An introduction to common PUF terminologies and performance metrics used throughout this chapter is provided in Section 2.2. Section 2.3 provides a brief overview of some of the most popular silicon-based PUF circuits along with their performance metrics. Next, the machine learning modeling method is reviewed as a threat in attacking PUFs in Section 2.4. Our discussion continues on applying machine learning techniques to help with understanding and mitigating the physical disorder of CMOS circuitry in Sections 2.5 and 2.6. More specifically, machine learning techniques can also be employed in a constructive way to improve the reliability of PUFs and the resolution of TDCs. Finally, we provide concluding remarks in Section 2.7.

2.1 Sources of CMOS Process Variations

The sources of process variations in ICs are summarized in this section. The interested reader is also referred to the book chapter by Kim et al. [3] for more details on the sources of CMOS variability. From the perspective of circuits, the sources of variations can be either *desirable* or *undesirable*. The desirable source of variations refers to process manufacturing variations (PMVs) as identified in [3]. Environmental variations and aging are undesirable for some functional circuits like PUFs.

2.1.1 Fabrication Variations

Due to the complex nature of the manufacturing process, the circuit parameters often deviate from their intended value. The various sources of variability include proximity effects, chemical-mechanical polishing (CMP), lithography system imperfections and so on. The PMVs consist of two components as identified in the literature, namely systematic and random variations [4].

2.1.1.1 Systematic Variations

The systematic component of process variations includes variations in the lithography system, the nature of the layout and CMP [4]. By performing a detailed analysis of the layout, the systematic sources of variations can be predicted in advance and accounted for in the design step. If the layout is not available for analysis, the variations can be assigned statistically [4].

2.1.1.2 Random Variations

Random variations refer to non-deterministic sources of variations. Some of the random variations include random dopant fluctuations (RDF), line edge roughness (LER) and oxide thickness variations. Random variations are often modeled using random variables for design and analysis purposes.

2.1.2 Environmental Variations and Aging

Environmental variations are detrimental to PUF circuits. Some of the common environmental sources of variations include power supply noise, temperature fluctuations and external noise. These variations must be minimized to improve the reliability of PUF circuits. Aging is a slow process and it reduces the frequency of operation of circuits by slowing them down. Circuits are also subjected to increased power consumption and functional errors due to aging [5].

2.2 Terminologies and Performance Metrics of PUF

Some terminologies and performance metrics like challenge–response pair (CPR), reliability, uniqueness and unpredictability used to evaluate PUFs are briefly summarized in the following sections.

2.2.1 Challenge–Response Pairs (CRPs)

The inputs to a PUF circuit are known as *challenges* and the outputs are referred to as *responses*. A challenge associated with its corresponding response is known as a *challenge–response pair* (CRP). In an application scenario, the responses of a PUF circuit are collected and stored in a database. This process is generally known as *enrollment*. Under a *verification* or *authentication* process, the PUF circuit is queried with a challenge from the database. The response is then compared against the one stored in the database. If the responses match, the device is authenticated.

2.2.2 Performance Metrics

Three important metrics are used to analyze a PUF circuit, namely uniqueness, reliability and unpredictability.

2.2.2.1 Uniqueness

One important application of PUF devices is to generate unique signatures for device authentication. Thus, it is desirable that any two PUF instances can be easily identified from another. Toward this end, a typical measure used to analyze uniqueness is known as *inter-distance* and is given by [3]

$$d_{\text{inter}}(C) = \frac{2}{k(k-1)} \sum_{i=1}^{i=k-1} \sum_{j=i+1}^{j=k} \frac{\text{HD}(R_i, R_j)}{m} \times 100\%. \tag{2.1}$$

In Equation 2.1, $\text{HD}(R_i, R_j)$ stands for the Hamming distance between two responses R_i and R_j of m bits long for challenge C, and k is the number of PUF instances under evaluation. For a group of PUF instances, the desired inter-distance (i.e., uniqueness) of them is 50%. By carefully looking at Equation 2.1, one can correspond the inter-distance $d_{\text{inter}}(C)$ to the mean of the Hamming distance distribution obtained over k chips for challenge C. While designing a PUF circuit, inter-distance is often measured through circuit simulations. A common practice is to perform Monte Carlo simulations over a large population of PUF instances. In simulations, care must be taken to efficiently model various sources of manufacturing variations in CMOS circuits, as they directly translate into uniqueness. During the simulations, manufacturing variations are modeled using a Gaussian distribution. In such cases, the mean and standard deviation of the Gaussian distribution under consideration must correspond to either inter-die or inter-wafer variations' statistics.

2.2.2.2 Reliability

As PUFs are built on microscopic process variations, a challenge applied to a PUF operating on an IC will not necessarily produce the same response under different operating conditions. The reliability of a PUF refers to its ability to produce the same responses under varying operating conditions. Reliability can be measured by averaging the number of flipped responses for the same challenge under different operating conditions. A common measure of reliability is *intra-distance*, given by [3,6]

$$d_{\text{intra}}(C) = \frac{1}{s} \sum_{j=1}^{s} \frac{\text{HD}(R_i, R'_{i,j})}{m} \times 100\%. \tag{2.2}$$

In Equation 2.2, R_i is the response of a PUF to challenge C under nominal conditions, s is the number of samples of response R_i obtained at different operating conditions, $R'_{i,j}$ corresponds to the jth sample of response R_i for challenge C and m is the number of bits in the response. Intra-distance is expected to be 0% for ideal PUFs, which corresponds to 100% reliability. The terms intra-distance (d_{intra}) and reliability have been used interchangeably further in this chapter. Given d_{intra}, reliability can always be computed (100 d_{intra}(%)). As an important feature of PUF performance, several contributions exist in the literature to improve the reliability of PUFs from circuit and system perspectives [7–13].

2.2.2.3 Unpredictability

For security purposes, the responses from a PUF circuit must be unpredictable. Unpredictability is a measurement that quantifies the randomness of the responses from the same PUF device. This metric can be evaluated using the National Institute of Standards and Technology (NIST) tests [6,14]. Silicon PUFs produce unique responses based on intrinsic process variations, which are very difficult to clone or duplicate by the manufacturer. However, by measuring the responses from a PUF device for a subset of challenges, it is possible to create a model that can mimic the PUF under consideration. Several modeling attacks on PUF circuits have been proposed in the literature [15]. The type of modeling attack depends on the PUF circuit. A successful modeling attack on a PUF implementation may not be effective for other PUF implementations. Modeling attacks can be made more difficult by employing some control logic surrounding the PUF block, which prevents direct read-out of its responses. One such technique is to use a secure one-way hash over PUF responses. However, if PUF responses are noisy and have significant intra-distance, this technique will require some sort of error correction on PUF responses prior to hashing [8].

2.3 CMOS PUFs

This section provides an overview of some PUF instantiations proposed in the literature. Based on the challenge–response behavior, PUFs can be classified into two categories, namely strong PUFs and weak PUFs. Strong PUFs have a complex challenge–response behavior and it is very difficult to physically clone them. They are often characterized by CRPs exponential to the number of challenge bits. So, it is not practical to read out all CRPs within a limited time using measurements. On the other hand, weak PUFs accept only a reduced set of challenges and in the extreme case, they accept just a single challenge. Such a construction requires that the response is never shared with the external world and is only used internally for security operations. We present an overview of some popular constructions of PUFs, namely arbiter PUFs in Sections 2.3.1, static random-access memory (SRAM) PUFs in Section 2.3.2 and one of its variants called a data retention voltage (DRV) PUF in Section 2.3.3.

2.3.1 Arbiter PUF

The concept of arbiter PUFs was introduced in [16,17]. An arbiter PUF architecture is shown in Figure 2.1. An n-bit arbiter PUF is composed of n stages, with each stage employing two 2:1 MUXs, as depicted in Figure 2.1. A challenge vector $\mathbf{C} = \{c_1, c_2,..., c_n\}$ is applied as the

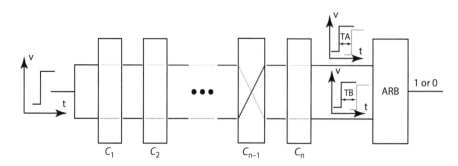

FIGURE 2.1

Schematic of an arbiter PUF. Challenge **C** controls the propagation paths of rising edges that gather the delay mismatch as they propagate toward the final arbiter.

control signals for all stages to configure two paths through the PUF toward the arbiter; at each stage, the paths are configured to be either straight or crossing. Thus, a rising edge applied at the input of the first stage gathers the delay mismatch from the paths through each stage while propagating toward the arbiter. The arbiter is a latch that digitizes the response into "1" or "0" by judging which rising edge is the first to arrive. It is important to note that the number of CRPs is exponential to the number of challenge bits. So, arbiter PUFs fall under the category of strong PUFs.

2.3.2 SRAM PUF

The concept of SRAM PUFs was first introduced in [18–20]. SRAM is a type of semiconductor memory, which is composed of CMOS transistors. SRAM is capable of storing a fixed written value "0" or "1", when the circuit is powered up. A SRAM cell capable of storing a single bit is shown in Figure 2.2. Each SRAM bit cell has two cross-coupled inverters and two access transistors. The inverters drive the nodes q and qb, as shown in Figure 2.2. When the circuit is not powered up, the nodes Q and $\bar{Q}$ are at logic low (00). When the power is applied, the nodes enter a period of metastability and settle down to either one

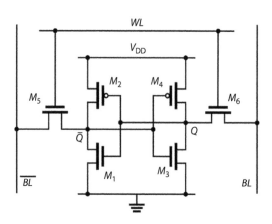

FIGURE 2.2

A SRAM cell using standard CMOS transistors. *BL* is the bit line, *WL* refers to word line, Q and $\bar{Q}$ are the state nodes for storing a single bit value. The transistors that make up inverters are shown in black and the access transistors are shown in gray.

of the states ($Q=0$, $\bar{Q}=1=1$ or $Q=1$, $\bar{Q}=0$). The final settling state is determined by the extent of the device mismatch in driving inverters and thermal noise. For PUF operation, it is desirable to have device mismatch dominant over thermal noise, as thermal noise is random in nature. The power-up states of SRAM cells were used for generating unique fingerprints in [19,20]. Experiments were conducted over 5120 64-bit SRAM cells from commercial chips and d_{inter} was found to be around 43%. Similarly, d_{intra} was found to be around 3.7%. Similar results were obtained in [18]. An extensive large-scale analysis over 96 application-specific integrated circuits (ASICs), with each ASIC having four instances of 8kB SRAM memory, was performed in [21]. The experimental results from [21] show that SRAM PUFs typically have high entropy, and the responses from different PUF instances were found to be highly uncorrelated.

2.3.3 DRV PUF

In the study of SRAM, DRV stands for the minimum supply voltage at which the written state is retained in a SRAM. In [22], DRV is proposed as the basis for a new SRAM-based PUF that is more informative than SRAM PUFs [23,24]. For example, DRV can uniquely identify circuit instances with 28% greater success than SRAM power-up states that are used in PUFs [22]. The physical characteristics responsible for DRV are imparted randomly to each cell during manufacturing, providing DRV with a natural resistance to cloning. It is shown that DRVs are not only random across chips, but also have relatively little spatial correlation within a single chip and can be treated in analysis as independent [25]. The proposed technique has the potential for wide application, as SRAM cells are among the most common building blocks of nearly all digital systems.

However, one major drawback of DRV is that it is highly sensitive to temperature, which makes it unreliable and unsuitable for use in a PUF. In [10], the idea of DRV fingerprinting is further extended to create a PUF based on DRV. Moreover, to overcome the temperature sensitivity of DRV, a DRV-based hashing scheme that is robust against temperature changes is proposed. The robustness of this hashing comes from the use of reliable DRV-ordering instead of less reliable DRV values. The use of DRV-ordering can be viewed as a differential mechanism at the logical level instead of the circuit level as in most PUFs. To help validate the DRV PUF, a machine learning technique is also proposed for simulation-free prediction of DRVs as a function of process variations and temperature. The machine learning model enables the rapid creation of large DRV data sets required for evaluating the DRV PUF approach.

2.4 Machine Learning Modeling Attacks on PUFs

In this section, we review some existing attacks on PUFs; more specifically, we focus on the machine learning–based modeling attacks. Recently, the emergence of unreliable CRP data as a tool for modeling attacks has introduced an important aspect to be addressed. Also, the vulnerabilities posed by strong PUFs toward modeling attacks will pave the way for immense competition between codemakers and codebreakers, with the hope that the process will converge on a PUF design that is highly resilient to known attacks.

2.4.1 Modeling Attacks on Arbiter PUFs

Because of the additive nature of path delay, several successful attempts have been made to attack arbiter PUFs using *modeling attacks* [15,26–28]. The basic idea is to observe a set of CRPs through physical measurements and use them to derive runtime delays using various machine learning algorithms. In [27], it was shown that arbiter PUFs composed of 64 and 128 stages can be attacked successfully using several machine learning techniques, achieving a prediction accuracy of 99.9% by observing around 18,000 and 39,200 CRPs, respectively. These attacks are possible only if the attacker can measure the response, i.e., the output of a PUF circuit is physically available through an input/output (I/O) pin in an IC. If the response of a PUF circuit is used internally for some security operations and is not available to the external world, modeling attacks cannot be carried out unless there is a mechanism to internally probe the output of the PUF circuit. However, probing itself can cause delay variations, thereby affecting the accuracy of the response measurement.

Several nonlinear versions have been proposed in the literature to improve the modeling attack resistance of arbiter PUFs. One version is feed-forward arbiter PUFs, in which some of the challenge bits are generated internally using an arbiter as a result of racing conditions at intermediate stages. However, such a construction has reliability issues because of the presence of more than one arbiter in the construction. This is evident from the test chip data provided in [16]. It was reported that feed-forward PUFs have $d_{intra} \approx 10\%$ and d_{inter} 38%. Modeling attacks have been attempted on feed-forward arbiter PUFs and the attacks used to model simple arbiter PUFs were found to be ineffective when applied to feed-forward arbiter PUFs. However, by using evolution strategies (ES), feed-forward arbiter PUFs have been shown to be vulnerable to modeling attacks [27]. Non-linearities in arbiter PUFs can also be introduced using several simple arbiter PUFs and using an XOR operation across the responses of simple arbiter PUFs to obtain the final response. This type of construction is referred to as an XOR arbiter PUF. Though XOR arbiter PUFs are tolerant to simple modeling attacks, they are vulnerable to advanced machine learning techniques. For example, a 64-stage XOR arbiter PUF with 6 XORs has been attacked using 200,000 CRPs to achieve a prediction accuracy of 99% [29–32]. All these modeling attacks demonstrate an urgent need to design a modeling attack–resistant arbiter PUF.

2.4.2 Attacks against SRAM PUFs

Since SRAM PUFs based on power-up states only have a single challenge, the response must be kept secret from the external world. Modeling attacks are not relevant for SRAM PUFs. However, other attacks such as side-channel and virus attacks can be employed, but they are not covered in this chapter.

2.5 Constructively Applying Machine Learning on PUFs

Reliability is an important feature of PUFs that reflects their ability to produce the same response for a particular challenge despite the existence of noise. So, it is possible that a PUF operating in different conditions generates a different response to the same challenge vector. The PUF output is therefore a function of not only the challenge and process variations, but also the transient environmental conditions. Therefore, to make a PUF highly reliable, unstable CRPs that are easily flipped by environmental noise and aging should be corrected or even not used. Generally, the reason that PUF responses are unreliable is because supply voltage and temperature variations can overcome the impacts of process variations and flip the responses. Besides the transient noise, device aging is also an important but rarely studied source of unreliability in PUFs. Unlike environmental noise that temporarily flips PUFs' CRPs (PUFs work more reliably when the supply voltage and temperature return to normal), device aging causes a permanent change in the behavior of a PUF. Device aging is usually caused by negative bias temperature instability, hot carrier injection, time-dependent dielectric breakdown and electro-migration [33,34]. In this section, we present some methodologies that use machine learning–related techniques to improve the reliability of PUFs.

2.5.1 Using Machine Learning to Improve the Reliability of Arbiter PUFs

2.5.1.1 Mechanism of Arbiter PUFs

In an n-bit arbiter PUF (Figure 2.1), the propagation delay from the input to the first stage to the top and bottom outputs of the ith stage can be defined as D_{top}^i and D_{bottom}^i, respectively. The delay mismatch between two delay paths is summed up as the timing difference between D_{top}^n and D_{bottom}^n (Figure 2.3). By mapping the original challenge $c_i \in \{0, 1\}$ into $c_i \in \{-1, 1\}$, the path delay can be formulated as

$$D_{top}^i = \frac{1+c_i}{2}\left(t_{top}^i + D_{top}^{i-1}\right)$$

$$+ \frac{1-c_i}{2}\left(t_{u\_across}^i + D_{bottom}^{i-1}\right)$$

$$D_{bottom}^i = \frac{1+c_i}{2}\left(t_{bottom}^i + D_{bottom}^{i-1}\right)$$

$$+ \frac{1-c_i}{2}\left(t_{d\_across}^i + D_{top}^{i-1}\right)$$

(2.3)

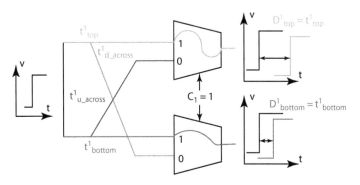

FIGURE 2.3
Propagation paths through the delay cells of an arbiter PUF.

where $D_{top}^0 = D_{bottom}^0 = 0$ and $t_{top}^i, t_{bottom}^i, t_{u\_across}^i, t_{d\_across}^i$ represent the four possible delays through the ith stage. Denoting the delay difference between the top and bottom arbiter inputs as $T_A - T_B$, following Equation 2.3, the delay difference between the two paths is $T_A - T_B = D_{top}^n - D_{bottom}^n$. The response ($r$) of an arbiter PUF of n-bit length is therefore determined by the sign of $T_A - T_B$ (Equation 2.4):

$$r = \begin{cases} 0, & \text{if} \quad \text{sgn}(T_A - T_B) > 0 \\ 1, & \text{if} \quad \text{sgn}(T_A - T_B) < 0 \end{cases} \tag{2.4}$$

From Equation 2.4, it is clear that a PUF response is flipped when the sign of $T_A - T_B$ changes, either from positive to negative or vice versa.

2.5.1.2 Modeling the $T_A - T_B$ of Arbiter PUFs

Because a PUF operating in different conditions can generate a different response to the same challenge vector, the PUF output is a function of not only the challenge and process variations, but also the transient environmental conditions. Therefore, to make a PUF highly reliable, unstable CRPs that are easily flipped by environmental noise and aging should not be used or corrected. Toward this end, a machine learning–based modeling method is proposed in [35], which helps with generating reliable CRPs without extra circuitry other than a normal PUF. Based on this technique, the unreliability source of a PUF is classified into two aspects: transient noise (e.g., temperature and supply voltage variations) and device aging. A machine learning model can be trained for PUF characterization and utilize the model for identifying and filtering out the unreliable challenge vectors for each PUF, allowing higher reliability to be achieved.

An n-bit arbiter PUF is composed of n stages, with each stage employing two 2:1 MUXs as depicted in Figures 2.1 and 2.3. A challenge vector $\mathbf{C} = \{c_1, c_2, ..., c_n\}$ is applied as the control signals for all stages to configure two paths through the PUF toward the arbiter; at each stage, the paths are configured to be either straight or crossing. To explore the impact of noise in more detail, simulations on a set of 64-bit arbiter PUFs show that the flipped responses are from challenges that correspond to $T_A - T_B$ in a small range:

$$DD_{umin} \leq T_A - T_B \leq DD_{umax} \tag{2.5}$$

where DD_{umin} and DD_{umax} represent the minimum/maximum delay difference between T_A and T_B of unreliable challenges. Based on this observation, it can be concluded that if only the challenge vectors satisfying either $T_A - T_B > DD_{umax}$ or $T_A - T_B < DD_{umin}$ are applied to the PUF, then the responses of the PUF will be reliable.

Since knowing the $T_A - T_B$ of each challenge vector makes it possible to get reliable CRPs by avoiding challenges with smaller delay difference, these challenges with smaller $T_A - T_B$ are likely to be unreliable and therefore can be discarded. However, as probing inside a PUF to measure $T_A - T_B$ is not practical, since it will introduce extra bias into the circuit operation and impact the original results, the machine learning modeling method is proposed to accomplish this job by building a PUF model. With a PUF model, the $T_A - T_B$ value for each challenge vector can be derived. Two parameters are defined as

$$\alpha_i = \left(t_{top}^i - t_{bottom}^i + t_{d\_across}^i - t_{u\_across}^i \right) / 2$$

$$\beta_i = \left(t_{top}^i - t_{bottom}^i - t_{d\_across}^i + t_{u\_across}^i \right) / 2 \tag{2.6}$$

Based on Equation 2.6, for a given challenge vector $\mathbf{C} = \{c_1, c_2,..., c_n\}$, a corresponding response generation can be modeled by accumulating the delay mismatches through delay stages as

$$T_A - T_B = \alpha_1 k_0 + \cdots + (\alpha_n + \beta_{n-1}) k_{n-1} + \beta_n k_n \tag{2.7}$$

where $k_n = 1$ and $k_i = \prod_{j=i+1}^{n} c_j$, reflecting the number of times that the rising edges will change tracks between the ith stage and the arbiter. Thus, knowing the challenge and α_i and β_i, $i \in (1,..., n)$ makes it possible to compute DD for any challenge. By denoting the delay parameters of an arbiter PUF with vector $\mathbf{p}_{model} = \{\alpha_1, \alpha_2 + \beta_1,..., \alpha_n + \beta_{n-1}, \beta_n\}$, and defining challenge features as $\mathbf{k} = \{k_0, k_1,..., k_n\}$, the model-predicted $T_A - T_B$ of each challenge vector can be denoted as

$$T_A - T_B = \langle \mathbf{p}_{model}, \mathbf{k} \rangle \tag{2.8}$$

The foregoing equations show that the machine learning modeling technique can be used to model $\mathbf{p}_{model}$ and predict $T_A - T_B$. To accomplish this, a set of known CRPs can be used to train a support vector machine (SVM) classifier. SVM models are powerful learning tools that can perform binary classification of data. Classification is achieved by a linear or nonlinear separating surface in the input space of the data set. Previously, SVMs have been widely used in attacking arbiter PUFs [16,29,36]. Note that finding the accurate delay difference is not an explicit objective of the SVM model, as the SVM model only seeks a value of $\mathbf{p}_{model}$ that can accurately predict responses. Fortunately, because the raw PUF responses are determined by the sign of $T_A - T_B$ (Equation 2.4), a model that is good at predicting unknown responses is also accurate in quantifying the values of $T_A - T_B$. Therefore, the flow of using a PUF model to enhance PUF reliability becomes: (1) train a binary classifier for a PUF and (2) use the trained PUF model to quantify the delay difference induced by each challenge. In other words, the model-quantified delay difference can be used to infer whether or not a challenge will generate a reliable PUF response. The proposed method can be evaluated by comparing the model-predicted $T_A - T_B$ with the golden value of $T_A - T_B$ that is extracted from a PUF. If the PUF model works well for response prediction, then it is expected that ρ in Equation 2.9 is close to 1 when the size of the CRP data set used to train the model is large enough.

$$\mathrm{corr}\left((T_A - T_B)_{golden}, (T_A - T_B)_{model} \right)$$
$$= \mathrm{corr}\left((T_A - T_B)_{golden}, \langle \mathbf{p}_{model}, \mathbf{k} \rangle \right) = \rho \tag{2.9}$$

The correlation coefficient (ρ) between $(T_A - T_B)_{golden}$ (from simulation) and $(T_A - T_B)_{model}$, which is calculated as $(\mathbf{p}_{model}, \mathbf{k})$, is as shown in Figure 2.4a.

As can be seen, although $(T_A - T_B)_{golden}$ and $(T_A - T_B)_{model}$ are different in scale, the correlation between them is very high. Note that it is not required to know the exact value of the delay differences to select the reliable challenges, but only the relative magnitudes of the delay differences are needed. From Figure 2.4a and b, it can be inferred that the model-predicted $T_A - T_B$ values from the model are in good agreement with that from the PUF circuit.

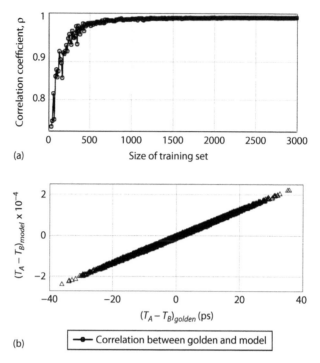

(a)

(b)

FIGURE 2.4
In Equation 2.4a, the correlation coefficient ρ between golden delay difference and model-predicted delay difference. While the PUF training size is increasing, higher ρ is achieved. In Equation 2.4b, based on the model trained with 3000 CRPs, there is good agreement between $(T_A - T_B)_{golden}$ and $(T_A - T_B)_{model}$ for 2000 random challenges. (a) Correlation coefficient increases with training size. (b) Good agreement between golden and model-predicted delay difference.

2.5.1.3 Improving PUF Reliability with PUF Model

As previously concluded, the flipped PUF responses are only those satisfying $DD_{umin} \leq (T_A - T_B)_{golden} \leq DD_{umax}$. However, due to the large size of a PUF CRP space, 2^n for an n-bit arbiter PUF, it is necessary to reconsider how to build the PUF model to predict the delay cutoffs that will rule out approximately fraction discarded ratio (dr) of the overall challenge space. For this purpose, there are still two questions that need to be answered:

Algorithm 1 Use the machine learning model to compute the range of model-predicted delay differences that are likely to be unreliable for a given PUF. Challenges predicted to have delay differences inside this range will not be applied to the PUF, and this will improve the overall PUF reliability. Reliability can be improved by using a larger value of dr to discard more challenges.
Input: A discard ratio dr and a set of challenges and corresponding responses obtained from a single PUF at nominal supply voltage and temperature.
Output: A range $[DD_{min}, DD_{max}]$ of delay differences to consider as unreliable for this PUF.

1: Let **k** be the challenges mapped to challenge features (Equations 2.7 and 2.8)

2: $\mathbf{p}_{\text{model}} \leftarrow$ **SVM**(**k**, responses) {train the PUF model}

3: $\mu_p = \text{avg}\langle \mathbf{p}_{\text{model}}, \mathbf{k} \rangle$ {mean predicted delay difference}

4: $\sigma_p^2 = \text{var}\langle \mathbf{p}_{\text{model}}, \mathbf{k} \rangle$ {variance of predicted delay difference}

5: the distribution of delay differences across all challenges is modeled to be N (μ_p, σ_p^2)

6: $\text{DD}_{\text{min}} = F^{-1}(0.5 - dr/2) = \mu_p + \sigma_p \Phi^{-1}(0.5 - dr/2)$

7: $\text{DD}_{\text{max}} = F^{-1}(0.5 + dr/2) = \mu_p + \sigma_p \Phi^{-1}(0.5 + dr/2)$ {delay difference cutoffs based on PUF model $\mathbf{p}_{\text{model}}$ and selected challenge features **k** (Equation 2.11)}

8: **return** $[\text{DD}_{\text{min}}, \text{DD}_{\text{max}}]$

1. How many CRPs are enough to train an accurate model to characterize the DD_{umax} and DD_{umin} cutoffs for each PUF?
2. For a given PUF model, what is the dr of CRPs that must be filtered to achieve an expected reliability level?

The first question can be answer by employing the techniques in Algorithm 1. The physical features of a PUF follow Gaussian distribution; therefore, the $T_A - T_B$ of each applied challenge vector will also follow a Gaussian distribution. A model can be firstly built to model this distribution, as shown in Equation 2.10. By selecting and applying the challenges randomly, the training set of PUF ensures that it covers a more unreliable range.

$$F\left(\text{DD}, \mu_p, \sigma_p\right) = \frac{1}{\sigma_p \sqrt{2\pi}} \exp^{-((\text{DD}-\mu p)^2/2\sigma_p^2)} \tag{2.10}$$

Because the unreliable responses are only related to challenges that are generating smaller $T_A - T_B$, even without knowing the exact range of the delay difference for these unreliable CRPS, it can be quantified by applying a quantile function. For example, by denoting the probit function of a standard normal distribution with $\Phi^{-1}(dr)(dr \in (0, 1))$, the range of the delay difference that should be discarded can be expressed in Equation 2.11 (as Steps 6 and 7 in Algorithm 1), where dr stands for the ratio of challenges that should be discarded from the CRP database of each PUF. Apparently, there exists a tradeoff between the value of dr and the number of usable challenges. For example, a larger dr implies that more challenges should be discarded, while a smaller dr means less challenges will be filtered out.

$$F^{-1}\left(dr\right) = \mu_p + \sigma_p \Phi^{-1}\left(dr\right) \tag{2.11}$$

By applying a challenge to the trained PUF model, if its delay difference satisfies $((T_A - T_B)_{\text{model}} \varepsilon [\text{DD}_{\text{min}}, \text{DD}_{\text{max}}])$, then it will be marked as reliable and applied on PUFs. By formulating a set of such reliable challenges and comparing the corresponding responses with the golden data set, it is found that as the training size of a PUF model increases, the characterized cutoffs DD_{min} and DD_{max} also become more accurate, as shown in Figure 2.5. As a result, fewer challenges need to be discarded to achieve the same reliability.

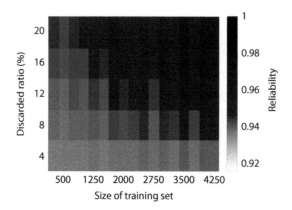

FIGURE 2.5
Validation under aging and environmental noise, across all of the simulated PUF instances. The tradeoff between training size and discarded ratio can be seen in the figure. A larger *dr* is conservative and can compensate for the lower-quality delay predictions of a model trained from a smaller training set.

2.5.2 Using Machine Learning to Model the Data Retention Voltage of SRAMs

DRV is a commonly studied feature of SRAMs in low-power research. A *simulator* program with integrated circuit emphasis (SPICE) simulator is usually employed to characterize the lowest DRV of different CMOS technology nodes. However, the time consumed by a SPICE simulator is relatively high to formulate the DRV of a SRAM cell. This is because multiple supply voltage values should be applied to find the maximum voltage that induces a data retention failure. In [10], it is reported that simulating a single test voltage on a single SRAM cell for 2 ms has a runtime of 0.17 s with an Intel Xeon E5-2690 processor running at 2.90 GHz with 64GB of RAM.

To save the DRV simulation time, an alternative method is to predict the DRV using a device model. According to [37], the DRV of a SRAM cell is determined by the environmental temperature and the process variations of the transistors. Therefore, the DRV value of a SRAM cell can be formulated as a function of physical features like temperature T and transistor width, length and threshold voltage (W, L and V_{th}, respectively). To study this relationship, Qin et al. [37] propose an analytical model as shown in Equation 2.12, where DRV_r is the DRV at room temperature, and DRV_f is defined in Equation 2.13 with ΔT representing the temperature difference from room temperature. Terms a_i, b_i and c in Equation 2.13 are fitting coefficients and their values are determined empirically for each CMOS technology process [37].

$$DRV = DRV_r + DRV_f \qquad (2.12)$$

$$DRV_f = \sum_{i=1}^{6} a_i \times \frac{\Delta\left(W_i/L_i\right)}{W_i/L_i} + \sum_{i=1}^{6} b_i \times \Delta\left(V_{thi}\right) + c \times \Delta T \qquad (2.13)$$

Although this model can be used to estimate the DRV of a SRAM cell, it has two weaknesses that create the need for a more advanced model:

1. To formulate a specified DRV value with Equation 2.13, firstly, the user needs to know the DRV_r for each SRAM cell. However, this physical feature cannot be

expressed as a function of transistor parameters and can only be characterized through hardware measurement or computationally expensive circuit simulation.

2. It is impractical to apply the same coefficients a_i, b_i and c to different SRAM cells. In reality, the DRV of different cells increases according to different coefficients depending on their unique process variations. This distinction is especially important in building the model.

2.5.2.1 Predicting DRV using Artificial Neural Networks

To address the two aforementioned weaknesses, Xu et al. propose to use machine learning for DRV prediction [10]. The proposed technique can predict the DRV value for a SRAM cell by feeding the physical parameters into a machine learning model. The basis of this technique is that for a certain CMOS technology node, the values of process variations only vary over a bounded range; therefore, the DRV values also fall into a bounded range $[\text{DRV}_{min}, \text{DRV}_{max}]$. Based on the modeling technique, the range of DRVs is firstly divided into K labels with each standing for a smaller range with size DRV (Equation 2.14). By training a machine learning model that maps the physical features into these K DRV classes, the DRV value for unknown SRAM cells can be predicted with the model, as shown in Figure 2.6.

$$\left[\text{DRV}_{min}, \text{DRV}_{max}\right] = \{[\text{DRV}_{min}, \text{DRV}_{min} + \Delta\text{DRV}) \cup$$

$$[\text{DRV}_{min} + \Delta\text{DRV}, \text{DRV}_{min} + 2 \times \Delta\text{DRV}) \cup \ldots \qquad (2.14)$$

$$\ldots \cup [\text{DRV}_{max} - \Delta\text{DRV}, \text{DRV}_{max}]\}$$

An artificial neural network (ANN) is a widely used machine learning method to solve so-called multi-classification problems. More specifically, to model the DRV values of SRAMs, the outputs of an ANN are used for different DRV classes. During the modeling

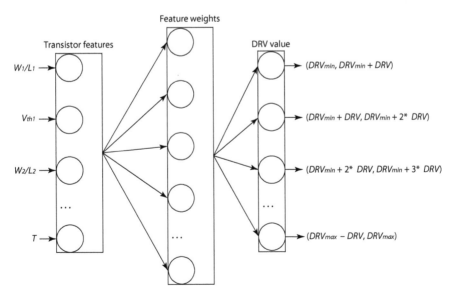

FIGURE 2.6
The layer composition of an artificial neural network for DRV classification and prediction.

training, the SRAM parameters are used to optimize the internal layers and the neuron nodes of an ANN and minimize the general prediction errors. A set of samples (including the DRV values and the corresponding physical characteristics of SRAM cells) are collected by SPICE simulation and used to train an ANN model.

2.5.3 Performance of DRV Model

The ANN model–predicted DRV values are compared with the SPICE simulated values in Figure 2.7, where the R value denotes the correlation between the model outputs and golden values. For the ANN-based DRV model, it is noted that there is a high correlation between prediction and output for all data sets. Before the ANN model proposed in [10], the linear model described in Equation 2.13 was widely used to model the DRV value of SRAM designs, which can be optimized with a linear regression (LR) method that fits a data set with linear coefficients. The most common type of LR is a *least-squares fit*, which can find an optimal line to represent the discrete data points. In an LR model, the same physical parameters of ANN models are defined as input training features $p = \{p_i, p_i \in \{W_1/L_1, W_2/L_2,..., T\}\}$. By denoting the linear coefficients with $\theta = \{\theta_0, \theta_1,..., \theta_n\}$:

$$h_\theta(p) = \theta_0 + \theta_1 \times p_1 + \cdots + \theta_n \times p_n \tag{2.15}$$

where θ stands for the set of coefficients (e.g., a_i and b_i as shown in Equation 2.13). Each training sample is composed of a transistor feature set p and the corresponding golden DRV value, DRV_{golden}, from SPICE simulation. Based on the least-squares fit rule, the cost function of m training examples can be expressed as

$$J(\theta) = \frac{1}{2m} \sum_{k=1}^{m} \left(h_\theta\left(p^{(k)}\right) - DRV_{golden}^{(k)} \right)^2 \tag{2.16}$$

where $p^{(k)}$ corresponds to the training features of a kth training sample, like the transistor sizes and temperature. To obtain the optimal θ, "gradient descent" can be applied simultaneously on each coefficient $\theta_j, j \in (1, 2,..., n)$:

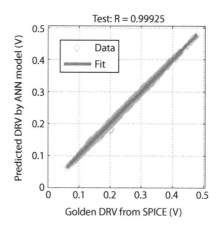

FIGURE 2.7
Training results based on the neural network model, across three data sets. R denotes the correlation between golden DRV data from SPICE simulation and the predicted DRV value from our model.

Repeat{

$$\theta_j := \theta_j - \alpha \frac{\partial J(\theta)}{\partial \theta_j}$$

$$= \theta_j - \alpha \frac{1}{m} \sum_{i=1}^{m} \left(h_\theta \left(p^{(k)} \right) - \mathrm{DRV}_{\mathrm{golden}}^{(k)} \right) p_i^{(k)}$$

(2.17)

}

α   is the learning rate of the LR model and $p^{(k)}$ is the ith feature of the kth training sample.

The experimental results demonstrate that the neural network model achieves smaller prediction errors than the LR model. According to [10], the mean μ and the standard deviation σ of the prediction error for the neural network model are 0.01 and 0.35 mV, respectively, while those of the LR model are 0.041 and 0.9 mV. Hence, it can be concluded that the neural network model outperforms the linear model in modeling the DRV of SRAMs. This is because varied weights and bias are employed in an ANN model for different feature patterns, whereas the linear model formulates all input features with the same optimized coefficients.

2.6 Using Machine Learning to Mitigate the Impact of Physical Disorder on TDC

This section presents a case study that uses machine learning to help mitigate the negative impact of physical disorder on TDC designs. Instead of discussing the traditional TDC schemes that are based on a delay line, the TDC design scheme presented in this section is the configurable compact algorithmic TDC (CCATDC) proposed in [38,39]. For brevity, this section does not cover the design scheme of this TDC architecture but focuses more on the use of machine learning for mitigating the process variations of this circuitry.

2.6.1 Background of TDC

High-resolution time measurement is a common need in modern scientific and engineering applications, such as time-of-flight measurement in remote sensing [40,41], nuclear science [42], biomedical imaging [43], frequency synthesizer and time jitter measurement for radio frequency transceivers in wireless communication [44]. To advance the post-processing of time signals, a TDC that bridges time measurements and digital electronic devices is proposed. As a measurement system, a TDC quantifies the time interval between two events by digitizing it, which greatly favors the post-processing in a digital way. Various TDC schemes have been proposed, such as the cyclic successive approximation–based TDC, pulse interpolation–based TDC and delta-sigma-based TDCs [45,46].

Most conventional TDC designs are implemented with delay lines, usually composed of two channels for two input signals, as shown in Figure 2.8a. The time difference (TD)

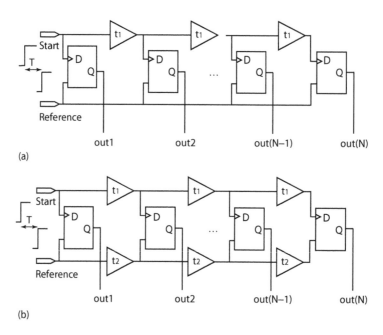

FIGURE 2.8
Schematic of traditional delay line–based TDCs. (a) Delay line–based TDC. (b) Vernier delay line–based TDC.

between the two input signals, "start" and "reference", is digitized into binary codes. To characterize the TD between two signals, the "start" signal is sequentially postponed by the delay elements. Theoretically, each delay element "slows down" the "start" signal by a constant timescale (td1 as in Figure 2.8b); the output of each delay element is then compared with a "reference" signal to generate a binary output. The highest resolution of a delay line–based TDC is the delay of each single element, which is limited by the CMOS fabrication technology node (a more advanced CMOS fabrication technology achieves higher resolution). To improve the resolution and conversion accuracy of such TDCs, several techniques have been proposed, such as a stretching pulse [47], using a tapped delay line [48] and employing a differential delay line [49].

As it becomes more difficult to control the process variations of CMOS fabrication with advanced technology nodes, the physical deviation from the designed value also becomes a big concern that limits the resolution of TDC design. For example, the fabricated delay elements in a TDC circuitry usually deviate from the designed (nominal) delay length, for example, T_{d1} in Figure 2.8a. Such process variations introduce design uncertainty into the time-to-digital conversion and decrease the conversion accuracy [2].

2.6.2 Mitigate the Process Variations by Reconfiguring the Delay Elements

The formulation of the physical disorder of CMOS devices comes from the fabrication stage, i.e., during which the length of the TDC delay element deviates from the designed value. Therefore, one solution to mitigate such impact is designing the delay element with an adjustable length and then regulating the length of the delay chain circuitry to improve its performance. This philosophy is based on the truth that: once an electronic circuit is fabricated, it is infeasible to remove the process variations. Fortunately, the actual in-path delay can be configured by using an adjustable delay line, thereby improving the

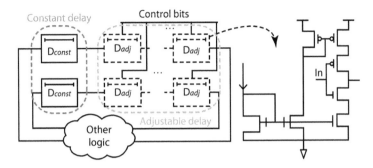

FIGURE 2.9
Schematic of the proposed configurable delay element, composed of two blocks, D_{const} and D_{adj}. There are many possible implementations of the adjustable delay element, and in this figure a current-controlled delay element is shown as an example.

robustness of CCATDC. There are many possible realizations of the adjustable delay element, such as voltage, current and digital controlled.

The schematic of an adjustable delay channel is shown in Figure 2.9, which is composed of two parts: the constant delay line, D_{const}, and the adjustable delay line, D_{adj}. Note that there still exist process variations in the constant delay line that deviate it from the designed delay length. The purpose of adding it is to roughly calibrate the reference time, T_{ref}, for the purposes of course tuning. A high-resolution adjustment is realized with the adjustable delay line, which fulfills the job of fine-tuning. To achieve a higher resolution, two parallel delay chains are utilized in two channels to form a vernier architecture. The control signals of the adjustable delay line can be supplied by an external calibration circuit.

2.6.3 Delay Chain Reconfiguration with Machine Learning

An important purpose of configuring the propagation delay of each element is to minimize the deviation of the process variations of CMOS transistors. However, since all components in the CCATDC design are of nanometer magnitude, it is again impractical to measure such process variations with an external instrument [35]. In [10,35], a machine learning technique was utilized to characterize the microscopic process variations and related physical features. This section continues this thread by employing machine learning techniques to characterize and configure the delay length of CCATDC. More specifically, the backward propagation of errors (backpropagation) algorithm is used. Backpropagation is a widely used technique in training ANNs, and is often used with other optimization methods such as gradient descent. Backpropagation training is usually composed of two phases: propagation and weight updating. Once an input vector is applied to the neural network model, it will be propagated from the input layer to the output layer. The output value of each input vector will be obtained and compared with the desired (golden) value; a loss function will be used to calculate the error for each neuron in the trained model and update the weight parameters correspondingly.

Within the framework, the controlling signals (e.g., the controlling current inputs) of the adjustable delay line (as shown in Figure 2.9) are fed into the backpropagation model as input vectors: $\mathbf{C} = (C_0, C_1, \dots, C_{n-1})$ in Algorithm 2. To comprehensively configure the delay chain, m different time inputs $\mathbf{T} = (t_0, t_1, \dots, t_{m-1})$ and corresponding digital conversion outputs $\mathbf{O} = (O_0, O_1, \dots, O_{m-1})$ are used to train the model. The accuracy of CCATDC is optimized by setting Δ_{err} as the threshold of conversion error. Before the model is trained, weight parameters are initialized as 0. During the model training, the controlling signals are fed into the HSPICE simulation and corresponding conversion outputs are obtained (Line 5); the simulated results are then compared with the golden values (Line 7). The conversion error is then backpropagated to the network (Line 8) and the controlling input vector is optimized to achieve a higher conversion accuracy (Line 9). The configuration procedure will terminate while the conversion error is below the pre-set threshold Δ_{err} (Line 10).

2.6.3.1 Performance of Configurable Compact Algorithmic TDC

The proposed CCATDC architecture is implemented with the predictive technology model (PTM) 45 nm standard cell libraries [50] and its performance is tested by applying random process variations on all transistors of the circuit; transient noise is also added in the simulation. The experimental result demonstrates that CCATDC is robust against process variations and transient noise, and the conversion accuracy is much higher if more conversion bits are utilized.

Algorithm 2 Given a CCATDC chip, configure the delay elements to improve the time digitization accuracy.

Input: a configurable compact algorithmic TDC
Input: n control signals $\mathbf{C} = (C_0, C_1, ..., C_{n-1})$
Input: weight vector with k parameters, $\mathbf{\Theta} = (\theta^0, \theta^1, ..., \theta^k)$
Input: m time input $\mathbf{T} = (t_0, t_1, ..., t_{m-1})$ and corresponding ideal conversion output
$\quad\quad O = (O_0, O_1, ..., O_{m-1})$ for the designed T_{ref} and amplification gain
Input: Δ_{err}, the threshold of conversion error

  1: Initialize weight parameters $\theta^i \leftarrow 0$

  2: Initialize error parameters $\Delta^i \leftarrow \Delta_{err}$

  3: **while** $\max(\Delta^i) >= \Delta_{err}$ **do**

  4:   **for** $i := 0$ **to** $m - 1$ **do**

  5:     $\hat{O}_i = \text{HSPICE}(t_i, \mathbf{C})$

  6:   **end for**

  7:   compute conversion error $\Delta = (\hat{O} - O)$

  8:   BackwardPropagateError(Δ)

  9:   UpdateWeights(Θ, $\mathbf{C}$)

  10: **end while**

  11: **return C**

2.7 Conclusion

In this chapter, an overview of some of the latest CMOS circuits like PUFs and TDCs in the literature is provided. Notably, the relationship between a physical disorder and these two circuit primitives is different: PUF is built on the physical disorder of CMOS fabrication, while the uncontrollable process variations of CMOS transistors degrade the resolution of TDCs. Apart from constructively leveraging these physical disorders, the modeling of them also provides a new way to study the process and environmental variations on the performance of CMOS circuitry. Based on the case studies presented in this chapter, we can see that machine learning is a good tool to study the microscopic physical disorder of CMOS circuitry and therefore provides a unique perspective to guide the design. The performance metrics used for analyzing these circuits are identified and several machine learning techniques are employed in a constructive way to benefit the performance of these designs.

References

1. S. R. Sarangi, B. Greskamp, R. Teodorescu, J. Nakano, A. Tiwari, and J. Torrellas. Varius: A model of process variation and resulting timing errors for microarchitects. *IEEE Transactions on Semiconductor Manufacturing*, 21(1):3–13, 2008.

2. J.-P. Jansson, V. Koskinen, A. Mantyniemi, and J. Kostamovaara. A multichannel high-precision CMOS time-to-digital converter for laser-scanner- based perception systems. *IEEE Transactions on Instrumentation and Measurement*, 61(9):2581–2590, 2012.

3. I. Kim, A. Maiti, L. Nazhandali, P. Schaumont, V. Vivekraja, and H. Zhang. From statistics to circuits: Foundations for future physical unclonable functions. In A.-R. Sadeghi and D. Naccache, editors, *Towards Hardware-Intrinsic Security*, Information Security and Cryptography, pages 55–78. Springer, 2010.

4. D. Blaauw, K. Chopra, A. Srivastava, L. Scheffer. Statistical timing analysis: From basic principles to state of the art. *IEEE Transactions on Computer-Aided Design of Integrated Circuits and Systems*, 27(4):589–607, 2008.

5. W. Wang, V. Reddy, B. Yang, V. Balakrishnan, S. Krishnan, and Y. Cao. Statistical prediction of circuit aging under process variations. In *IEEE Custom Integrated Circuits Conference*, pages 13–16. IEEE, 2008.

6. D. Forte and A. Srivastava. On improving the uniqueness of silicon-based physically unclonable functions via optical proximity correction. In *IEEE/ACM Design Automation Conference*, pages 96–105. June 2012.

7. V. Vivekraja and L. Nazhandali. Feedback-based supply voltage control for temperature variation tolerant PUFs. In *IEEE International Conference on VLSI Design*, VLSID '11, Washington, DC, USA, pages 214–219. IEEE Computer Society, 2011.

8. M.-D. Yu and S. Devadas. Secure and robust error correction for physical unclonable functions. *IEEE Design Test of Computers*, 27(1):48–65, 2010.

9. C. Bösch, J. Guajardo, A.-R. Sadeghi, J. Shokrollahi, and P. Tuyls. Efficient helper data key extractor on FPGAs. In *Cryptographic Hardware and Embedded Systems*, volume 5154 of *Lecture Notes in Computer Science*, pages 181–197. Springer, 2008.

10. X. Xu, A. Rahmati, D. E. Holcomb, K. Fu, and W. Burleson. Reliable physical unclonable functions using data retention voltage of SRAM cells. *IEEE Transactions on Computer-Aided Design of Integrated Circuits and Systems*, 34(6):903–914, 2015.

11. X. Xu and D. Holcomb. A clockless sequential PUF with autonomous majority voting. In *Great Lakes Symposium on VLSI, 2016 International*, pages 27–32. IEEE, 2016.

12. X. Xu and D. E. Holcomb. Reliable PUF design using failure patterns from time-controlled power gating. In *Defect and Fault Tolerance in VLSI and Nanotechnology Systems (DFT), 2016 IEEE International Symposium on*, pages 135–140. IEEE, 2016.

13. X. Xu, V. Suresh, R. Kumar, and W. Burleson. Post-silicon validation and calibration of hardware security primitives. In *VLSI (ISVLSI), 2014 IEEE Computer Society Annual Symposium on*, pages 29–34. IEEE, 2014.

14. R. Maes and I. Verbauwhede. Physically unclonable functions: A study on the state of the art and future research directions. In A.-R. Sadeghi and D. Naccache, editors, *Towards Hardware-Intrinsic Security*, Information Security and Cryptography, pages 3–37. Springer Berlin Heidelberg, 2010.

15. G. Hospodar, R. Maes, and I. Verbauwhede. Machine learning attacks on 65 nm arbiter PUFs: Accurate modeling poses strict bounds on usability. In *IEEE International Workshop on Information Forensics and Security*, pp. 37–42. IEEE, 2012.

16. D. Lim. Extracting secret keys from integrated circuits. Master's thesis, Massachusetts Institute of Technology, Dept. of Electrical Engineering and Computer Science, 2004.

17. G. Suh and S. Devadas. Physical unclonable functions for device authentication and secret key generation. In *IEEE/ACM Design Automation Conference*, pages 9–14, June 2007.

18. J. Guajardo, S. S. Kumar, G.-J. Schrijen, and P. Tuyls. FPGA intrinsic PUFs and their use for IP protection. In *International Workshop on Cryptographic Hardware and Embedded Systems*, CHES'07, pages 63–80. Springer-Verlag, 2007.

19. D. E. Holcomb, W. Burleson, and K. Fu. Initial SRAM state as a fingerprint and source of true random numbers for RFID tags. In *Proceedings of the Conference on RFID Security*, 2007.

20. D. E. Holcomb, W. P. Burleson, and K. Fu. Power-up SRAM state as an identifying fingerprint and source of true random numbers. *IEEE Transactions on Computers*, 58(9):1198–1210, 2009.

21. S. Katzenbeisser, U. Kocabaş, V. Rožić, A.-R. Sadeghi, I. Verbauwhede, and C. Wachsmann. PUFs: Myth, fact or busted? A security evaluation of physically unclonable functions (PUFs) cast in silicon. In *International Conference on Cryptographic Hardware and Embedded Systems*, CHES'12, pages 283–301. Springer-Verlag, 2012.

22. D. E. Holcomb, A. Rahmati, M. Salajegheh, W. P. Burleson, and K. Fu. DRV-fingerprinting: Using data retention voltage of SRAM cells for chip identification. In *International Workshop on Radio Frequency Identification. Security and Privacy Issues*, pages 165–179. Springer, 2013.

23. J. Guajardo, S. Kumar, G. Schrijen, and P. Tuyls. FPGA intrinsic PUFs and their use for IP protection. *Cryptographic Hardware and Embedded Systems*, 2007.

24. D. E. Holcomb, W. P. Burleson, and K. Fu. Power-up SRAM state as an identifying fingerprint and source of true random numbers. *IEEE Transactions on Computers*, 2009.

25. A. Kumar, H. Qin, P. Ishwar, J. Rabaey, and K. Ramchandran. Fundamental data retention limits in SRAM standby: Experimental results. In *Quality Electronic Design, 2008. ISQED 2008. 9th International Symposium on*, pages 92–97. IEEE, 2008.

26. M. Majzoobi, F. Koushanfar, and M. Potkonjak. Testing techniques for hardware security. In *IEEE International Test Conference*, pages 1–10. Oct. 2008.

27. U. Rührmair, F. Sehnke, J. Sölter, G. Dror, S. Devadas, and J. Schmidhuber. Modeling attacks on physical unclonable functions. In *ACM Conference on Computer and Communications Security*, CCS'10, New York, USA, pages 237–249. ACM, 2010.

28. X. Xu, U. Rührmair, D. E. Holcomb, and W. Burleson. Security evaluation and enhancement of bistable ring PUFs. In *International Workshop on Radio Frequency Identification: Security and Privacy Issues*, pages 3–16. Springer, 2015.

29. U. Rührmair, J. Sölter, F. Sehnke, X. Xu, A. Mahmoud, V. Stoyanova, G. Dror, J. Schmidhuber, W. Burleson, and S. Devadas. PUF modeling attacks on simulated and silicon data. *Information Forensics and Security, IEEE Transactions on*, 2013.

30. X. Xu and W. Burleson. Hybrid side-channel/machine-learning attacks on PUFs: A new threat? In *Proceedings of the Conference on Design, Automation & Test in Europe*, page 349. European Design and Automation Association, 2014.

31. U. Rührmair, X. Xu, J. Sölter, A. Mahmoud, M. Majzoobi, F. Koushanfar, and W. Burleson. Efficient power and timing side channels for physical unclonable functions. In *International Workshop on Cryptographic Hardware and Embedded Systems*, pages 476–492. Springer, 2014.

32. U. Rührmair, X. Xu, J. Sölter, A. Mahmoud, F. Koushanfar, and W. Burleson. Power and timing side channels for PUFs and their efficient exploitation. *IACR Cryptology ePrint Archive*, 2013:851, 2013.
33. S. Khan, S. Hamdioui, H. Kukner, P. Raghavan, and F. Catthoor. Incorporating parameter variations in BTI impact on nano-scale logical gates analysis. In *Defect and Fault Tolerance in VLSI and Nanotechnology Systems (DFT), 2012 IEEE International Symposium on*, pages 158–163. IEEE, 2012.
34. D. Lorenz, G. Georgakos, and U. Schlichtmann. Aging analysis of circuit timing considering NBTI and HCI. In *On-Line Testing Symposium, 2009. IOLTS 2009. 15th IEEE International*, pages 3–8. IEEE, 2009.
35. X. Xu, W. Burleson, and D. E. Holcomb. Using statistical models to improve the reliability of delay-based PUFs. In *VLSI (ISVLSI), 2016 IEEE Computer Society Annual Symposium on*, pages 547–552. IEEE, 2016.
36. S. S. Avvaru, C. Zhou, S. Satapathy, Y. Lao, C. H. Kim, and K. K. Parhi. Estimating delay differences of arbiter PUFs using silicon data. In *2016 Design, Automation & Test in Europe Conference & Exhibition (DATE)*, pages 543–546. IEEE, 2016.
37. H. Qin, Y. Cao, D. Markovic, A. Vladimirescu, and J. Rabaey. SRAM leakage suppression by minimizing standby supply voltage. In *5th International Symposium on Quality Electronic Design*, pages 55–60. IEEE, 2004.
38. S. Li and C. D. Salthouse. Compact algorithmic time-to-digital converter. *Electronics Letters*, 51(3):213–215, 2015.
39. S. Li, X. Xu, and W. Burleson. CCATDC: A configurable compact algorithmic time-to-digital converter. In *VLSI (ISVLSI), 2017 IEEE Computer Society Annual Symposium on*, pages 501–506. IEEE, 2017.
40. D. Marioli, C. Narduzzi, C. Offelli, D. Petri, E. Sardini, and A. Taroni. Digital time-of-flight measurement for ultrasonic sensors. *IEEE Transactions on Instrumentation and Measurement*, pages 93–97. IEEE, 1992.
41. S. Li and C. Salthouse. Digital-to-time converter for fluorescence lifetime imaging. In *Instrumentation and Measurement Technology Conference (I2MTC), 2013 IEEE International*, pages 894–897. IEEE, 2013.
42. N. Bar-Gill, L. M. Pham, A. Jarmola, D. Budker, and R. L. Walsworth. Solid-state electronic spin coherence time approaching one second. *Nature Communications*, 4:1743, 2013.
43. A. S. Yousif and J. W. Haslett. A fine resolution TDC architecture for next generation PET imaging. *IEEE Transactions on Nuclear Science*, pages 1574–1582. IEEE, 2007.
44. J.-P. Jansson, A. Mantyniemi, and J. Kostamovaara. A CMOS time-to-digital converter with better than 10 ps single-shot precision. *IEEE Journal of Solid-State Circuits*, 41(6):1286–1296, 2006.
45. A. Mantyniemi, T. Rahkonen, and J. Kostamovaara. A CMOS time-to-digital converter (TDC) based on a cyclic time domain successive approximation interpolation method. *IEEE Journal of Solid-State Circuits*, 44(11):3067–3078, 2009.
46. W.-Z. Chen and P.-I. Kuo. A ΔΣ TDC with sub-ps resolution for PLL built-in phase noise measurement. In *European Solid-State Circuits Conference, ESSCIRC Conference 2016: 42nd*, pages 347–350. IEEE, 2016.
47. S. Tisa, A. Lotito, A. Giudice, and F. Zappa. Monolithic time-to-digital converter with 20ps resolution. In *Solid-State Circuits Conference, 2003. ESSCIRC'03. Proceedings of the 29th European*, pages 465–468. IEEE, 2003.
48. R. B. Staszewski, S. Vemulapalli, P. Vallur, J. Wallberg, and P. T. Balsara. 1.3 v 20 ps time-to-digital converter for frequency synthesis in 90-nm CMOS. *IEEE Transactions on Circuits and Systems II: Express Briefs*, 53(3):220–224, 2006.
49. P. Dudek, S. Szczepanski, and J. V. Hatfield. A high-resolution CMOS time-to-digital converter utilizing a vernier delay line. *IEEE Journal of Solid-State Circuits*, 35(2):240–247, 2000.
50. W. Zhao and Y. Cao. New generation of predictive technology model for sub-45 nm early design exploration. *IEEE Transactions on Electron Devices*, 53(11):2816–2823, 2006.

3

Design of Alkali-Metal-Based Half-Heusler Alloys Having Maximum Magnetic Moments from First Principles

Lin H. Yang, R.L. Zhang, Y.J. Zeng, and C.Y. Fong

CONTENTS

3.1 Introduction

Half-Heusler alloys have a $C1_b$ structure, which is a variation of the full-Heusler crystalized in the $L2_1$ structure, as schematically shown in Figure 3.1. The Bravais lattice is a face-centered-cubic (fcc) lattice shown as the outer cube in Figure 3.1. The notations of the atomic positions in a unit cell are adopted from Wyckoff (1963). For the $C1_b$ structure, three elements occupy 4a, 4b, and 4c sites, respectively, while 4d sites are vacant. The first $C1_b$ alloy studied was composed of two transition-metal (TM) elements denoted by X (Ni) and Y (Mn) and one pnictogen (Sb) designated as Z (de Groot *et al.*, 1983). Later, alloys with X, a simple metal, were considered (Kieven *et al.*, 2010; Jungwirth *et al.*, 2011; Roy *et al.*, 2012).

To search for other potential half-metallic candidates in half-Heusler alloys, Damewood *et al.* (2015) replaced one of the TM elements, the said X, by an alkali-metal element and the pnictogen (Z) by a Group IV element. The arrangements of the three atoms (X, Y, and Z) at three different sites – 4a, 4b, and 4c (Figure 3.1) – in the $C1_b$ structure define three phases, the alpha (α), beta (β), and gamma (γ) phases (Larson *et al.*, 2000) (Table 3.1).

The unique feature of these half-Heusler alloys is to exhibit half-metallicity (de Groot *et al.*, 1983) in their electronic and magnetic properties. The half-metallicity is characterized by showing metallic properties in one spin channel and insulating behaviors in the opposite spin states. Consequently, the conduction current in these alloys is expected to have a 100% spin polarization according to the Julliére (1975) formula:

$$P = \left(N_m - N_i\right)\Big/\left(N_m + N_i\right)$$

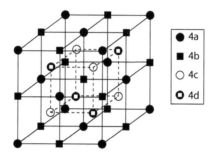

FIGURE 3.1

The L2$_1$ structure. Wyckoff's notations of atomic positions are expressed in rectangular coordinates: 4a=(0,0,0)a_0, 4b = (1/2,1/2,1/2)a_0, 4c = (1/4,1/4,1/4)a_0, and 4d = (3/4,3/4,3/4)a_0, where a_0 is the lattice parameter of the large cube. The C1$_b$ structure shares the same picture except 4d sites are vacant.

TABLE 3.1

Arrangements of X, Y, and Z in α, β, and γ Phases

Phase	4a	4b	4c
α	Z	Y	X
β	X	Y	Z
γ	Z	X	Y

Note: The 4d sites are vacant (not listed here).

where P is the spin polarization at the Fermi energy, E$_F$, so that the spin carrying mobile carriers can be easily transported. In this representation, Julliére defined P in terms of the density of states (DOS) at E$_F$; N_m represents the DOS of the metallic channel; and N_i is the corresponding DOS of the insulating channel. For half-metals, N_m is finite while N_i is zero because E$_F$ falls in the insulating gap. The spin polarization at E$_F$, P, is therefore 1. In principle, they can be ideal materials for spintronic devices.

To predict whether a half-Heusler alloy can be a half-metal, two criteria related to its electronic and magnetic properties must be satisfied. The first one concerns the DOS at E$_F$. The E$_F$ should intersect at least one band in the metallic states and fall in the gap of the insulating spin channel. The second one is the integer value of the magnetic moment/unit cell. Because the valence states in the insulating spin channel are completely occupied, the occupation number should be an integer. Adding to the fact that the total number of electrons is an integer, the remaining number of electrons in the other spin channel should also be integers. The resulting net electrons contribute their spins to the moment. With the electronic spin g-*factor*, g_s to be 2, the predicted value of the magnetic moment/unit cell should strictly be an integer in units of μ_B, where μ_B is the Bohr magneton.

After the seminal publication by de Groot *et al.* (1983), tremendous efforts have been devoted to realizing spintronic devices fabricated by half-Heusler alloys and to searching for new alloys based on theoretical predictions and experimental measurements. However, devices designed to be made of half-Heusler alloys are still illusive due to the complications of growing flawless samples (Otto *et al.*, 1987) and other factors discussed in chapter 3 of the monograph by Fong *et al.* (2013). Although polycrystalline samples of NiMnSb have been reported (Gardelis *et al.*, 2004), a quality single crystal of NiMnSb, on the other hand, has not appeared in the literature. Earlier predictions listed in the monograph (Fong *et al.*, 2013) were not helpful because the stability issue of the half-Heusler alloys was not addressed at the time. Recently, this issue has been studied by Damewood *et al.* (2015) and

Zhang *et al.* (2017). By comparing the acoustic phonon spectra of half-Heusler alloys to the ones in half-metals with a zinc blende structure, they have concluded that half-Heusler alloys are stable even when not at their optimized lattice constants at $T = 0$ K. One should expect, however, that in such samples there will be some stresses that can affect the life-time of the devices made from these alloys. To reduce the stress, it will be more effective to first carry out a theoretical search of alloys exhibiting half-metallic properties at or near their optimized lattice constants instead of using a trial-error approach.

We can now summarize the criteria for spintronic applications using half-Heusler. They are (a) to have half-metallic properties at lattice constants at or near equilibrium at $T = 0$ K; (b) to have the largest magnetic moment possible for a 3d TM element; and (c) to have the largest moment exhibiting at or near the optimized lattice constant at $T = 0$ K. Among the three conditions, the last one is the most challenging.

In this review, we focus on the following issue: How do we overcome the challenge? This is accomplished by (a) choosing the alkali-metal atoms for the non-TM element in the $C1_b$ structure; and (b) applying the Pauli principle to half-Heusler alloys to have the largest possible magnetic moment ($5\,\mu_B$) for 3d TM elements. We restrict our study of half-Heulser alloys to have Cr as the TM element in the β and γ phases because these two phases have lower free energies than the α phase. We will present our findings of 9 out of 30 β- and γ-phase alloys that show favorable half-metallicity and have the largest magnetic moments at their respective equilibrium lattice constants.

In Section 3.2, we present guidelines deduced from physics for how to choose the three elements and how to attain the largest magnetic moment for the alloys. In Section 3.3, a method of calculation will be presented. Results and discussion will be given in Section 3.4. Finally, a summary will be given in Section 3.5.

3.2 Guiding Principles of Designing the Half-Heusler Alloys

Our choices of the three elements in the primitive cell of the half-Heusler alloys are guided by the following physics: (a) the nearest-neighbor configurations of the non-metal atom with the 3d TM element under the tetrahedral environment governing the charge transfer between the two atoms to form bonds; (b) the requirement of the strong electro-negativity of the non-metal atoms on the alkali-metal elements to force a complete charge transfer from the metal atom to the non-metal atom; and (c) the Pauli principle dictates the volume required to align the spins at the TM site.

To address the first guiding principle, we first realize that the arrangement of atoms in the III-V compound semiconductors is closely related to the one in the half-Heusler alloys. A III-V compound, e.g. GaAs, has a zinc blende structure as shown in Figure 3.2. Each atom in the zinc blende structure is surrounded by four neighbors of different species form-ing the tetrahedral environment. The s- and p-states of the cation under the tetrahedral environment hybridize with the s- and p-states of a possibly different principal quantum number of its neighbors to form the sp^3 covalent bonds. It is clearly seen that the $C1_b$ struc-ture (Figure 3.1) and the zinc blende structure (Figure 3.2) should share some similarity in their bonding features. In Figure 3.3, we show the charge distributions of the β-phase LiMnSi and the zinc blende MnSi; both crystals are half-metals. The sections are defined by the solid lines cutting through the (110) plane in Figure 3.2. The bond charges between Mn and Si are shown and labeled. Their orientations and locations show close similarity

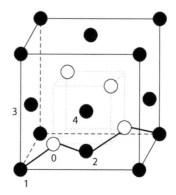

FIGURE 3.2
The zinc blende structure. The filled circles correspond to 4a atoms and the open circles are equivalent to 4c atoms in Figure 3.1. The solid lines indicate where the bonds formed between the atoms. The nearest neighbors of the atom 0 (open circle) are indicated by numbers 1–4.

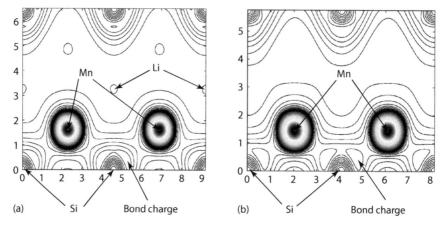

FIGURE 3.3
(a) Charge distribution of β-LiMnSi in a section defined by the solid lines shown in Figure 3.2. (b) Charge distribution of MnSi in the same section. The bond charges for both cases are seen. The shaded regions are the high densities from the Mn d-states.

for the two compounds. In Figure 3.3a, the label "Li" indicates the location of the Li atom in LiMnSi. There is no charge around the Li atom. Its electron is completely transferred to the non-metallic atom.

To further understand the guiding principles, we put forward the notion of electro-negativity. Physically, an element having a strong electro-negativity attracts electrons from its nearest neighbors in a solid to form bonds. Most of these elements are located at the right of the Group IV elements (C, Si, and Ge), such as pnictides, chalcogens, and halides in the periodic table. With respect to the alkali-metal and the TM elements, these species including Group IV elements are called "non-metal elements," which have the strongest electro-negativities among the three species in the compound. We therefore concluded that the important role played by "electro-negativity" in full-Heusler alloys composed of two TM elements – the bonding to hold the crystals together – is by virtue of the nearest-neighbor configuration of one of the TM elements with the non-metal element (Shaughnessy *et al.*, 2013).

In a half-Heusler alloy, the consequences of transferring electrons from the TM element to its nearest-neighbor non-metal element are twofold: (a) to fill the outer electronic shell of the non-metal element; and (b) to determine the magnetic moment of the alloy by the remaining electrons at the TM site. The transferred electrons do not contribute to any magnetic moments of the alloys because their spins are paired due to bond formations. These results suggest that the magnetic moment of a half-Heusler alloy can be simply accounted for by counting the number of electrons being transferred from the TM element and the number of remaining electrons at the site where the TM element is located. This fact has been recognized by Schwarz (1986) who proposed the so-called "ionic model" and has been used by Damewood *et al.* (2015). Based on the physics governed by "electro-negativity," the desired half-Heusler alloys should have the TM atom as the nearest neighbor to the non-metal atom with strong electro-negativity and should have the values of the magnetic moment determined by the ionic model.

In this review, we decided to take Cr as the TM element. Its valence is 6. To have the largest magnetic moment of the alloy, only one electron should be transferred to its neighboring non-metal atom. We then chose an alkali element as the nearest neighbor to the non-metal element to stabilize the crystalline structure. This alkali-metal offers one electron to the most electro-negative element. Thus, we chose Group VI atoms as the non-metal element in the alloys, which takes one electron each from the alkali-metal and TM element to attain the largest magnetic moment/unit cell.

We now turn to the third guiding principle. Depending on the open space or volume available around the TM element, the value of the magnetic moment can vary. The reason is the Pauli principle – electrons with parallel spins need a large volume to avoid each other – they should not occupy the same spatial region. We recognized this physical fact in the study of MnC, a half-metal with the zinc blende structure (Qian *et al.*, 2004). In MnC, its moment/unit cell is 3 μ_B at a larger lattice constant (5.0 Å), while at a smaller lattice constant (4.0 Å) the moment reduces to only 1 μ_B. The larger value is consistent with the prediction of the ionic model but not the one for the smaller lattice constant. The smaller volume causes one of the spins to flip due to the Pauli principle. In order to achieve the maximum magnetic moment for an alloy, we should take the Pauli principle into consideration. However, one should be aware that the free energy of each compound can be influenced by the bonding strength and the exchange interaction as a function of lattice constants. In order to have the optimized lattice constant (at $T = 0$ K) and the largest magnetic moments, this has become a challenging optimization problem.

To address the optimization problem, we have applied the guiding principles in α-LiCrS alloy (Zhang *et al.*, 2016). It is the first alloy to demonstrate the power of the guiding principles and to meet the optimization challenge. In this alloy, the S atom has the strongest electro-negativity and thus needs two electrons to fill its $n = 3$ electronic shell. The TM atom, Cr, is located at $(1/2, 1/2, 1/2)a_0$, where $a_0 = 6.0$ Å is the lattice constant of the corresponding fcc cell. With the S atom at $(1/4, 1/4, 1/4)a_0$, the nearest-neighbor configuration causes the Cr atom in principle to lose two of its electrons. That would result in four d-electrons remaining at the Cr site to give the magnetic moment of 4 μ_B. In order to increase the magnetic moment, we then chose Li, the least electro-negative element, as the other neighbor to the S atom, so the two required transferred electrons are equally contributed by Li and Cr atoms – one electron for each element.

Consequently, we found that α-LiCrS has 5 μ_B at its optimized lattice constant at $T = 0$ K (Zhang *et al.*, 2017). Unfortunately, compared to the other two phases, this α phase has the highest free energy at $T = 0$ K and may not be favorable to grow. Therefore, it is worth the effort to find other half-Heusler alloys in the β and γ phases because they have comparable lower free energies than the α phase at the same lattice constants. These two phases have the potential to be grown in polycrystalline thin-film forms with minimum lattice mismatch at the interface.

3.3 Method of Calculation

To find the desired half-metallic half-Heusler alloys, the fcc primitive cell was used for the two phases (β and γ alloys). There is one TM atom per unit cell. The spin-polarized version of the Vienna ab-initio simulation package (VASP) (Kresse and Furthmüller, 1996) with projector-augmented-wave (PAW) pseudopotentials (Blöchl, 1994) for the elements was used in this review. The generalized gradient approximation (GGA) of Perdew *et al.* (1996) was used to treat the electron–electron exchange–correlation interactions in the PAW pseudopotentials and crystalline calculations. We used plane-wave basis functions with a 1200 eV kinetic energy cutoff, E_{cut}, for all calculations. The Monkhorst and Pack (1976) mesh of (15, 15, 15) was used to calculate the total charge density. The convergences of the total energy and the magnetic moment for each studied alloy are better than 1.0 meV and 1.0 mμ_B, respectively. The values of the E_{cut} and the mesh points should be checked for the convergences of the two quantities. Otherwise, interesting physics will be missed (Zhang *et al.*, 2017).

3.4 Results and Discussion

In Table 3.2, we tabulate first our results according to the order of the β and γ phases of the half-Heusler alloys exhibiting the desired properties – maximum magnetic moment per unit cell, $M = 5\ \mu_B$, at the optimized lattice constant. In addition, the following information is also given in Table 3.2: (a) the range of lattice constants and the corresponding optimized

TABLE 3.2

Range of Lattice Constants (a), Equilibrium Lattice Constant (a_0), Magnetic Moment (M), Free Energy (F), and Energy Gap (E_g) in the Insulating Channel for the Alloys with Integer Magnetic Moments

Alloy	a (Å)	a_0 (Å)	M (μ_B)	F (eV)	E_g (eV)	E_c (eV)	E_v (eV)
β-CsCrS	6.50–8.00	7.35	5	−13.492	1.926	0.107	−1.819
β-CsCrSe	6.40–7.68	7.58	5	−13.062	1.758	0.003	−1.755
γ-KCrS	6.50–7.00	6.65	5	−14.030	3.025	0.251	−2.774
γ-KCrSe	6.40–6.96	6.86	5	−13.692	2.619	0.026	−2.593
γ-CsCrO	6.15–7.70	6.49	5	−13.500	2.153	0.621	−1.532
γ-CsCrS	6.70–7.50	7.06	5	−13.461	2.519	0.582	−1.937
γ-CsCrSe	6.98–7.80	7.27	5	−13.150	2.398	0.593	−1.805
γ-RbCrS	6.40–8.00	6.83	5	−13.710	2.866	0.414	−2.452
γ-RbCrSe	6.60–7.40	7.04	5	−13.390	2.382	0.139	−2.243

Note: The bottom of conduction band energy (E_c) and the top of valence band energy (E_v) relative to Fermi energy (set to be zero) are also listed in the last two columns.

one at $T=0$ K for each alloy that shows the largest integer magnetic moments; (b) the magnetic moment, M; (c) the energy gap, E_g, in the insulating channel; and (d) the energies at the bottom of the conduction bands and the top of the valence bands measured with respect to E_F. We list E_c and E_v relative to E_F to examine whether the spin–orbit interaction can be neglected to maintain their half-metallicity in these alloys.

For the β phase, only two Cs-based half-Heusler alloys, β-CsCrS and β-CsCrSe, are predicted to have maximum magnetic moments at their corresponding equilibrium lattices at $T=0$ K. Their lattice constants are larger than 7.0 Å as a result of the Pauli principle. This means that the Cr–S bond length is more than 3.031 Å, so there is enough space around Cr to align its spin moments. There are, however, seven favorable γ-phase K-, Cs- and Rb-based half-Heusler alloys. Among the seven cases, two of them (γ-CsCrS and γ-CsCrSe) can compete with their counterparts in the β phase. The β-phase CsCrS is lower in free energy by 0.031 eV than the γ phase, while the β-phase CsCrSe is higher in energy than the γ phase by 0.088 eV. We therefore expect that the growth of these two half-Heusler alloys can have mixed phases.

In each phase, the optimized lattice constants, a_0, for the S-composed alloys are smaller than the Se ones. This demonstrates a combination of stronger electro-negativity and smaller ionic radius of the S atom. Except the β-CsCrS and β-CsCrSe as well as the γ-KCrS and γ-KCrSe, we can roughly state that the range of lattice constants for the alloys to have the largest magnetic moment in a unit cell is consistent with the ionic radii of the Group VI elements. Another general statement can be made: the ionic radii of three alkali-elements, K, Cs, and Rb, are larger than Na and Li to form half-Heusler alloys having the most favored half-metallic properties. This is another manifestation of the role played by the Pauli principle. To examine whether the spin–orbit interactions have an effect on the desired half-metallic properties, Table 3.2 lists the values at the top of the valence bands (E_v) and the bottom of the conduction bands (E_c) relative to the Fermi energies, which are set to zero. Based on the conclusion given by Zhang *et al.* (2017): the spin–orbit interaction is negligible when both E_c and E_v are more than 0.1 eV from E_F. We expect that, except β-CsCrSe and γ-KCrSe, the spin–orbit effect on these half-Heusler alloys is negligible (Zhang *et al.*, 2016).

Table 3.3 lists those alloys that have integer magnetic moments in the range not covering their respective equilibrium lattice constants at $T=0$ K. For comparison, their magnetic moments at their equilibrium lattice constants are also listed. They are all in the β phase. We suggest that these alloys can be grown in layered forms.

The remaining 17 alloys predicted not to have half-metallic properties are listed in Table 3.4. In general, these alloys have smaller equilibrium volumes due to the presence of Li and Na, respectively. The unfavorable predictions are understandable based on the Pauli principle.

TABLE 3.3

Equilibrium Lattice Constant (a_0), Magnetic Moment (M), Free Energy (F), and the Range of Lattice Constants (a) Have Integer Magnetic Moments for Half-Heusler Alloys

Alloy	a_0 (Å)	M (μ_B)	F (eV)	a (Å)	M (μ_B)
β-CsCrO	6.50	4.9996	−14.353	5.50–6.40	5
β-KCrS	6.80	4.9982	−14.388	6.50	5
β-RbCrSe	7.30	4.9952	−13.472	6.30–6.60	5
β-RbCrS	7.00	4.9996	−13.936	6.40–6.60	5

TABLE 3.4

Equilibrium Lattice Constant (a_0), Magnetic Moment (M), and Free Energy (F) for Alloys that Do Not Have the Desired Half-Metallic Properties

Alloy	a_0 (Å)	M (μ_B)	F (eV)
β-RbCrO	6.40	4.9716	−14.699
β-LiCrO	5.50	4.8268	−17.666
β-NaCrS	6.30	4.9819	−15.310
β-LiCrSe	6.10	4.7110	−15.691
β-KCrO	6.50	4.9463	−14.931
β-NaCrO	6.40	4.8481	−15.374
β-LiCrS	6.40	4.9817	−16.064
β-KCrSe	7.00	4.9934	−13.885
β-NaCrSe	6.50	4.9763	−14.670
γ-LiCrO	5.50	4.8892	−16.156
γ-RbCrO	6.40	4.9973	−13.633
γ-NaCrS	6.30	4.9817	−14.676
γ-LiCrSe	6.20	4.8972	−15.217
γ-KCrO	6.50	4.9761	−13.763
γ-NaCrO	6.40	4.8642	−14.169
γ-LiCrS	6.40	4.9809	−15.405
γ-NaCrSe	6.50	4.9729	−14.256

3.5 Summary

We have carried out systematic studies in an attempt to search for alkali-metal-based and Cr-related half-Heusler alloys with half-metallic properties desirable for spintronic applications. In order to have their half-metallicity and the maximum magnetic moments happening at their optimized lattice constants, the guiding principles of selecting the elements based on physics were given. A total of 30 cases were studied, 9 of which show the desired properties. Among them, β-CsCrS is best and is worth growing. Two of them, β-CsCrSe and γ-KCrSe, may have a large spin–orbit effect due to E_F and the bottom of the conduction bands is close in energy to mix the up-and-down spin states. The growth of the remaining six alloys may have mixed phases. Four β-phase alloys showing the largest magnetic moment/unit cell at lattice constants away from the respective optimized value should be grown in thin-film forms. The remaining 17 cases may not be worth trying. They mostly involve smaller alkali-metal elements causing smaller lattice constants and do not exhibit half-metallic properties attributed to the role played by the Pauli principle.

Acknowledgments

Work at Lawrence Livermore National Laboratory was performed under the auspices of the U.S. Department of Energy by Lawrence Livermore National Laboratory under Contract DE-AC52-07NA27344. RLZ was supported by grants from the National Natural Science Foundation of China (Grant No. 10904061) and China Scholarship Council. Work at UC Davis was supported in part by the National Science Foundation (Grant No. ECCS-0725902).

References

Blöchl, P. E., 1994. Projector augmented-wave method. *Phys. Rev. B*, Volume 50, p. 17953.

Damewood, L. *et al.*, 2015. Stabilizing and increasing the magnetic moment of half-metals: The role of Li in half-Heusler LiMn Z (Z = N, P, Si). *Phys. Rev. B*, Volume 91, p. 064409.

de Groot, R. A., Mueller, F. M., van Engen, P. G. & Buschow, K. H. J., 1983. New class of materials: Half-metallic ferromagnets. *Phys. Rev. Lett.*, Volume 50, p. 2024.

Fong, C. Y., Pask, J. E. & Yang, L. H., 2013. *Half-Metallic Materials and Their Properties.* London: Imperial College Press.

Gardelis, S. *et al.*, 2004. Synthesis and physical properties of arc melted NiMnSb. *J. Appl. Phys.*, Volume 95, p. 8063.

Julliére, M., 1975. Tunneling between ferromagnetic films. *Phys. Lett. A*, Volume 54, p. 225.

Jungwirth, T. *et al.*, 2011. Demonstration of molecular beam epitaxy and a semiconducting band structure for I-Mn-V compounds. *Phys. Rev. B*, Volume 83, p. 035321.

Kieven, D. *et al.*, 2010. I-II-V half-Heusler compounds for optoelectronics: Ab initio calculations. *Phys. Rev. B*, Volume 81, p. 075208.

Kresse, G. & Furthmüller, J., 1996. Efficient iterative schemes for ab initio total-energy calculations using a plane-wave basis set. *Phys. Rev. B*, Volume 54, p. 11169.

Larson, P., Mahanti, S. D. & Kanatzidis, M. G., 2000. Structural stability of Ni-containing half-Heusler compounds. *Phys. Rev. B*, Volume 62, p. 12754.

Monkhorst, H. J. & Pack, J. D., 1976. Special points for Brillouin-zone integrations. *Phys. Rev. B*, Volume 13, p. 5188.

Otto, M. J. *et al.*, 1987. Electronic structure and magnetic, electrical and optical properties of ferromagnetic Heusler alloys. *J. Mag. and Mag. Mat.*, Volume 70, p. 33.

Perdew, J. P., Burke, K. & Ernzerhof, M., 1996. Generalized gradient approximation made simple. *Phys. Rev. Lett.*, Volume 77, p. 3865.

Qian, M. C., Fong, C. Y. & Yang, L. H., 2004. Coexistence of localized magnetic moment and opposite-spin itinerant electrons in MnC. *Phys. Rev. B*, Volume 70, p. 052404.

Roy, A., Bennett, J. W., Rabe, K. M. & Vanderbilt, D., 2012. Half-Heusler semiconductors as piezoelectrics. *Phys. Rev. Lett.*, Volume 109, p. 037602.

Schwarz, K., 1986. CrO_2 predicted as a half-metallic ferromagnet. *J. Phys. F: Met. Phys.*, Volume 16, p. L211.

Shaughnessy, M. *et al.*, 2013. Structural variants and the modified Slater-Pauling curve for transition-metal-based half-Heusler alloys. *J. Appl. Phys.*, Volume 113, p. 043709.

Wyckoff, R. W. G., 1963. *Crystal Structures.* 2nd ed. San Francisco: John Wiley & Sons.

Zhang, R. L. *et al.*, 2016. A half-metallic half-Heusler alloy having the largest atomic-like magnetic moment at optimized lattice constant. *AIP Advances*, Volume 6, p. 115209.

Zhang, R. L. *et al.*, 2017. Two prospective Li-based half-Heusler alloys for spintronic applications based on structural stability and spin-orbit effect. *J. Appl. Phys.*, Volume 122, p. 013901.

Zhang, R. L., Fong, C. Y. & Yang, L. H., 2017. unpublished.

4

Defect-Induced Magnetism in Fullerenes and MoS$_2$

Sinu Mathew, Taw Kuei Chan, Bhupendra Nath Dev, John T.L. Thong,
Mark B.H. Breese, and T. Venkatesan

CONTENTS

4.1 Magnetism in Allotropes of Carbon

The observation of magnetism in materials made purely of carbon origin [1–5] has created enormous interest both in basic and applied fields of research. The difference in the magnetic susceptibilities of various carbon allotropes and the influence of particle size on susceptibility were first reported by Sir C.V. Raman and P. Krishnamurthy in 1929 [6]. Bulk crystalline graphite is a strong diamagnet, with magnetic susceptibility second only to superconductors. However, graphite containing certain defects can exhibit long-range magnetic ordering [7]. C.V. Raman reported that *"Diamagnetic susceptibility of graphite is markedly a function of particle size, diminishing steadily with increase in sub-division of the substance"* [6]. Later in 1995, Ishi et al. found that microcrystalline graphite shows ferromagnetic ordering [8,9]. The first organic ferromagnet was reported by Korshak et al. in 1987 [10]. Later in 1991, Ovchinnikov and Shamvosky theoretically predicted the possibility of a ferromagnetically ordered phase of mixed sp^2 and sp^3 pure carbon [11]. Interestingly, the calculated saturation magnetic moment in the sp^2–sp^3 mixed phase is higher than that of pure iron (Fe). Makarova et al. reported ferromagnetism in a two-dimensionally polymerized rhombohedral phase of C$_{60}$ [1]. Highly oriented pyrolitic graphite (HOPG) samples were found to show ferromagnetic (or ferrimagnetic) ordering when irradiated with MeV protons [12,13]. Low energy nitrogen ion (a few keV energy)-implanted nano-diamonds have also been reported to be ferromagnetic [14]. According to a theoretical investigation by Vozmediano et al., proton irradiation can produce large local defects that give rise to the appearance of local moments whose interaction can induce ferromagnetism in a large portion of the graphite sample [7].

Ohldag et al. demonstrated that magnetic order in proton-irradiated metal-free carbon films originates from a carbon π-electron system using x-ray dichroism spectromicroscopy [15]. Ferromagnetic ordering has also been observed in doped C$_{60}$ and microcrystalline carbon samples [16–18]. The discovery of ferromagnetism in rhombohedral C_{60} polymers [2,14] has opened up the possibility of a whole new family of magnetic fullerenes and fullerides.

We have found that magnetism is induced in proton-irradiated C_{60} films [19]. The use of ion beams to produce magnetically active areas points toward the possibility of tailoring the size of such magnets and has potential applications in spin electronics [14,20,21].

4.2 Soft Ferromagnetism in C_{60} Thin Films

Magnetization measurements were performed on pristine and 2 MeV proton-irradiated C_{60} films deposited on hydrogen-passivated Si(111) surfaces [19,22,23]. Magnetic measurements were carried out using a superconducting quantum interference device (SQUID) magnetometer.

The results of magnetization vs field (M–H) measurements on a pristine sample and samples irradiated at a fluence of 6×10^{15} H$^+$/cm^2 are shown in Figure 4.1.

The magnetization vs field (M–H) measurements at 5 K for the irradiated film show a marked increase in magnetization and a tendency toward saturation (Figure 4.1). A weak remnant magnetization of the order of a few tens of μemu is observed (data not shown).

At 300 K, both the as-deposited and the irradiated sample show diamagnetic behavior. Magnetization data in the temperature range of 2 K–300 K, in a 1 T applied field, for the irradiated film shows much stronger temperature dependence compared with that of the pristine film (Figure 4.2).

The magnetization (M) vs field (H) isotherm of irradiated C_{60} obtained at 5 K (Figure 4.1) clearly shows a tendency toward saturation, which is a signature of ferromagnetism. The hysteresis associated with the M–H curve is feeble, as expected for a soft ferromagnetic material. Similar soft-ferromagnetism was indeed observed earlier in an organic fullerene C_{60} material, namely C_{60} TDAE$_{0.86}$, where the ferromagnetic state showed no remanence [17]. An alternative explanation of the M–H curve for the irradiated sample in Figure 4.1 is that it arises from superparamagnetism. In order to demonstrate that the observed M–H curve is not due to superparamagnetism and the observed magnetic behavior in the irradiated C_{60} film is stable over a long period, we present the results of measurements made

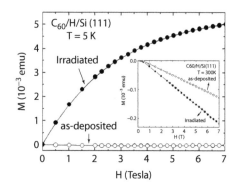

FIGURE 4.1

M vs H at 5 K for an as-deposited and an irradiated C_{60} film after subtracting the substrate [H-Si(111)] contribution (substrate data not shown here). (Reprinted with permission from S. Mathew et al., *Phys. Rev. B* 75, 75426, 2007 copyright (2017) by the American Physical Society.)

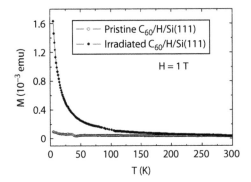

FIGURE 4.2
M vs T for as-deposited and irradiated C$_{60}$ film in an applied field of 1 T (substrate [H-Si(111)] contribution subtracted). (Reprinted with permission from S. Mathew et al., *Phys. Rev. B.* 75, 75426, 2007 copyright (2017) by the American Physical Society.)

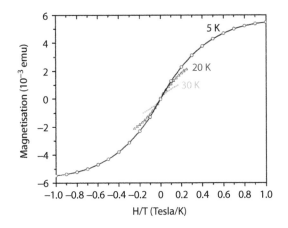

FIGURE 4.3
M vs H/T at 5 K, 20 K and 30 K for the same irradiated C$_{60}$ film as in Figure 4.2 measured again one year after the first measurement (Figure 4.1). (Reprinted with permission from S. Mathew et al., *Phys. Rev. B* 75, 75426, 2007 copyright (2017) by the American Physical Society.)

on the same sample one year after the original measurements (Figure 4.3) had been made. Measurements at three different temperatures (5 K, 20 K and 30 K) have been made to test one criterion for superparamagnetism, that is M–H isotherms should scale as H/T [24]. The magnetization vs H/T curves for different temperatures are shown in Figure 4.3 and from the figure it is clear that they do not superimpose. These isotherms should superimpose for paramagnetic or superparamagnetic materials. From these results, in conjunction with the observed feeble remnant magnetization, we conclude that the irradiated C$_{60}$ films are soft ferromagnets. The results also show that the ferromagnetic behavior in the irradiated film is stable over a long period of time.

The effects of the foregoing irradiation on the nature of the fullerene cage structure and the atomic arrangements can be understood using Raman spectra and high-resolution transmission electron microscope (HRTEM) images. Raman spectroscopy results for a pristine and an irradiated sample at a fluence of 6×10^{15} ions/cm^2 are given in Figure 4.4. An enlarged part of the spectrum in the range of 1400–1550 cm^{-1} is given in Figure 4.4 (b) and (c). In the pristine sample, two dominant peaks are seen at 1458 and 1467 cm^{-1}. In the irradiated sample,

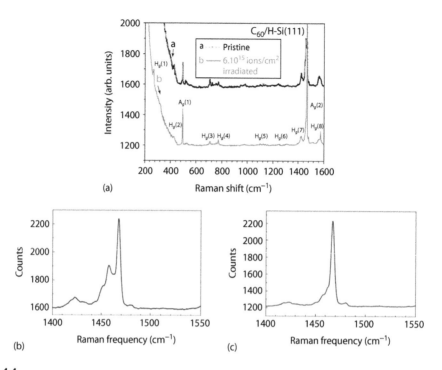

FIGURE 4.4

Raman spectrum of (a) pristine and C_{60}/H-Si(111) film irradiated with 2 MeV H^+ at a fluence of 6×10^{15}/cm². An enlarged part of $A_g(1)$ mode of pristine and irradiated spectra in (a) is given in (b) and (c). (Reprinted with permission from S. Mathew et al., *Phys. Rev. B.* 75, 75426, 2007 copyright (2017) by the American Physical Society., and S. Mathew et al., 2007, Two mega-electron volt proton-irradiation effects on fullerene films, *Radiat Eff Defect S*, Nov. 2007, Taylor & Francis.)

the first peak height is reduced while the second is enhanced. Sauvajol et al. [25] showed that the appearance of a mode at 1458 cm^{-1} was due to phototransformation (partial polymerization) of pure C_{60}. However, in the irradiated film, the intensity of this mode is reduced along with an enhancement of the 1467 cm^{-1} vibrational mode. The reduction in the intensity of the 1458 cm^{-1} mode in the irradiated sample in Figure 4.4 may be due to the fact that irradiation has caused enough disorder so that the fraction of the phototransformed component is reduced in comparison with the pristine film. Another explanation for the appearance of the mode at 1458 cm^{-1} is given by Duclos et al. [26]. They showed that a C_{60} film grown and kept under vacuum shows only a broad peak at 1458 cm^{-1}. Exposure of the film to O_2 or air gives rise to the peak at 1467 cm^{-1}. Although the authors could not confirm the origin of this peak, they tend to believe that it lies in the structural perturbation of the C_{60} molecule itself. This conclusion was based on, besides other reasons, the fact that a total of 16 modes were observed, while the icosahedral symmetry of C_{60} predicts only 10, signifying a lowering of the symmetry of the C_{60} molecule in the O_2- or air-exposed film. With ion irradiation, a fraction of C atoms is displaced from the C_{60} cage and thus contributes to symmetry lowering. The displacement of C atoms might be responsible for the enhancement of the intensity of the peak at 1467 cm^{-1} in the irradiated film. We believe this is why the intensity of the peak at 1467 cm^{-1} increases in the irradiated film.

The perturbation of the C_{60} molecular cage can be inferred from the HRTEM image in Figure 4.5 (b), although the overall crystalline structure, such as the crystalline planes and the diffraction spots (Figure 4.5), remain that of the fcc C_{60} crystals.

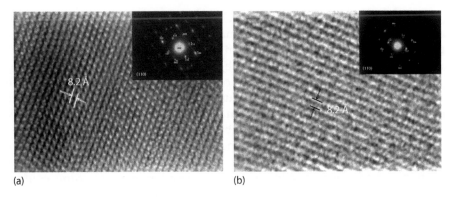

(a)   (b)

FIGURE 4.5

A HRTEM lattice image from (a) an as-deposited C_{60} film and (b) an irradiated C_{60} film. The corresponding selected area transmission electron diffraction (TED) patterns are given in the insets, respectively. The (111) planar spacing of fcc fullerene is marked in (a) and (b). (Reprinted with permission from S. Mathew et al., *Phys. Rev. B 75*, 75426, 2007 copyright (2017) by the American Physical Society.)

HRTEM lattice images of as-deposited and irradiated samples with corresponding diffraction patterns are shown in Figure 4.5. The transmission electron diffraction (TED) and HRTEM images confirm the crystalline nature of the C_{60} film in the fcc structure. Twin structures are seen in Figure 4.5 (a). The presence of defect structures in the irradiated film is evident from Figure 4.5 (b). In the case of the irradiated film (Figure 4.5 (b)), although the overall crystalline structure, such as the crystalline planes and the diffraction spots, remain that of fcc C_{60} crystals, one can clearly observe the perturbation of the C_{60} molecular cages.

The magnetic moment observed for the irradiated sample in an applied field of 7 T is about 5×10^{-3} emu with a tendency toward saturation. The magnetization curve of the irradiated film in high fields at 5 K compared with that of the as-deposited film and the empty substrate (the latter not shown here) gives clear evidence for the irradiation-induced magnetism in C_{60} films. The total amount of magnetic impurities (Fe, Cr, Ni) was determined by post-irradiation proton-induced x-ray emission (PIXE) experiments and was estimated to be ~50 ppm; the maximum magnetic moment contribution due to all of these impurities in our film will be less than 5×10^{-7} emu. Thus, the contribution to the observed magnetization due to these impurities is negligible.

The range of 2 MeV protons, calculated using the *stopping and range of ions in matter* (SRIM) [27] simulation code for an amorphous carbon target having the density of C_{60}, is found to be ~50 μm. Since our C_{60} film is only ~1.9 μm thick, the protons pass through the film and are buried deep in the Si substrate. The total energy loss of the proton beam in the present 1.9 μm thick C_{60} film is ~45 keV. The energy loss of protons at the top and bottom of the C_{60} film are 24.4 and 24.8 eV/nm, respectively. As the proton energy loss is nearly uniform over the whole thickness of the film, it is reasonable to assume that the irradiation damage is uniformly produced throughout the whole thickness of the film. We have used the total thickness of the film in order to determine the magnetization value in emu/g. The magnetization curve of the irradiated sample in Figure 4.1 has a tendency toward saturation at high fields. The magnetization at 7 T is about 200 emu/g. In the proton-irradiated C_{60} films, although defects are created, the observation of ordered periodic lattice fringes in the irradiated sample and the corresponding TED pattern indicate that irradiation did not cause the disintegration of the C_{60} cage leading to amorphization. Regarding the mechanism for the formation of the magnetic state in all carbon systems, among others, the defect-mediated mechanism appears to be the most general one. The defect-mediated mechanism has been addressed in a number of publications [1,2,7]. The possible origin of magnetism in these irradiated C_{60} films could be (i) irradiation-induced carbon vacancy in the system could lead to a singly occupied dangling sp^2 orbital that can give rise to a magnetic moment; (ii) the presence of nanographitic fragments that have zigzag edges could lead to splitting up of the flat energy bands and lowering the energy of the spin-up band than the spin-down band and hence the appearance of ferromagnetism in the material. Recent theoretical studies have predicted that magnetic ground states are stable in the edges of isolated graphene sheets whether they are hydrogen passivated or not [28]. In the present work, 2 MeV protons (range 50 μm) were used to irradiate a film of thickness 2 μm and hence the possibility of a H-terminated edge is negligible. The proton irradiation on C_{60} films could lead to the formation of nanographitic fragments leading to magnetism. (iii) A broken cage-like structure of polymerized fullerene and consequent broken interfragment C–C bonds could also lead to a magnetic moment. However, the Raman spectrum of irradiated film does not reveal any significant damage to the cage-like structure of C_{60}. The possible presence of magnetic moments in a hexagonal polymeric C_{60} layer has been predicted [1]. In this prediction, the magnetic moments would apparently arise from the formation of radical centers in polymerized C_{60}, with partially broken intermolecular bonds, without damage to the fullerene cages [29].

Although the details may be different for different carbon systems (graphite, polymeric fullerene, nanotubes, etc.), the common feature is the presence of undercoordinated atoms, such as atoms near vacancies [30] and atoms in the edges of graphene-like nanofragments [29–32]. Ohldag et al. have performed x-ray magnetic circular dichroism (XMCD) measurements on 2.25 MeV proton-irradiated graphitized carbon samples of thickness 200 nm and obtained a ferromagnetic signal [15]. They also demonstrated that the intrinsic origin of magnetism is from the π-electrons of carbon [15]. The clear observation of magnetism in graphitized carbon films of thickness 200 nm indicates that the effects are mainly due to the defects produced during the passage of protons through the film. Ion irradiation of any materials generates vacancies. The enhancement in magnetization observed in H^+-irradiated C_{60} samples may be due to defect moments from vacancies and/or the deformation and partial destruction of the fullerene cage. The HRTEM image in Figure 4.5 points to this possibility. Raman spectroscopy and HRTEM measurements on the irradiated sample show the stability of the fullerene crystal structure under the present irradiation condition. Further studies, such as estimating the presence of nanographitic fragments and carbon vacancies in such systems, will provide deeper insight into the origin of magnetism in the ion-irradiated C_{60} films.

According to a density functional study [31] of magnetism in proton-irradiated graphite [13], it is shown that H-vacancy complex plays a dominant role in the observed magnetic signal. For a fluence of 10 μC, the predicted signal is 0.8 μemu, which is in agreement with the experimental signal [3]. The implanted proton fluence in our sample is 77 μC (6×10^{15} ions/cm²) and all the protons are buried in silicon. So far, we have not come across any report showing magnetic ordering in proton-irradiated silicon. Even if we assume the same kind of magnetism due to H-vacancy complex in proton-irradiated silicon as in proton-irradiated graphite, the expected magnetic signal would be three orders of magnitude smaller than our observed result. Considering this fact, we can safely ignore the contribution of implanted protons in the Si substrate to the observed magnetism, which is predominantly due to atomic displacements caused by energetic protons while passing through the film.

4.3 Magnetism in MoS$_2$

4.3.1 Introduction

MoS$_2$ is one of the central members in transition metaldichalcogenide compounds [33]. Its layered structure, held together with van der Waals forces between the layers, along with its remarkable electronic properties, such as charge density wave transitions in transition metal dichalcogenides, make the material interesting from both fundamental and applied research perspectives [33–35]. Furthermore, monolayer MoS$_2$ is a semiconducting analog of graphene and has been fabricated [36,37].

An interesting weak ferromagnetism phenomenon in nanosheets of MoS$_2$ had previously been reported by Zhang et al. who attributed the observed magnetic signal to the presence of unsaturated edge atoms [38]. There have been several theoretical efforts in understanding ferromagnetic ordering in MoS$_2$. Li et al. predicted ferromagnetism in zigzag nanoribbons of MoS$_2$ using density functional theory [39]. The formation of magnetic moments was also reported in Mo_nS_{2n} clusters [40], nanoparticles [41] and nanoribbons [42–44] of MoS$_2$ from first-principle studies. The discovery of ferromagnetism in MoS$_2$ nano-sheets along with various simulation studies and the above reports on magnetism in carbon allotropes raise the possibility of magnetism in ion-irradiated MoS$_2$ system.

Here, we present the results of magnetization measurements on pristine and 2 MeV proton-irradiated MoS$_2$ samples [45].We find that magnetism is induced in proton-irradiated MoS$_2$ samples. The observation of long-range magnetic ordering in ion-irradiated MoS$_2$ points toward the possibility of selectively fabricating magnetic regions in a diamagnetic matrix, which may enable the design of unique spintronic devices.

4.3.2 Ferrimagnetism in MoS$_2$

Samples for irradiation were prepared in the following way. MoS$_2$ flakes, 2 mm in diameter and ~200 μm in thickness, were glued with diamagnetic varnish onto a high purity silicon substrate. Ion irradiations were carried out at room temperature. Magnetic measurements were performed using a SQUID system (MPMS SQUID-VSM) with a sensitivity of 8×10^{-8} emu.

The results of magnetization vs field (M–H) measurements at 300 K and 10 K for the sample irradiated at a fluence of 1×10^{18} ions/cm^2 and the pristine sample are shown in Figure 4.6. The pristine sample is diamagnetic in nature. The appearance of hysteresis along with a clear remanence and coercivity (~700 Oe at 10 K) and its decrease with increasing temperature in the irradiated sample (Figure 4.6 (A) and (B)) clearly indicate that MeV proton irradiation has induced ferro- or ferrimagnetic ordering in the MoS$_2$. The observed magnetic ordering can be due to the presence of defects such as atomic vacancies, displacements and saturation of a vacancy by the implanted protons. The same sample was subsequently irradiated at cumulative fluences of 2×10^{18} and 5×10^{18} ions/cm^2 to probe the evolution of induced magnetism with ion fluence.

Magnetizations as a function of field isotherms at a fluence of 5×10^{18} ions/cm^2 are shown in Figure 4.7 (a). An enlarged view of the M–H curves near the origin is given in the inset and the decrease of coercivity with increasing temperature is clear from the plot. A plot of coercivity vs ion fluence at various temperatures is given in Figure 4.7 (b). The value of coercivity is found to increase with ion fluence at all temperatures used

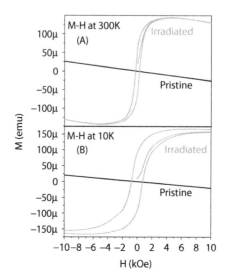

FIGURE 4.6

M vs H curve (A) at 300 K and (B) at 10 K for a pristine and an irradiated MoS_2 after subtracting the substrate Si contribution. (Reproduced from S. Mathew et al., *Appl. Phys. Lett.* 101, 102103, 2012 with the permission of AIP Publishing.)

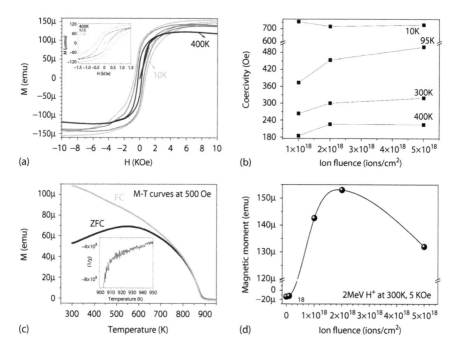

FIGURE 4.7

(a) M vs H curve at various temperatures from 400 K (black), 95 K (blue) and 10 K (red) for a pristine and an irradiated MoS_2 at a fluence of 5×10^{18} ions/cm². An enlarged view of the M–H isotherms near the origin is shown in the inset. The variation of coercivity vs ion fluence is shown in (b). (c) Zero field cooled (ZFC) and field cooled (FC) magnetization vs temperature measurements in an applied field of 500 Oe for an irradiated MoS_2 sample at a fluence of 5×10^{18} ions/cm². The inverse of the estimated magnetic susceptibility vs temperature plot near T_c (900–950 K) is shown in the inset. (d) The value of magnetization at 300 K with H = 5 kOe as a function of ion fluence from 1×10^{17} ions/cm² to 5×10^{18} ions/cm². (Reproduced from S. Mathew et al., *Appl. Phys. Lett.* 101, 102103, 2012 with the permission of AIP Publishing.)

in this study, although at 10 K it is almost constant. A large variation of coercivity with temperature indicates the presence of long-range magnetic ordering in the irradiated samples. The zero field cooled (ZFC) and field cooled (FC) magnetizations at an applied field of 500 Oe are given in Figure 4.7 (c). An insight into the nature of magnetic ordering (ferro- or ferrimagnetic) can be gained by analyzing the variation of susceptibility with temperature. The inverse of susceptibility plot is shown as an inset of Figure 4.7 (c). The value of Curie temperature (T$_c$) is estimated to be ~895 K. The nature of the curve near T$_c$, a concave curvature with respect to the temperature axis, is characteristic of ferrimagnetic ordering, whereas for a ferromagnetic material this curvature near T$_c$ would be convex [46]. The variation of magnetization at 300 K with an applied field of 5 kOe is plotted in Figure 4.7 (d) for different fluences. The magnetization of the pristine sample and the sample irradiated at a fluence of 1×10^{17} ions/cm^2 is negative, at a fluence of 1×10^{18} ions/cm^2 the sample magnetization becomes positive and at 2×10^{18} ions/cm^2 the magnetism in the sample increases further, and at a fluence of 5×10^{18} ions/cm^2 it has decreased. The dependence of magnetization on the irradiation fluence observed in Figure 4.7 (d) (the bell-shaped curve) indicates that the role of the implanted protons and thus the effect of end-of-range defects for the observed magnetism is minimal, as had been shown in the case of proton-irradiated HOPG [15,47,48]. It was demonstrated that 80% of the measured magnetic signal in the 2 MeV H$^+$-irradiated HOPG originates from the top 10 nm of the surface [48]. To probe the H$^+$-irradiated radiation-induced modification in our samples near the surface region, we used x-ray photoelectron spectroscopy (XPS) and Raman spectroscopy.

The modifications in the atomic bonding and core-level electronic structure can be probed using XPS. The XPS spectra of a pristine sample and the sample irradiated at a fluence of 5×10^{18} ions/cm^2 are shown in Figure 4.8(a)–(d). Fitting of the spectra was done by a chi-square iteration program using a convolution of Lorentzian–Gaussian functions with a Shirley background. For fitting the Mo 3d doublet, the peak separation and the relative area ratio for 5/2 and 3/2 spin–orbit components were constrained to be 3.17 eV and 1.5 eV, respectively, while the corresponding constraints for the S 2p 3/2 and 1/2 levels were 1.15 eV and 2 eV, respectively [49,50]. The peaks at 228.5 eV and 231.7 eV observed in the pristine spectrum of Mo are identified as Mo 3d$_{5/2}$ and 3d$_{3/2}$, while the small shoulder at 226 eV in Figure 4.8 (a) is the sulfur 2s peak [49]. In the irradiated spectrum in Figure 4.8 (b), apart from the pristine Mo peaks, two additional peaks at 229.6 eV and 232.8 eV are visible. The peak observed at 229.6 eV in Figure 4.8 (b) has 18% intensity of the total Mo signal, which could be due to a Mo valence higher than +4. The binding energy positions of 3d levels in Mo (V) have been reported to be 2 eV higher than those of Mo (IV) [51]. We found a peak at 229.6 eV in the irradiated sample that is only 1.0 eV above that of the Mo (IV) level. The pristine spectrum of S consists of S 2p$_{3/2}$ and 2p$_{1/2}$ peaks at 161.4 and 162.5 eV and another two peaks at 163 eV and 164.2 eV. The peak at 163 eV in the irradiated spectrum of S in Figure 4.8 (d) had increased by 6% in intensity compared to that in the pristine sample.

An indication of the nature of the induced defects and crystalline quality can be gained using Raman spectroscopy, which was a major tool for characterizing ion irradiation–induced defects in graphene and graphite in a recent study [51]. The Raman spectra of the pristine and irradiated samples at a fluence of 5×10^{18} ions/cm^2 are shown in Figure 4.9. The E_{2g}^1 mode at 385 cm^{-1} and A$_{1g}$ mode at 411 cm^{-1} are clearly seen in Figure 4.9 [52]. In the low frequency sides of E_{2g}^1 and A$_{1g}$ phonon modes, the peaks observed in the deconvoluted spectra are the Raman-inactive E_{1u}^2 and B$_{1u}$ phonons; these modes become Raman active due to the resonance effect, as observed by Sekine et al. [53]. These extra phonons are

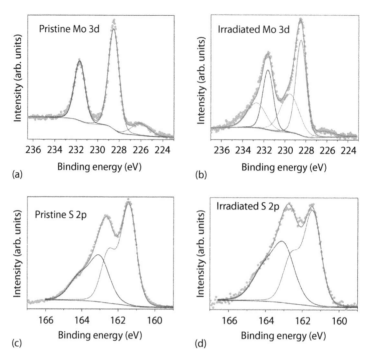

FIGURE 4.8

XPS spectra from pristine and irradiated MoS_2 at a fluence of 5×10^{18} ions/cm². The Mo peak is given in (a) and (b) and the S peak in (c) and (d). The fitted spectra along with the constituent peaks and experimental points are also shown. (Reproduced from S. Mathew et al., *Appl. Phys. Lett.* 101, 102103, 2012 with the permission of AIP Publishing.)

Davydov pairs of the E_{2g}^1 and A_{1g} modes [54]. The broad peak observed at ~452 cm⁻¹ can be the second order of LA(M) phonon [54].

In the irradiated sample, a peak at 483 cm⁻¹ is clearly visible. Frey et al. reported a peak at 495 cm⁻¹ in chemically synthesized fullerene-like and platelet-like nanoparticles of MoS_2 and assigned this to the second-order mode of the zone-edge phonon at 247 cm⁻¹ [54]. Phonon dispersion and the density of the phonon state calculations have shown a peak ~250 cm⁻¹ due to a TO branch phonon [56]. The mode at 483 cm⁻¹ is close to the above zone-edge phonon observed in nanoparticles of MoS_2 [54]. The appearances of a mode at 483 cm⁻¹ along with the broadening of the mode at 452 cm⁻¹ indicate the presence of lattice defects due to proton irradiation of the sample. The FWHM of the E_{2g}^1 and A_{1g} modes has not increased in the irradiated MoS_2, and this shows that the lattice structure has been preserved in the near-surface region of the irradiated sample. The A_{1g} mode couples strongly to the electronic structure compared to E_{2g} as observed in resonance Raman studies [55]. The ratio of the intensity of A_{1g} to E_{2g} modes (hereafter R) can be a measure of Raman cross section as discussed in high pressure Raman spectroscopy studies [56]. The intensity ratio R is found to enhance by 16% in the irradiated sample compared to the pristine sample as shown in Figure 4.9. This increase in the intensity ratio R in the irradiated sample can be attributed to the induced changes in the electronic band structure, which enhances the interaction of electrons with A_{1g} phonons [56]. The intensities of the E_{1u}^2 and B_{1u} phonon modes were also found to be enhanced in the irradiated sample. This enhancement of the Davydov pairs and second-order LA(M) peak along with the appearance of a defect mode indicate a deviation from the perfect symmetry of the system.

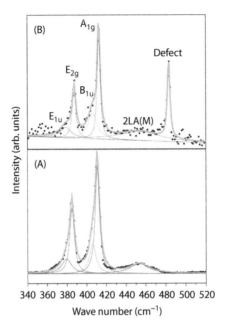

FIGURE 4.9

Raman spectra of (A) pristine MoS$_2$ and (B) irradiated MoS$_2$ at a fluence of 5×10^{18} ions/cm^2. The deconvoluted modes are labeled in the spectrum, and the fitted curve with constituent peaks and experimental points are also given. (Reproduced from S. Mathew et al., *Appl. Phys. Lett.* 101, 102103, 2012 with the permission of AIP Publishing.)

The magnetic moment observed for the irradiated sample at an applied field of 2000 Oe is as high as ~150 µemu along with a clear hysteresis and coercive field of 700 Oe at 10 K. The magnetization curve of the irradiated flake compared with that of the pristine sample and the blank substrate (the latter not shown here) gives clear evidence for the irradiation-induced magnetism in MoS$_2$. The total amount of magnetic impurities present in the sample was determined by post-irradiation PIXE experiments. PIXE results show ~54 ppm of Fe in the sample and if we assume that all of these Fe impurities became ferromagnetic, which is a rather unrealistic assumption, the maximum signal would be only 39 µemu, far short of the observed magnetization that is at least a factor of 4 larger. Thus, the observed magnetism cannot be explained by magnetic impurities. The fact that we observe diamagnetism in the pristine sample and clear magnetic hysteresis loops in the same sample after exposure to MeV protons provides irrefutable evidence for the intrinsic nature of the observed magnetic signal.

Regarding the mechanism for the formation of the magnetic state in MoS$_2$, among others, the defect-mediated mechanism appears to be the most general one. Possible origins of magnetism in an irradiated MoS$_2$ system could include the following: (i) irradiation-induced point defects that can give rise to a magnetic moment; and (ii) the presence of edge states as fragments that have zigzag or armchair edges could lead to splitting the flat energy bands and lowering the energy of the spin-up band compared to the spin-down band, leading to ferromagnetism in the material. The Raman spectrum in Figure 4.9 (b) showed a defect mode at 483 cm^{-1} due to the presence of zone-edge phonons in the irradiated sample. The appearance of well-defined E_{2g}^1 and A$_{1g}$ modes in Figure 4.9 (b) does not reveal any significant damage to the crystal structure of MoS$_2$ lattice. The presence of atomic vacancies, predominantly S in our case, can create a loss of symmetry and hence the coordination number of Mo atom would not remain as 6 in the irradiated sample. Tiwari et al. observed sulfur vacancy–induced surface reconstruction of MoS_2 under high-temperature treatment (above 1330 K) [57]. The higher valence of Mo observed in the irradiated sample could be due to a reconstruction of the lattice, apart from the presence of sulfur vacancies [57–59].

An estimate of the defect density created by the proton beam in MoS_2 can be determined from Monte Carlo simulations (SRIM 2008) using full damage cascade. The displacement energy of Mo and S for the creation of a Frenkel pair used for the calculation is 20 eV and 6.9 eV, respectively, as reported by Komsa et al. in a recent study of electron irradiation hardness of transition metal dichalcogenides [60]. According to this calculation, the 2 MeV proton comes to rest at a depth of 31 μm from the surface. The distance between vacancies estimated at the surface and at the end of the range after irradiating with 1×10^{18} ions/cm^2 is 6.3 and 2.7 Å, respectively. These calculated values are overestimates because the annealing of defects and the crystalline nature of the target have not been incorporated in SRIM simulations.

We carried out another experiment involving low-energy proton irradiation where we subjected the MoS_2 sample to 0.5 MeV H$^+$ irradiation at an ion fluence of 1×10^{18} ions/cm^2 (the fluence at which magnetic ordering was observed using 2 MeV protons) and at a lower fluence of 2×10^{17} ions/cm^2. Ferromagnetism was not observed in the former case, whereas a weak magnetic signal with a clear variation of coercivity with temperature was observed at the lower fluence in the latter case as evident from Figure 4.10. Also, we have observed a weak magnetic signal with a clear hysteresis loop in the former case (the sample irradiated at a fluence of 1×10^{18} ions/cm^2) after annealing (350°C, 1 hour in Ar gas flow). If ion damage were solely responsible for magnetization, then at the end of the range of the 2 MeV ion and of the 0.5 MeV ion there should be very little difference. The electronic energy loss of the ion helps recrystallization while the nuclear energy loss is only responsible for defect creation [61,62]. The ratio between electronic and nuclear energy loss is 42% greater

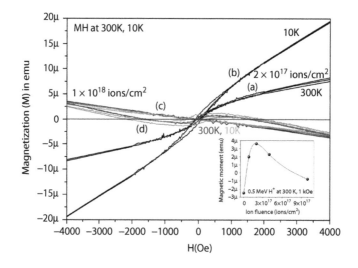

FIGURE 4.10

M vs H curves for an irradiated MoS_2 sample using 0.5 MeV protons (a) at 300 K, (b) at 10 K for a fluence of 2×10^{17} ions/cm^2, (c) for a fluence of 1×10^{18} ions/cm^2 at 300 K (blue curve) and at 10 K (red curve) and (d) the same sample in (c) after annealing in an Ar flow in a tube furnace (350°C, 1 hour). The value of the magnetization at 300 K with H = 1 kOe as a function of ion fluence is shown in the inset. The background contribution from the substrate Si is subtracted in all the curves. (Reproduced from S. Mathew et al., *Appl. Phys. Lett.* 101, 102103, 2012 with the permission of AIP Publishing.)

for a 2 MeV proton compared to a 0.5 MeV proton. This enhanced electronic energy loss component will allow the defective lattice to recover albeit with atomic displacements in the case of 2 MeV ion irradiation [61,62]. In the case of 0.5 MeV ions, the electronic energy loss appears to be insufficient for the required reconstruction of the lattice together with the atomic displacements to induce a strong ferromagnetic signal and hence it is not effective in creating magnetic ordering in MoS$_2$.

The appearance of magnetism observed in proton-irradiated MoS$_2$ samples can be due to a combination of defect moments arising from vacancies, interstitials, deformation and partial destruction of the lattice structure, i.e., the formation of edge states and reconstructions of the lattice. To identify the relative contributions of the ion beam–induced defects toward the observed magnetism in MoS$_2$, first-principle simulations incorporating atomic vacancies, edge states and lattice reconstructions, such as the ones performed in the case of various carbon allotropes, are required.

4.4 Conclusions

We have shown that soft-ferromagnetism can be induced in C$_{60}$ films with 2 MeV proton irradiation, while MoS$_2$ displays ferrimagnetic behavior with a Curie temperature of 895 K. Magnetism in this irradiated C$_{60}$ film arises due to atomic displacements caused by the energetic protons as they pass through the film. Possible sources of magnetization in the above systems are isolated vacancies, vacancy clusters, formation of edge states and reconstructions of the lattice. The discovery of ion irradiation–induced magnetism in MoS$_2$ and C$_{60}$ films sheds light on tailoring its properties by engineering the defects using energetic ions and provides a route for future applications of these materials.

Acknowledgments

S. Mathew would like to dedicate this article to the memory of Professor S. N. Behera (IOP Bhubaneswar, India) who had been an inspiration to look for magnetism in unconventional magnetic materials such as in modified carbon allotropes and other layered solids.

I acknowledge the support and care received from my mentors Professors T. Venkatesan and M.B.H. Breese from NUSNNI-NanoCore NUS Singapore, Professor J.T.L. Thong (Dept. ECE, NUS Singapore) and Professor B.N. Dev (IOP Bhubaneswar and IACS Kolkata India and now in IIT Kharagpur). I also thank all my collaborators in NUS Singapore and IOP Bhubaneswar, without their support this work would not have been possible.

References

1. T.L. Makorova, B. Sundqvist, R. Hohne, P. Esquinazi, Y. Kopelevich, P. Scharff, V.A. Davydov, L.S. Kashevarova and A.V. Rakhmania, *Nature* 413 (2001) 716; *Nature* 440 (2006) 707.
2. R.A. Wood, M.H. Lewis, M.R. Lees, S.M. Bennington, M.G. Cain and N. Kita-mura, *J. Phys. Cond. Matter.* 14 (2002) L385.

3. P. Esquinazi and R. Hohne, *J. Magn. Magn. Mater* 290–291 (2005) 20.
4. A.V. Rode, E.G. Gamaly, A.G. Christy, J.G.F. Gerald, S.T. Hyde, R.G. Elliman, B. Luther-Davies, et al., *Phys. Rev. B* 70 (2004) 54407.
5. J. Tuček, K. Holá, A. B. Bourlinos, P. Błoński, A. Bakandritsos, J. Ugolotti, M. Dubecký, et al., *Nat. Commun.* 8 (2017) 14525.
6. C.V. Raman and P. Krishnamurthy, *Nature* 124 (1929) 53.
7. M.A.H. Vozmediano, F. Guinea and M.P. Lopez-Sancho, *J. Phys. Chem. Solids* 67 (2006) 562.
8. C. Ishii, Y. Matsumura and K. Kaneko, *J. Phys. Chem.* 99 (1995) 5743.
9. C. Ishii, N. Shindo and K. Kaneko, *Chem. Phys. Lett.* 242 (1995) 196.
10. Yu.V. Korshak, T.V. Medvedeva, A.A. Ovchnnikov and V.N. Spector, *Nature* 326 (1987) 370.
11. A.A. Ovchinnikov and I.L. Shamovsky, *J. Mol. Struct. (Theochem.)* 251 (1991) 133.
12. P. Esquinazi, D. Spemann, R. Hohne, A. Setzer, K.H. Han and T. Butz, *Phys. Rev. Lett.* 91 (2003) 227201.
13. P. Esquinazi, D. Spemann, K. Schindler, R. Höhne, M. Ziese, A. Setzer, K.-H. Han, S. Petriconi, et al., *Thin Solid Films* 505 (2006) 85.
14. S. Talapatra, P.G. Ganesan, T. Kim, R. Vajtai, M. Huang, M. Shima, G. Ramanath, et al., *Phys. Rev. Lett.* 95 (2005) 97201.
15. H. Ohldag, T. Tyliszczak, R. Höhne, D. Spemann, P. Esquinazi, M. Ungureanu and T. Butz, *Phys. Rev. Lett.* 98 (2007) 187204.
16. B. Narymbetov, A. Omerzu, V.V. Kabanov, M. Tokumoto, H. Kobayashi and D. Mihallovic, *Nature* 407 (2000) 883.
17. P.M. Allemand, K.C. Khemani, A. Koch, F. Wudl, K. Holczer, S. Donovan, G. Gruner, et al., *Science* 253 (1991) 301.
18. T.L. Makarova and F. Palacio (Eds.), *Carbon-Based Magnetism*, Elsevier, Ams-terdam, 2006.
19. S. Mathew, B. Satpati, B. Joseph, B.N. Dev, R. Nirmala, S.K. Malik and R. Kesavamoorthy, *Phys. Rev. B* 75 (2007) 75426.
20. D. Spemannn, P. Esquinazi, R. Höhne, A. Setzer, M. Diaconu, H. Schmid and T. Butz, *Nucl. Instr. Meth. Phys. B* 231 (2005) 433.
21. O.V. Yazev and M.I. Katsnelson, *Phys. Rev. Lett.* 100 (2008) 47209.
22. A.F. Hebard, O. Zhou, Q. Zhong, R.M. Fleming and R.C. Haddon, *Thin Solid Films* 257 (1995) 147.
23. S. Mathew, B. Satpati, B. Joseph and B.N. Dev, *Appl. Surf. Sci.* 249 (2005) 31.
24. A.E. Berkowitz and E. Kneller, *Magnetism and Metallurgy*, Academic Press, New York, 1969 p. 393; W.E. Henry. *Rev. Mod. Phys.* 25 (1953) 163.
25. J.L. Sauvajol, F. Brocard, Z. Hircha and A. Zahab, *Phys. Rev. B* 52 (1995) 14839.
26. S.J. Duclos, R.C. Haddon, S.H. Glarum, A.F. Hebard and K.B. Lyons, *Solid State Comm.* 80 (1991) 481.
27. J.F. Ziegler, J.P. Biersack and U. Littmark, *SRIM 2003—A Version of the TRIM Program: The Stopping and Range of Ions in Matter*, Pergamon Press, New York, 1995.
28. H. Lee, N. Park, Y.W. Son, S. Han and J. Yu, *Chem. Phys. Lett.* 398 (2004) 207.
29. V.V. Belavin, L.G. Bulusheva, A.V. Okotrub and T.L. Makarova, *Phys. Rev. B* 70 (2004) 155402.
30. P.O. Lehtinen, A.S. Foster, Y. Ma, A.V. Krasheninnikov and R.M. Nieminen, *Phys. Rev. Lett.* 93 (2004) 187202.
31. P. Esquinazi, A. Setzer, R. Hohne, C. Semmelhak, Y. Kopelevich, D. Spemann, T. Butz, et al., *Phys. Rev. B* 66 (2002) 24429.
32. K. Nakada, M. Fujita, G. Dresselhaus and M.S. Dresselhaus, *Phys. Rev. B* 54 (1996) 17954.
33. J.A. Wilson and A.D. Yoffe, *Adv. Phys.* 18 (1969) 193.
34. A.H. Castro Neto and K. Novoselov, *Rep. Prog. Phys.* 74 (2011) 082501.
35. H.S.S.R. Matte, A. Gomathi, A.K. Manna, D.J. Late, R. Datta, S.K. Pati and C.N.R. Rao, *Angew. Chem.* 122 (2010) 4153.
36. K.F. Mak, C. Lee, J. Hone, J. Shan and T.F. Heinz, *Phys. Rev. Lett.* 105 (2010) 136805.
37. Y. Zhan, Z. Liu, S. Najmaei, P.M. Ajayan and J. Lou, *Small* 8 (2012) 966.
38. J. Zhang, J.M. Soon, K.P. Loh, J. Yin, J. Ding, M.B. Sullivian and P. Wu, *Nano Lett.* 7 (2007) 2370.
39. Y. Li, Z. Zhou, S. Zhang and Z. Chen, *J. Am. Chem. Soc.* 130 (2008) 16739.

40. P. Murugan, V. Kumar, Y. Kawazoe and N. Ota, *Phys. Rev. A* 71 (2005) 063203.
41. K. Zberecki, *J. Supercond. Novel Magn.* 25 (2012) 2533.
42. A. Vojvodic, B. Hinnemann and J.K. Nørskov, *Phys. Rev. B* 80 (2009) 125416.
43. A.R. Botello-Mendez, F. Lopez-Urias, M. Terrones and H. Terrones, *Nanotechnology* 20 (2009) 325703.
44. R. Shidpour and M. Manteghian, *Nanoscale* 2 (2010) 1429.
45. S. Mathew, K. Gopinadhan, T.K. Chan, D. Zhan, L. Cao, A. Rusdi, M.B.H. Breese, et al., *Appl. Phys. Lett.* 101 (2012) 102103.
46. H.P. Myers, *Introductory Solid State Physics*, Taylor & Francis Group, Oxford, 1997.
47. T.L. Makarova, A.L. Shelankov, I.T. Serenkov, V.I. Sakharov and D.W. Boukhvalov, *Phys. Rev. B* 83 (2011) 085417.
48. H. Ohldag, P. Esquinazi, E. Arenholz, D. Spemann, M. Rothermel and A. Setzer, *New J. Phys.* 12 (2010) 123012.
49. J.R. Lince, T.B. Stewart, M.M. Hills, P.D. Fleischauer, J.A. Yarmoff and A. Taleb-Ibrahimi, *Surf. Sci.* 210 (1989) 387.
50. T.A. Patterson, J.D. Carver, D.E. Leyden and D.M. Hercules, *J. Phys. Chem.* 80 (1976) 1700.
51. S. Mathew, T.K. Chan, D. Zhan, K. Gopinadhan, A.R. Barman, M.B.H. Breese, S. Dhar, et al., *J. Appl. Phys.* 110 (2011) 84309.
52. T.J. Wieting and J.L. Verble, *Phys. Rev. B* 3 (1971) 4286.
53. T. Sekine, K. Uchinokura, T. Nakashizu, E. Matsuura and R. Yoshizaki, *J. Phys. Soc. Jpn.* 53 (1984) 811.
54. G.L. Frey, R. Tenne, M.J. Matthews, M.S. Dresselhaus and G. Dresselhaus, *Phys. Rev. B* 60 (1999) 2883.
55. N. Wakabayashi, H.G. Smith and R.M. Nicklow, *Phys. Rev. B* 12 (1975) 659.
56. T. Livneh and E. Sterer, *Phys. Rev. B* 81 (2010) 195209.
57. R.K. Tiwari, J. Yang, M. Saeys and C. Joachim, *Surf. Sci.* 602 (2008) 2628.
58. R.St.C. Smart, W.M. Skinner and A.R. Gerson, *Surf. Interface Anal.* 28 (1999) 101–105.
59. J.R. Lince, D.J. Carre and P.D. Fleischauer, *Langmuir* 2 (1986) 805.
60. H.P. Komsa, J. Kotakoski, S. Kurasch, O. Lehtinen, U. Kaiser and V. Krasheninnikov, *Phys. Rev. Lett.* 109 (2012) 035503.
61. T. Venkatesan, R. Levi, T.C. Banwell, T. Tomberllo, M. Nicolet, R. Hamm and E. Mexixner, *Mat. Res. Soc. Symp. Proc.* 45 (1985) 189.
62. A. Benyagoub and A. Audren, *J. Appl. Phys.* 106 (2009) 83516.

5

Hot Electron Transistors with Graphene Base for THz Electronics

Filippo Giannazzo, Giuseppe Greco, Fabrizio Roccaforte,
Roy Dagher, Adrien Michon, and Yvon Cordier

CONTENTS

5.1 Introduction

The terahertz (THz) frequency region of the electromagnetic spectrum (from 0.1 to 10 THz, corresponding to millimeter and sub-millimeter wavelengths) is the spectral range that separates electronics from photonics. Although access to this spectral range is strategic for application areas like communications, medical diagnostics and security, this has been historically difficult because of the technical challenges associated with generating, detecting, processing and radiating such high-frequency signals. From the electronic devices point of view, controlling and manipulating signals in this portion of the radio frequency (RF) spectrum requires solid-state transistors with a cut-off frequency (f_T) and maximum oscillation frequencies (f_{max}) well above 1 THz.

To date, the high electron mobility transistors (HEMTs) and heterojunction bipolar transistors (HBTs) have been the two main device architectures employed for ultrahigh-frequency applications. In particular, indium phosphide–based HEMTs and HBTs have been demonstrated with f_{max} exceeding 1 THz [1–5]. These record RF figures of merit have been achieved by coupling the advantageous physical properties of the InGaAs/InP material system (i.e. the large heterojunction offset for carrier confinement, the high electron mobilities in quantum well channels for HEMTs and p-doped bases for HBTs and the high achievable doping levels for low Ohmic contact resistivities) with aggressive lateral and vertical scaling of transistors geometries. However, further improvements of HEMTs' performances will ultimately be limited by the saturation velocity of electrons in the channel, whereas the diffusion of the minority carriers (electrons) across the p-type base will limit the performances of HBTs.

In this context, the hot electron transistor (HET) can represent a promising device concept with the potential to overcome the fundamental limitations of HEMTs and HBTs in ultrahigh-frequency applications. The HET is a unipolar and majority carrier vertical device where the base-to-emitter voltage controls the injection and the transport of ballistic hot electrons through an ultrathin transit layer (base) to the collector terminal. Due to its working principles, the HET does not suffer from the intrinsic limitations of HBTs (i.e. the minority carrier diffusion and recombination in the base) and has the potential to reach superior performances in the THz frequency range. However, the practical implementation of this device concept relies primarily on the possibility of achieving ballistic transport in the base, as well as on the efficiency of hot electrons injection from the emitter and finally on the filtering efficiency of the base-collector (B-C) barrier.

The HET device concept was introduced more than 50 years ago by Mead [6]. Since then, several material systems have been used for HET development including metal thin films [6–9], complex oxides [10], superconducting materials [11], III-V semiconductor heterostructures [12] and, more recently, Group III-nitride semiconductors heterostructures [13]. However, the successful demonstration of high-performance HETs has been limited by some technological issues, such as the difficulty to scale the base thickness below the electron mean free path of the carriers. In this context, single-layer, two-dimensional (2D) materials are naturally suitable for applications requiring ultrathin, defect-free films. In particular, monolayer graphene (Gr) with its excellent transport properties (high mobility, from ~10^3 up to ~10^5 cm^2 V^{-1} s^{-1}, and micrometer electron mean free path) [14–16] and a dangling-bond-free inert surface is an ideal candidate as a low resistance, scattering-free base material in HETs. Theoretical studies have predicted that with an optimized structure, f_T up to several terahertz [17], I_{on}/I_{off} over 10^5 and high-current gain can be achieved with Gr base HETs (GBHET). The initial experimental demonstrators of GBHETs, reported in 2013, showed successful direct current (DC) operation in terms of current modulation ($I_{on}/I_{off} > 10^5$) but suffered from a low output current density (~μA/cm^2), low current gain, low injection efficiency and high threshold voltage. These limitations were not intrinsic to the use of Gr as a base material and significant improvements have been obtained more recently by the careful choice of the barrier layers' materials and the improvement of the interfaces quality.

In this chapter, the operating principles of an ideal HET will be introduced, illustrating the device's DC characteristics and discussing the impact of the main physical parameters on the DC figures of merit (current transit ratio and current gain) and on the alternating current (AC) figures of merit (f_{max} and f_T). Therefore, an historical perspective on the attempted implementations of this device concept will be provided, starting from the first proposal of a metal base HET [6] to more recent implementations, such as the nitride semiconductors–based HETs and the GBHETs. The theoretical DC and AC performances of GBHETs will be discussed and the state-of-the-art GBHETs will be presented. The last section of the chapter will present open issues and new ideas to improve the performances of GBHETs.

5.2 Hot Electron Transistor (HET): Device Concept and Operating Principles

A general schematic illustration of a HET device and the related energy band diagram under equilibrium conditions are reported in Figure 5.1a and b, respectively. The HET

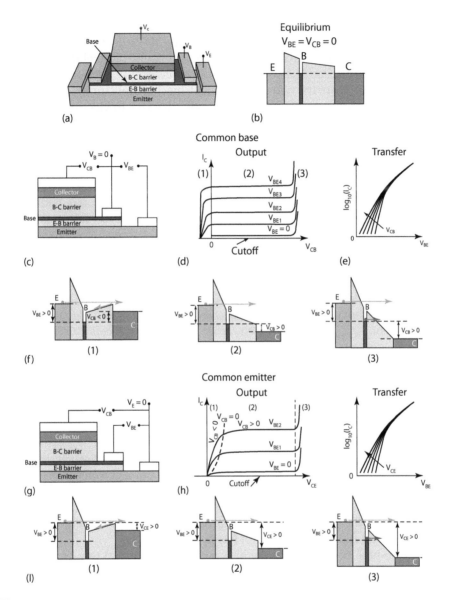

FIGURE 5.1
(a) Schematic illustration of a hot electron transistor HET and (b) energy band diagram of the device under equilibrium conditions ($V_{BE} = V_{CB} = 0\,V$). Cross-sectional schematic of a HET biased in the common-base configuration (c), output characteristics $I_C - V_{CB}$ (d), transfer characteristics $I_C - V_{BE}$ (e) and band diagrams for different V_{CB} values (f). Cross-sectional schematic of a HET biased in the common-base configuration (g), output characteristics $I_C - V_{CB}$ (h), transfer characteristics $I_C - V_{BE}$ (i) and band diagrams for different V_{CB} values (l).

consists of an emitter (E), a base (B) and a collector (C) region, which are separated by two barrier layers, i.e. the emitter-base (E-B) barrier with barrier energy height ϕ_{EB} and thickness d_{EB}, and the B-C barrier with barrier energy height ϕ_{BC} and thickness d_{BC}.

For a sufficiently high forward bias applied between the base and the emitter, electrons are injected into the base. A key aspect of the HET operation is that the injected electrons (hot electrons) have a higher energy compared to the Fermi energy of the electrons thermal population (cold electrons) in the base. Depending on the values of ϕ_{BE} and d_{BE}, as well as on the barrier's material quality, several potential mechanisms can rule the injection of electrons across the E-B barrier, i.e. (i) direct tunneling (DT), (ii) Fowler–Nordheim (FN) tunneling through the thinned triangular barrier, (iii) Poole–Frenkel (PF) traps' assisted emission through the barrier and (iv) thermionic emission (TE) above the barrier. As a matter of fact, FN and TE are the two most favorable mechanisms for hot electron injection into the base. On the contrary, DT and PF emission give rise to electron injection at any energy in the range from the emitter conduction band to the top of the E-B barrier, so that most of the injected carriers will keep part of the cold electron population in the base. As discussed later in this section, high values of the injected hot electrons current density, J_E, are fundamental to achieving high-frequency performances of hot electrons transistors. Hence, ϕ_{BE} and d_{BE} must be properly chosen to maximize J_E.

Ideally, for a base thickness $d_B < \lambda_{mfp}$ (with λ_{mfp} the scattering mean free path of hot electrons), a large fraction of the injected electrons can traverse the base ballistically, i.e. without losing energy, and finally reach the edge of the B-C barrier. This second barrier is aimed to act as an energy filter, which allows the hot electrons to reach the collector and reflects back the electrons with insufficient energy. These reflected electrons eventually become part of the cold electrons population in the base and contribute to the base current (I_B), whereas the hot electrons reaching the collector give rise to the collector current (I_C). Besides transmitting hot electrons, the B-C barrier must be thick and high enough to block the leakage current, I_{BCleak}, of cold electrons from the base to the collector. For a thick B-C barrier, the transmission probability T of hot electrons with energy $E > \phi_{BC}$ is ultimately ruled by quantum mechanics and it is related to the energy difference $E - \phi_{BC}$ as follows:

$$T = \left[\frac{1}{2} + \frac{1}{4}\left(\sqrt{\frac{E}{E - \phi_{BC}}} + \frac{E - \phi_{BC}}{E} \right) \right]^{-1} \tag{5.1}$$

Equation 5.1 implies that, for finite values of ϕ_{BC}, a certain fraction of hot electrons is always reflected back at the B-C barrier. Furthermore, since the hot electrons' energy is comparable to the E-B barrier height, ϕ_{BC} should be significantly lower than ϕ_{EB}.

Figure 5.1c–f further illustrates the principles of operation and the DC electrical characteristics of a HET in the common-base configuration, whereas the operation in the common-emitter configuration is illustrated in Figure 5.1g–l.

In the common-base configuration, the base contact is grounded (see Figure 5.1c), and a potential difference, $V_{BE} = (V_B - V_E) > 0$, is applied between the base and the emitter contacts, in order to allow hot electrons injection in the base. Depending on the values of the potential difference, $V_{CB} = V_C - V_B$, between the collector and base contacts, three current transport regimes can be observed in the common-base output characteristics, $I_C - V_{CB}$, for different values of V_{BE} (Figure 5.1d). For $V_{CB} > 0$ (region 2 in Figure 5.1d), I_C is almost independent of V_{CB}, i.e. all the injected hot electrons are transmitted above the B-C barrier (panel 2 of Figure 5.1f). For $V_{CB} < 0$, the collector edge of the B-C barrier is raised up and part of the hot electrons are reflected back in the base (panel 1 of Figure 5.1f), resulting in a decrease of I_C with increasing negative values of V_{CB}, up to device switch-off (region 1 of Figure 5.1d). For large positive values of V_{CB}, the leakage current (I_{BCleak}) contribution of cold electrons injected by FN tunneling through the B-C barrier becomes large (panel 3 of Figure 5.1f) and this leads to a rapid increase of I_C as a function of V_{CB} (region 3 of Figure 5.1d). The typical shape of the transfer characteristics ($I_D - V_{BE}$) for different values of V_{CB} are also reported in Figure 5.1e.

In the common-emitter configuration (Figure 5.1g), the emitter contact is grounded ($V_E = 0$), and electrons injection from the emitter to the base is achieved due to the potential difference, $V_{BE} = (V_B - V_E) > 0$. The common-emitter output characteristics (Figure 5.1h) are obtained by sweeping $V_{CE} = V_C - V_E$ from 0 to positive values, for different values of V_{BE}. Three current transport regimes can be observed in the $I_C - V_{CE}$ characteristics. For a fixed V_{BE}, low V_{CE} values (region 1 of Figure 5.1h) correspond to a $V_{CB} < 0$, i.e. to an upward bending of the B-C barrier at the collector edge (panel 1 of Figure 5.1l), resulting in a partial back-reflection of hot electrons from the base. In this transport regime, I_C grows with increasing V_{CE} up to the condition corresponding to $V_{CB} = 0$. For larger V_{CE} values (region 2 of Figure 5.1h), I_C is independent of V_{CE}, i.e. all the injected hot electrons are transmitted to the collector (panel 2 of Figure 5.1l). Finally, for very large V_{CE} values (region 3 of Figure 5.1h), the cold electrons leakage current through the B-C triangular barrier (panel 3 of Figure 5.1l) becomes dominant, leading to a rapid increase of I_C. The typical shape of the transfer characteristics ($I_D - V_{BE}$) for different values of V_{CE} are also reported in Figure 5.1i.

The main figures of merits for the DC operation of a HET are

1. The common-base current transfer ratio $\alpha = I_C/I_E$
2. The common-emitter current gain β defined as $\beta = I_C/I_B$

For good DC performances, $\alpha \approx 1$ and β as large as possible are needed. This ensures that most of the electrons injected from the emitter reach the collector and only a very small fraction of them contribute to the base current. We can express α as $\alpha = \alpha_B \alpha_{BC} \alpha_C$, where $\alpha_B = \exp(-d_B/\lambda_{mfp})$ is the base efficiency, α_{BC} is the B-C barrier filtering efficiency and α_C is the collector efficiency, respectively. An ultrathin base with a high electron mean free path is required to achieve $\alpha_B \approx 1$.

The high-frequency figures of merit for a transistor are the current gain cut-off frequency, f_T, and the power gain maximum oscillation frequency, f_{max}. Both of these are defined in terms of the parameters of the small-signal (SS) equivalent circuit, which is very similar to that of the HBT.

In particular, f_T is defined as follows:

$$\frac{1}{2\pi f_T} = \tau_d + \frac{C_{tot}}{g_m} \tag{5.2}$$

τ_d is the sum of the delay times associated with electrons transit in the E-B barrier layer, in the base and in the B-C filtering layer, i.e.

$$\tau_d = \tau_{EB} + \tau_B + \tau_{BC} \tag{5.3}$$

In particular, τ_B can be expressed as $\tau_B = d_B/v_B$, where d_B is the base thickness and v_B is the velocity of hot electrons in the base. In the case of an ultrathin base, ensuring ballistic transport, this term is very small and can be neglected. The delay τ_{EB} due to electronic transport in the B-E barrier strongly depends on the current injection mechanisms, i.e. it can be very small for FN tunneling through the barrier and higher for TE over the barrier. The delay τ_{BC} associated with electrons transit in the B-C barrier is typically the largest term in Equation 5.3. It is proportional to d_{BC}/v_s, where d_{BC} is the B-C barrier thickness and v_s is the saturated electron velocity in this region. A thin B-C barrier would be necessary to reduce τ_{BC}. On the other hand, d_{BC} cannot be reduced too much in order to avoid an increase of J_{BCleak} and a low collector breakdown voltage, V_{Cbr}.

The second term in Equation 5.2 is the base charging delay, i.e. the time spent charging the base. It is directly proportional to the total capacitance, $C_{tot} = C_{EB} + C_{BC}$, where $C_{EB} = \varepsilon_0 \varepsilon_{EB}/d_{EB}$ and $C_{BC} = \varepsilon_0 \varepsilon_{BC}/d_{BC}$ are the capacitances of the E-B barrier layer and the B-C filtering layer connected in parallel, ε_0 is the vacuum permittivity and ε_{EB} and ε_{BC} are the relative permittivities of the E-B and B-C barrier layer materials. Furthermore, it is inversely proportional to the device transconductance $g_m = dJ_C/dV_{BE}$. Clearly, the most effective way to minimize this charging delay is represented by the increase of g_m. In fact, a reduction of the barrier layer capacitances would imply an increase of the thickness d_{BE} and d_{BC}, with a consequent impact on the transit delay times. Under saturation conditions, when all the hot electrons injected from the emitter reach the collector ($J_C \approx J_E$), the transconductance $g_m \approx dJ_E/dV_{BE}$. As the emitter current is injected over a barrier, it exhibits an exponential dependence on V_{BE}, i.e. $J_E \propto \exp(qV_{BE}/kT)$. As a result, $g_m \propto qJ_E/kT$. This means that a high injection current density is one of the main requirements to achieve a high cut-off frequency, f_T.

The maximum oscillation frequency f_{max} can be expressed as

$$f_{max} = \sqrt{\frac{f_T}{2\pi R_B C_{BC}}} \qquad (5.4)$$

where R_B is the resistance associated with "lateral" current transport in the base layer from the device active area to the base contact. Hence, R_B is the sum of different contributions, i.e. the "intrinsic" base resistance $R_{B\,int} \propto \rho/d_B$ (with ρ the base resistivity and d_B the base thickness), the resistance of the Ohmic metal contact with the base and the access resistance from this contact to the device active area. All these contributions should be minimized to achieve a low R_B. Of course, the most challenging issue to obtain high f_{max} is to fabricate an ultrathin base (allowing ballistic transport of hot electrons in the vertical direction) while maintaining low enough intrinsic and extrinsic base resistances. However, for most of the bulk materials, reducing the film thickness to the nanometer or sub-nanometer range implies an increase of the resistivity, due to the dominance of surface roughness and/or grain boundaries scattering, as well as to the presence of pinholes and other structural defects in the film. In this context, 2D materials, such as graphene and transition metal dichalcogenides (TMDs), can represent ideal candidates to fabricate the base of HETs, since they maintain excellent conduction properties and structural integrity down to single atomic layer thickness. The state of the art and challenges in the use of graphene as a base material for HETs will be discussed in the final section of this chapter.

5.3 HET Implementations: Historical Perspective

This section presents an historical overview on the attempts made to implement the HET device concept using different materials systems, starting from the metal base transistor originally proposed by Mead [6] in the 1960s, to recent demonstrations of HETs fabricated by bandgap engineering of epitaxial nitride semiconductors [18–20]. Section 5.4 will be devoted to illustrating the very recent efforts to employ graphene or other 2D materials as ultimately thin base films of HETs [21–24].

5.3.1 Metal Base HETs

The first proposal of a three-terminal HET consisted of a metal-oxide-metal-oxide-metal (MOMOM) structure, whose band structure is illustrated in Figure 5.2a. The first

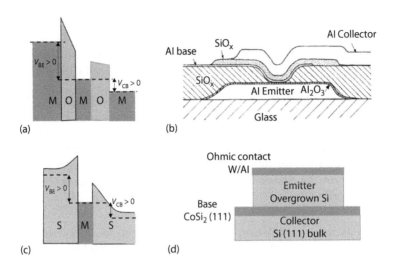

FIGURE 5.2
(a) Energy band diagram of a metal-oxide-metal-oxide-metal (MOMOM) hot electrons transistor and (b) experimental implementation of this device using an Al/Al$_2$O$_3$/Al/SiO$_x$/Al stack. (c) Energy band diagram of a semiconductor-metal-semiconductor (SMS) hot electron transistor and (d) its implementation with a Si/CoSi$_2$/Si stack. (Panel (b) adapted with permission from Ref. [6] Copyright (1961), American Institute of Physics. Panel (d) adapted with permission from Ref. [8] Copyright (1985), American Institute of Physics.)

prototype was practically implemented using an Al/Al$_2$O$_3$/Al/SiO$_x$/Al stack [6], as shown in the cross-sectional schematic of Figure 5.2b. The emitter terminal was represented by the thick Al stripes on the bottom, the Al$_2$O$_3$ E-B barrier (5–10 nm thick) was formed by oxidation of the Al surface and the base was obtained by the evaporation of an Al thin film (~30 nm). Finally, the B-C barrier was formed by the deposition of an SiO$_x$ thin film (≈10 nm), with the collector terminal obtained by evaporating thick Al stripes on top. Electron injection in the base was based on electron tunneling through the thin Al$_2$O$_3$ film into a high-energy state in the metal base, and a common-base current transfer ratio α ≈ 0.1 was reported for this very first HET prototype. This device structure suffered many drawbacks: (i) the difficulty of realizing a pinhole-free metal base thinner than the electron mean free path in the metal; and (ii) the presence of traps in the E-B and B-C barrier layers, resulting in low injection and collection efficiencies. To address this latter issue, the MOMOM device structure was later replaced by the semiconductor-metal-semiconductor (SMS) transistor (see schematic in Figure 5.2c), where the metal base layer was sandwiched between two semiconductors and the injection of hot electrons was primarily due to TE over the E-B Schottky barrier [7]. These devices were practically realized by an Si/CoSi$_2$/Si stack (see schematic in Figure 5.2d), where the CoSi$_2$ base (with thickness of 5–30 nm) was obtained by a solid-state reaction (silicidation) of Co films deposited on Si, resulting in an epitaxial interface [9]. This epitaxial interface between the emitter and base layers was beneficial to improve the hot electron injection. However, the limitation in the base efficiency still remained, resulting in a poor common-base transfer ratio α.

5.3.2 HETs Based on Semiconductor Heterostructures

Progress in the heteroepitaxial growth of semiconductors also opened the way to the implementation of HETs fully based on semiconductor heterostructures. III-V materials were first employed to fabricate HETs with the E-B and B-C barriers realized by bandgap engineering. As an example, Levi and Chiu [25] employed an $AlSb_{0.92}As_{0.08}/InAs/GaSb$ heterostructure to demonstrate for the first time a double heterojunction HET operating at room temperature with common-emitter gain $\beta > 10$ and high collector current density $>1200\,A/cm^2$.

Later on, Moise et al. [26] used an InAlGaAs/AlAs/InGaAs/InAlGaAs/InGaAs heterostructure to demonstrate a HET based on tunneling injection of hot electrons through the AlAls barrier. This device, originally named tunneling hot electron transfer amplifier (THETA), showed current gain over the entire voltage range from turn-on to breakdown.

More recently, Group III-nitride semiconductors (III-N) have been considered for epitaxial HETs fabrication. As compared to III-V, III-N materials offer several advantages for HETs implementation:

1. The wider bandgap modulation range and higher conduction band discontinuities (with ΔE_c between GaN and AlN being 1.8 eV and that between AlN and InN being ~3 eV) enable the design of HETs with higher injection energy and, hence, higher injection velocity.

2. Polarization charges in the III-N systems can be engineered to obtain a high-density and high-mobility 2DEG in the base that can be used to reduce the base resistance R_B in the HET.

3. The III-N systems possess high intervalley separations (~1.3 eV for GaN as compared to ~0.3 eV for GaAs), and hence, a wider range of injection energies is possible in the nitride systems prior to the onset of intervalley scattering events.

In 2000, Shur et al. [13] first simulated a unipolar HET based on a GaN/AlGaN/GaN/AlInGaN heterostructure, where hot electron injection from the emitter (n$^+$-GaN) to the thin (5 nm) base (n$^-$-GaN) occurred by FN tunneling through the AlGaN barrier layer. Those simulations predicted a high emitter current density ($J_E = 10^4 - 10^5\,A/cm^2$), high common-base current gain $\beta = I_C/I_B$ up to 60 and low base resistance R_B, thanks to the high mobility (~1500 cm^2/V_s) and high concentration ($n \approx 10^{13}\,cm^{-2}$) polarization induced 2DEG.

The first demonstrators of III-N-based HETs have been reported more recently. As an example, Dasgupta et al. [18] demonstrated a HET with an $Al_{0.24}Ga_{0.76}N$ barrier emitter, a GaN base (10 nm in thickness) and an $Al_{0.08}Ga_{0.92}N$ barrier collector structure. The values of the E-B barrier ($\phi_{EB} = 0.27\,eV$) and the B-C barrier ($\phi_{BC} = 0.13\,eV$) were obtained, respectively, for the used Al mole fractions. TE was the main current injection mechanism from the emitter to the base, resulting in the high drive current capability of this device. A common-base transfer ratio $\alpha \approx 0.97-0.98$ was obtained for this first transistor. More recently, III-N HETs based on tunneling current injection through an ultrathin AlN emitter barrier into a GaN base with sub-10 nm thickness [19,20] have been demonstrated. As an example, Yang et al. [19] fabricated a HET with the heterostructure illustrated in Figure 5.3a and b, where the E-B barrier was represented by 3 nm AlN (capped by 1 nm UID GaN) and the base was formed by 8 nm n$^+$-GaN. A polarization engineered B-C barrier was obtained by stacking several layers, i.e. a 4 nm UID GaN spacer, 6 nm $Al_{0.25}Ga_{0.75}N$, 5 nm $AlGa_{1-x}N$ (with x ranging from 0.2 to 0.25 moving from top to bottom), 42 nm $Al_{0.3}Ga_{0.7}N$ and 28 nm UID GaN. Finally, the sub-collector was represented by 210 nm n$^+$-GaN. The output characteristics of this HET in the common-emitter configuration are reported in Figure 5.3c. The base current density (J_B vs V_{BE}) and the transfer characteristics (J_C vs V_{BE}) are shown in Figure 5.3d, from which a current gain $\beta \approx 1.5$ (at $V_{BE} = 8\,V$) was deduced (see Figure 5.3d, right axis).

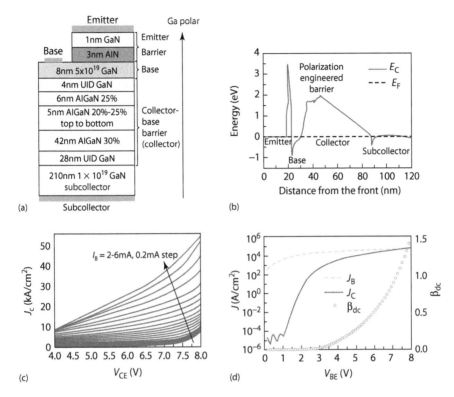

FIGURE 5.3
Schematic cross section (a) and energy band diagram (b) of a HET with a GaN emitter, an AlN (3 nm) tunneling barrier, an n$^+$-GaN (8 nm) base, a graded base-collector barrier and an n$^+$-GaN collector. Common-emitter output characteristics (c) and transfer characteristics (d) of this device. (Figures adapted with permission from Ref. [19] Copyright (2015), American Institute of Physics.)

Despite these very promising performances, these devices suffer from limitation in the base thickness scalability. This is an inherent limitation of the conventional semiconductor growth process, which increases the interface roughness and degrades the transport properties in very thin (<5 nm) base layers. Furthermore, the large polarization in III-N materials causes enhanced scattering in alloys and at interfaces. In this context, 2D materials, such as Gr [27] and semiconductor TMDs (MoS$_2$, MoSe$_2$, WS$_2$, WSe$_2$, etc.) [28,29], have the natural advantage of being stable materials down to single-layer thickness, which makes them ideal candidates to overcome the issue of scalability in the base of HETs. During the last few years, several groups have proposed different versions of devices with the base made with 2D materials [30,31]. In the following section, an overview of current strategies followed to implement graphene-base HETs (GBHETs) will be presented.

5.4 HETs with a Graphene Base

5.4.1 Theoretical Properties

A first theoretical estimation of the high-frequency performances of a GBHET was performed by Mehr et al. [32], by solving the 1D Schrödinger equation for the system formed by a metal emitter and collector, a Gr base, a thin insulator working as the E-B tunneling barrier and a thicker insulator working as the B-C filtering barrier. Figure 5.4a reports the schematic band diagrams of the GBHET under equilibrium conditions, in the off-state and in the on-state configurations. Figure 5.4b and c show the simulated output characteristics ($J_C - V_{CE}$) and transfer characteristics ($J_C - V_{BE}$) in the common-emitter configuration for a device with an E-B barrier layer with 3 nm thickness and barrier height $\phi_{EB} = 0.2$ eV, and a B-C filtering barrier with 70 nm thickness and $\phi_{BC} = 0.2$ eV. These curves illustrate a J_C modulation over several orders of magnitude as a function of the V_{BE} (Figure 5.4c) and

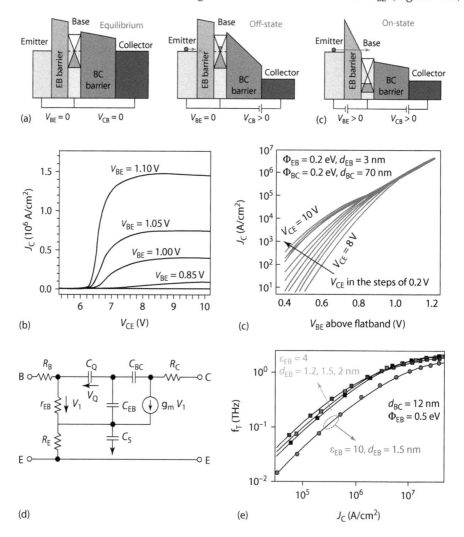

FIGURE 5.4

(a) Schematic band diagrams of a graphene-base hot electron transistor (GBHET) under equilibrium conditions, in the off-state and in the on-state. (b) Simulated common-emitter output characteristic $I_C - V_{CE}$ (for different V_{BE} values) and (c) transfer characteristics $I_C - V_{BE}$ (for different V_{CE} values). (d) Small-signal equivalent circuit of the GBHET and (e) evaluated cut-off frequency as a function of the collector current density J_C for different values of the E-B barrier thickness and permittivity. (Panels (b) and (c) adapted with permission from *Solid State Commun.*, 224, Vaziri, S., et al., Going ballistic: Graphene hot electron transistors, 64–75, Copyright (2015) from Elsevier. Panel (e) adapted with permission from *Microelectron. Eng.*, 109, Driussi, F., et al., Modeling, simulation and design of the vertical graphene-base transistor, 338–341, Copyright (2013) from Elsevier.)

current saturation in the output characteristics (Figure 5.4b). The cut-off frequency, f_T, was evaluated according to Equation 5.2 using the small-signal equivalent circuit illustrated in Figure 5.4d.

The main peculiar aspect of using a Gr base (instead of an ideal metal base) is that the applied bias $V_{CB} > 0$ and $V_{BE} > 0$ produce not only an increase in the electric field across the B-C and E-B barriers, but they also cause a Gr Fermi level shift with respect to the Dirac point, resulting in a change of the Gr charge density, Q_{gr}. This electrostatic behavior of Gr is expressed by its quantum capacitance $C_Q = |dQ_{gr}/dV_Q|$, where $V_Q = (E_F - E_D)/q$ is the potential drop on Gr.

C_Q is proportional to the density of states $D(E_F - E_D) = 2|E_F - E_D|/(\pi \hbar^2 v_F^2)$, where $\hbar$ is the reduced Planck's constant and v_F is the graphene Fermi velocity [33]. As compared to an ideal metal contact, which is able to accommodate large charge changes in response to small potential variations ($C_Q \longrightarrow \infty$), in the case of Gr C_Q has a finite value due to the finite density of states at the Fermi level. These peculiar properties of the Gr base clearly have an effect on both the DC and AC performances of the GBHET.

As illustrated in the equivalent circuit in Figure 5.4d, C_Q is connected in series to the parallel combination of the E-B capacitance (C_{EB}) and of the B-C capacitance (C_{BC}). As a result, the total capacitance in Equation 5.2 must be expressed as $C_{tot} = C_Q(C_{EB} + C_{BC})/(C_Q + C_{EB} + C_{BC})$. Hence, the finite value of C_Q for Gr results in a reduced C_{tot} (compared to the case of an ideal metal contact) and could be beneficial to the increase of f_T, according to Equation 5.2. On the other hand, since the bias V_{BE} partially falls on Gr, the effective electric field across the E-B barrier is lower with respect to the case of an ideal metal base. This results in a reduced injected current density J_E and, ultimately, in a lower transconductance g_m. Hence, the final effect of the quantum capacitance is a reduction of the estimated maximum f_T value with respect to the case when C_Q is not considered [32]. It should be noted that, notwithstanding this detrimental effect of the finite and bias dependent C_Q of Gr, f_T well above 1 THz have been predicted by these calculations. Figure 5.4e illustrates the behavior of f_T vs the collector current density J_C evaluated considering GBHETs with different values of the E-B barrier thickness and permittivity [34].

It should be noted that these early models did not include scattering effects and charge trapping at interfaces. More refined models for GBHETs including these effects have been subsequently proposed [35–37].

5.4.2 GBHETs Implementation

The experimental implementation of a GBHET firstly requires a selection of insulating or semiconducting materials with proper band alignment with respect to Gr to work as E-B and B-C barriers.

Figure 5.5 shows a summary of the reported literature values for the bandgap and electron affinity of Gr and common semiconductors and insulators [38]. In order to maximize the current injection efficiency from the emitter to the Gr base, the conduction band offset between Gr and the E-B barrier material should be as low as possible. Furthermore, to minimize back-scattering of hot electrons at the B-C filtering barrier, the conduction band offset of this layer with Gr should be lower than the E-B barrier one. Besides these criteria on the theoretical band alignment, high material quality is required for both the E-B and B-C barriers, in order to avoid leakage current through these layers.

The first experimental prototypes of GBHETs were reported by Vaziri et al. [21] and Zeng et al. [22] in 2013. Those demonstrations were fabricated on Si wafers using a complementary metal-oxide-semiconductor (CMOS) compatible technology and were based on metal/insulator/Gr/SiO$_2$/n$^+$-Si stacks, where n$^+$-doped Si substrate worked as the emitter, a thin (5 nm) SiO$_2$ as the E-B barrier, a thicker high-k insulator (Al$_2$O$_3$ or HfO$_2$) as the B-C barrier and the topmost metal layer (e.g. Ti) as the collector.

Figure 5.6 illustrates the band diagrams of this GBHET prototype in different operating conditions, i.e. the flatband (a), the off-state (b) and the on-state (c) conditions. The measured common-emitter output and transfer characteristics of this device are reported in Figure 5.6d and e, respectively. Although the DC characteristics showed several orders of magnitude modulation of J_C as a function of V_{BE}, these first prototypes suffered high threshold voltage and a very poor injected current density (in the order of μA/cm^2) due

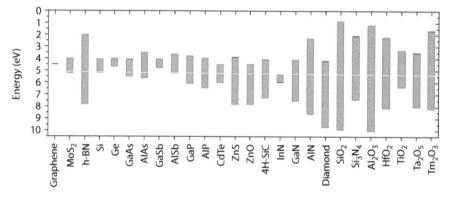

FIGURE 5.5
Band alignment of Gr with common semiconductors or insulators.

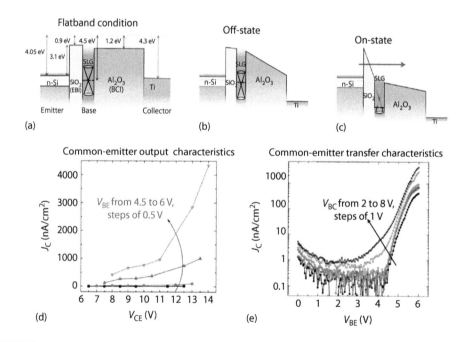

FIGURE 5.6
Band diagram illustrations of the Si/SiO$_2$/Gr/Al$_2$O$_3$/Ti GHET in the flatband (a), off-state (b) and on-state (c) conditions. Common-emitter output characteristics J_C–V_{CE} (d) and transfer characteristics J_C–V_{BE} (e) of the device. (Figures adapted with permission from Ref. [21] Copyright (2013) American Chemical Society.)

to the high Si/SiO$_2$ barrier, hindering their application at high frequencies. In order to improve the current injection efficiency, other materials have been investigated as E-B barrier layers in replacement of SiO$_2$ [23].

Figure 5.7 shows the comparison of the injected current density J_{BE} vs V_{BE} characteristics measured on n$^+$-Si/insulator/Gr junctions with different insulator barrier layers, i.e. a single layer of SiO$_2$ (5 nm) or HfO$_2$ (6 nm) (see Figure 5.7a) and a TmSiO (1 nm)/TiO$_2$ (5 nm) double layer (see Figure 5.7b). As an example, using a 6 nm thick HfO$_2$ (including a 0.5 nm interfacial SiO$_2$) deposited by atomic layer deposition (ALD) results in an improved threshold voltage and a higher J_{BE} with respect to the case of an SiO$_2$ barrier with similar thickness (see Figure 5.7a). Further improvements have been obtained using the TmSiO/ TiO$_2$ (1/5 nm) bilayer, as shown in Figure 5.7b. The role played by the two insulating layers with different electron affinities in the current injection is schematically illustrated in the insert of Figure 5.7b. The thin TmSiO layer (with low electron affinity) in contact with the Si emitter allows high current injection by step tunneling, while the thicker TiO$_2$ layer (with higher electron affinity) serves to block the leakage current from the Si valence band. A GBHET with this TmSiO/TiO$_2$ E-B barrier and the B-C barrier made of a 60 nm Si film deposited on Gr has also been fabricated [23]. A collector current density $J_C \approx 4$ A/cm^2 (more than five orders of magnitude higher than in the first prototypes) was obtained at $V_{BE} = 5$ V and for $V_{BC} = 0$ V (to avoid the contribution of cold electrons leakage current from the base to the collector). However, the device still suffered from low values of $\alpha \approx 0.28$ and $\beta \approx 0.4$. This can be due to the insufficient quality of the interface between Gr and the deposited B-C barrier, which is a common issue for all insulator or semiconductor thin films deposited on the chemically inert Gr surface.

The above-discussed attempts to implement the GBHET device using Si as the emitter material have been mainly motivated from the perspective of integrating this new technology with the state-of-the-art CMOS fabrication platform. More recently, some research groups in both the United States and Europe have investigated the possibility of demonstrating GBHETs by the integration of Gr with nitride semiconductors. As already discussed in Section 5.3.2, GaN/AlGaN or GaN/AlN heterostructures are excellent systems for use as emitter/emitter-base barriers, due the presence of high-density 2DEG at the interface and to the high structural quality of the barrier layer compared to current oxides.

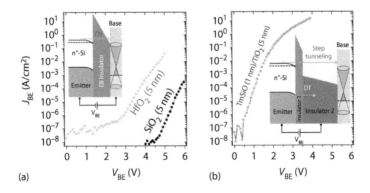

FIGURE 5.7

Injected current density J_{BE} vs V_{BE} measured on n$^+$-Si/insulator/Gr junctions with a single-layer insulator barrier: SiO$_2$ (5 nm), HfO$_2$ (6 nm) (a), and with a bilayer insulator barrier TmSiO (1 nm)/TiO$_2$ (5 nm) (b). Mechanisms of current injection through the single layer and the bilayer insulator are illustrated in the inserts. (Figures adapted with permission from *Solid State Commun.*, 224, Vaziri, S., et al., Going ballistic: Graphene hot electron transistors, 64–75, Copyright (2015) Elsevier.)

Very efficient current injection by FN tunneling has been shown in the case GaN/AlN/Gr heterojunctions with ultrathin (3 nm) AlN barrier [24], whereas TE has been demonstrated as the main current transport mechanisms in GaN/AlGaN/Gr systems with a thicker (~20 nm) AlGaN barrier layer [39–41].

Figure 5.8a and b illustrates a cross-section schematic and the band diagram of a recently demonstrated GBHET based on a GaN/AlN/Gr/WSe$_2$/Au stack [24]. A bulk n$^+$-doped GaN ($N_D \approx 10^{19}$ cm^{-3}) substrate with very low threading dislocation defect density (<10^5 cm^{-2}) was used as the emitter, and a 3 nm AlN tunneling barrier was grown on top of it by plasma-assisted molecular beam epitaxy. The use of a low threading dislocation density substrate ensured a high AlN film quality and minimal leakage current through the dislocations. A 3 nm GaN layer was used as a capping layer between the AlN and the Gr base. In order to circumvent the problems related to the poor interface quality between Gr and conventional insulators or semiconductors deposited on top of it, a layered semiconductor (WSe$_2$) from the family of TMDs was adopted as a B-C barrier layer. Thin films of WSe$_2$ were obtained by mechanical exfoliation from the bulk crystal and then transferred onto Gr to form a van der Waals (vdW) heterojunction with a defect-free, sharp interface. The resulting Gr/WSe$_2$ Schottky junction is characterized by a low barrier height due the small band offset (~0.54 eV) between Gr and WSe$_2$.

Figure 5.8c shows the common-base output characteristics ($I_C - V_{CB}$) for different values of the emitter injection current, I_E, in the case of a GBHET with a 2.6 nm thick WSe$_2$ barrier (corresponding to 4 WSe$_2$ monolayers). Furthermore, Figure 5.8d plots $\alpha = I_C/I_E$ as a function of V_{CB} in the same bias range. Three current transport regimes can be identified

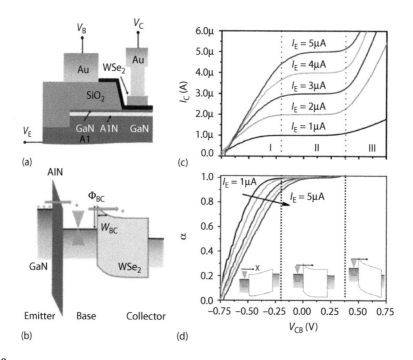

FIGURE 5.8

(a) Cross-section schematic and (b) band diagram of a recently demonstrated GBHET based on a GaN/AlN/Gr/WSe$_2$/Au stack. (c) Common-base output characteristics ($I_C - V_{CB}$) for different values of the emitter injection current I_E in the case of a GBHET with 2.6 nm thick WSe$_2$ barrier. (d) Plots of $\alpha = I_C/I_E$ as a function of V_{CB} in the same bias range. (Images adapted with permission from Ref. [24] Copyright (2017) American Chemical Society.)

in the output characteristics. At intermediate V_{CB} bias (region II), I_C is almost independent from the V_{CB} and $\alpha \approx 1$, indicating that almost all the injected hot electrons are able to overcome the Gr/WSe$_2$ Schottky barrier and reach the collector (as illustrated in the band diagram in the central insert of Figure 5.8c). For $V_{CB} < 0$ (region I), the injected electrons from the emitter are reflected back by the elevated B-C potential barrier (as shown in the band diagram in the left insert of Figure 5.8c), resulting in a reduced I_C and $\alpha < 1$. Finally, at higher positive V_{CB}, the current starts to increase and α becomes superior to 1 due to the increasing contribution of cold electrons leakage current from the base, as illustrated in the band diagram in the right insert of Figure 5.8c. Although this device shows excellent DC characteristics in terms of α, it exhibits a very limited operating V_{BC} window (~0.3 V), as a consequence of the poor blocking capability of the B-C junction with an ultrathin WSe$_2$ barrier. Increasing the number of layers was demonstrated to improve the blocking capability of the B-C barrier, as the larger interlayer resistance between layers of WSe$_2$ suppresses the cold electron transport between base and collector. On the other hand, a thicker filtering barrier results in a reduced value of α. As an example, $\alpha = 0.75$ was evaluated for a GaN/AlN/Gr/WSe$_2$/Au GBHET with 10 nm thick WSe$_2$ barrier [24].

5.4.3 Open Challenges for the Fabrication of GBHETs

Table 5.1 reports a comparison of the main DC electrical parameters (i.e. the collector current density J_C, the common-base current transfer ratio α and the common-emitter current gain β) for the state-of-the-art GBHETs and nitride semiconductors HETs with sub-10 nm base thickness.

This comparison shows that, in spite of their theoretically predicted superior performances (related to ballistic transport in the atomically thin Gr base), GBHETs still suffer reduced values of J_C, α and β with respect to HETs fabricated by bandgap engineering of III-N semiconductors (even with a thicker GaN base). This degradation of GBHET performances can be due to the not-ideal quality of Gr interfaces with the emitter and collector barriers. Hence, the first challenging task to achieve the optimal behavior of GBHETs is the deposition of a defects-free B-C barrier onto the chemically inert Gr surface. In addition, a suitable barrier height/thickness is required for this layer to minimize at the same time the quantum mechanical reflection of hot electrons and the leakage current of cold electrons from the Gr base. Uniform, i.e. pinhole-free, oxide layers (such as Al$_2$O$_3$ and HfO$_2$) can be deposited on Gr by ALD after proper surface functionalization, with minimal degradation of the Gr electronic/structural properties [42]. Although these oxide films typically exhibit a very low leakage current and a high breakdown field [43], their electron affinity is too low (as shown in Figure 5.5), resulting in a high B-C barrier, ϕ_{BC}. Furthermore, interfaces with ALD-deposited high-k dielectrics typically suffer significant electron trapping.

Semiconductor films are much more suitable as B-C barrier layers for two reasons: (i) the higher electron affinity as compared to oxides (resulting in a lower ϕ_{BC}), and (ii) the possibility of tailoring the barrier width (d_{BC}) by controlling the doping of the semiconductor. As an example, by simple band-alignment consideration, GaN would be the most suitable B-C barrier for a GBHET with a GaN emitter. However, the deposition of GaN thin films with suitable crystalline quality on Gr by MBE or MOCVD currently represents a challenging task [44,45]. The transfer of semiconducting TMDs thin films exfoliated from the parent bulk crystals onto Gr can represent an alternative solution to obtain a high-quality B-C barrier [24]. However, current methods to obtain such vdW heterostructures are not scalable and can be used for fabricating only a few small-size devices, whereas large-scale vdW epitaxy is still in its infancy [46].

TABLE 5.1

Comparison of J_C, α and β for the-State-of-the-Art GBHETs and Nitride-HETs with Sub-10nm Base Thickness

Emitter/Emitter-Base Barrier	Base (Thickness)	Base-Collector Barrier	JC (A/cm²)	α	β	References
Si/SiO$_2$	Gr (0.35nm)	Al$_2$O$_3$	~1×10^{-5}	~0.06	~0.06	[21]
Si/SiO$_2$	Gr (0.35nm)	Al$_2$O$_3$, HfO$_2$	~5×10^{-5}	~0.44	~0.78	[22]
Si/TmSiO/TiO$_2$	Gr (0.35nm)	Si	~4	~0.28	~0.4	[23]
GaN/AlN	Gr (0.35nm)	WSe$_2$ (10nm)	~50	~0.75	4–6	[24]
Si/SiO$_2$	MoS$_2$ (0.7nm)	HfO$_2$	~1×10^{-6}	~0.95	~4	[30]
GaN/Al$_{0.24}$Ga$_{0.76}$N	GaN (10nm)	Al$_{0.08}$Ga$_{0.92}$N	~5×10^3	~0.97	—	[18]
GaN/AlN	GaN/InGaN (7nm)	GaN	~2.5×10^3	>0.5	>1	[20]
GaN/AlN	GaN (8nm)	AlGaN/GaN	~46×10^3	~0.93	~14.5	[19]

The second critical task for the future implementation of the GBHET technology is the development of scalable methods for the fabrication of the Gr base. To date, the most used approach to fabricate the Gr base has been the transfer of Gr grown by chemical vapor deposition (CVD) on catalytic metals (such as Cu) to the surface of the E-B barrier layer. Although this is a very versatile and widely used method, it suffers from some drawbacks related to Gr damage and polymer contaminations during the transfer procedure, as well as from possible adhesion problems between Gr and the substrate. Furthermore, it typically introduces undesired metal (Cu, Fe) contaminations [47] originating from the growth substrate and the typically used Cu etchants. An intense research activity is still in progress to optimize Gr transfer procedures to minimize Gr defectivity and contaminations associated with transfer [48–51].

In many respects, the direct growth/deposition of Gr on the E-B barrier would be highly desirable to avoid some of the above-mentioned issues related to Gr transfer. However, to date, high-quality Gr growth has been demonstrated on few semiconducting or semi-insulating materials, such as silicon-carbide [52–56] and, more recently, germanium [57]. Single or a few layers of Gr can be obtained on the Si face (0001) of hexagonal SiC, either by controlled sublimation of Si at high temperatures (typically > 1650°C) in Ar at atmospheric pressure or by direct CVD at lower temperatures (~1450°C) using an external carbon source (such as C$_3$H$_8$) with H$_2$ or H$_2$/Ar carrier gases [55]. Gr grown on SiC(0001), commonly named epitaxial graphene (EG), generally exhibits a precise epitaxial orientation with respect to the substrate, which originates from the peculiar nature of the interface, i.e. the presence of a carbon buffer layer with mixed sp^2/sp^3 hybridization sharing covalent bonds with the Si face of SiC [58,59]. This buffer layer has a strong impact both on the lateral (i.e. in plane) current transport in EG, causing a reduced carrier mobility, and on the vertical current transport at the EG/SiC interface [60,61]. Hydrogen intercalation at the interface between the buffer layer and the Si face has been demonstrated to be efficient in increasing Gr carrier mobility and tuning the Schottky barrier and, hence, the vertical current transport across the Gr/SiC interface [62]. Recently, a GBHET system based on a metal/AlN/Gr/SiC stack has been presented [63], where SiC worked as the collector, the epitaxial Gr/SiC Schottky barrier worked as the B-C barrier and a thin AlN films deposited on Gr by atomic layer epitaxy worked as the E-B barrier layer [64].

Recently, CVD growth of Gr from carbon precursors on III-N substrates/templates has been investigated. Gr deposition on these non-catalytic surfaces represents a challenging task, as it requires significantly higher temperatures as compared to conventional

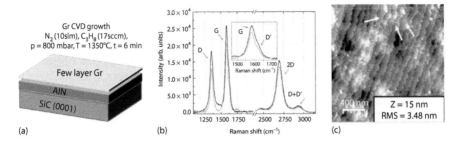

FIGURE 5.9
(a) Schematic illustration of the direct (non-catalytic) CVD growth of Gr at high temperature ($T = 1350°C$) on the surface of an AlN/SiC template using propane (C_3H_8) as the carbon precursor. (b) Raman spectra collected at two different positions of the AlN surface after Gr deposition. (c) Surface morphology of AlN with deposited Gr. Wrinkles on the Gr surface are highlighted by white arrows.

deposition on metals. The first experimental works addressing this issue showed the possibility of depositing a few layers of Gr both on bulk AlN (Al and N face) and on AlN templates grown on different substrates, such as Si(111) and SiC, at temperatures >1250°C using propane (C_3H_8) as the carbon source, without significantly degrading the morphology of AlN substrates/templates [65,66]. Figure 5.9a schematically illustrates the CVD growth conditions of Gr on the surface of an AlN/SiC template, i.e. the C_3H_8 and N_2 gas fluxes, the pressure $p = 800$ mbar and the temperature $T = 1350°C$. Figure 5.9b shows Raman spectra collected at different positions on the AlN surface after Gr deposition, demonstrating the deposition of a few layers of Gr, with Gr domain size in the order of 30 nm. Figure 5.9c illustrates the typical surface morphology of AlN with deposited Gr. Wrinkles, i.e. typical corrugations of the Gr membrane, are highlighted by white arrows. In spite of the very promising results of these experiments, further work will be required to evaluate the feasibility and the effects of CVD Gr growth on AlN/GaN or AlGaN/GaN heterostructures. Moreover, the possibility of integrating these high-temperature processes in the fabrication flow of GBHETs needs to be investigated.

5.5 Summary

In summary, we have discussed the HET device concept for THz applications, starting from the operating principles, to the physical and technological challenges related to its practical implementation. The most recent developments of this technology have been discussed, in particular nitride semiconductors–based HETs and the graphene-base HET, with open issues and new ideas to exploit the expected ultrahigh-frequency performances of this device.

Acknowledgments

The authors want to acknowledge the following colleagues for useful discussions: E. Frayssinet (CNRS-CRHEA, France); M. Leszczynski, P. Kruszewski and P. Prystawko (TopGaN, Poland); S. Ravesi, S. Lo Verso and F. Iucolano (STMicroelectronics, Catania,

Italy); G. Fisichella, E. Schilirò, S. Di Franco, P. Fiorenza, R. Lo Nigro, I. Deretzis, A. La Magna, G. Nicotra, C. Bongiorno and C. Spinella (CNR-IMM, Catania, Italy); R. Yakimova and A. Kakanakova (Linkoping University, Sweden); B. Pecz (HAS-MFA, Budapest, Hungary). This chapter has been supported, in part, by the Flag-ERA project "GraNitE: Graphene heterostructures with Nitrides for high frequency Electronics".

References

1. R. Lai, X. B. Mei, W. R. Deal, W. Yoshida, Y. M. Kim, P. H. Liu, J. Lee, J. Uyeda, V. Radisic, M. Lange, et al., Sub 50 nm InP HEMT device with f_{max} greater than 1 THz, *Proceeding of IEEE Electron Devices Meeting*, Washington, DC, December (2007), pp. 609–611.

2. D. H. Kim, J. A. del Alamo, P. Chen, W. Ha, M. Urteaga, and B. Brar, 50-nm E-mode In0.7Ga0.3As PHEMTs on 100-mm InP substrate with >1 THz, *Proceeding of IEEE Electron Devices Meeting*, San Francisco, CA, December (2010), pp. 30.6.1–30.6.4.

3. M. Urteaga, R. Pierson, P. Rowell, V. Jain, E. Lobisser, and M. J. W. Rodwell, 130 nm InP DHBTs with f_t > 0.52 THz and f_{max} > 1.1 THz, *Proceeding of 69th Annual Device Research Conference*, Santa Barbara, CA, June (2011), pp. 281–282.

4. V. Jain, J. C. Rode, H.-W. Chiang, A. Baraskar, E. Lobisser, B. J. Thibeault, M. Rodwell, M. Urteaga, D. Loubychev, A. Snyder, et al., 1.0 THz f_{max} InP DHBTs in a refractory emitter and self-aligned base process for reduced base access resistance, *Proceeding of 69th Annual Device Research Conference*, Santa Barbara, CA, June (2011), pp. 271–272.

5. J. C. Rode, H.-W. Chiang, P. Choudhary, V. Jain, B. J. Thibeault, W. J. Mitchell, M. J. Rodwell, M. Urteaga, D. Loubychev, A. Snyder, et al., Indium phosphide heterobipolar transistor technology beyond 1-THz bandwidth, *IEEE Trans. Electron. Devices* 62 (9), 2779–2785 (2015).

6. C. A. Mead, Operation of tunnel-emission devices, *J. Appl. Phys.* 32, 646–652 (1961).

7. M. M. Atalla and R.W. Soshea, Hot-carrier triodes with thin-film metal base, *Solid-State Electron.* 6, 245–250 (1963).

8. J. C. Hensel, A. F. J. Levi, R. T. Tung, and J. M. Gibson, Transistor action in Si/CoSi$_2$/Si heterostructures, *Appl. Phys. Lett.* 47, 151–153 (1985).

9. E. Rosencher, P. A. Badoz, J. C. Pfister, F. Arnaud d'Avitaya, G. Vincent, and S. Delage, Study of ballistic transport in Si-CoSi$_2$-Si metal base transistors, *Appl. Phys. Lett.* 49, 271–273 (1986).

10. T. Yajima, Y. Hikita, and H. Y. Hwang, A heteroepitaxial perovskite metal-base transistor, *Nat. Mater.* 10 (3), 198–201 (2011).

11. M. Tonouchi, H. Sakai, T. Kobayashi, and K. Fujisawa, A novel hot-electron transistor employing superconductor base, *IEEE Trans. Magn.* 23, 1674–1677 (1987).

12. M. Heiblum, D. C. Thomas, C. M. Knoedler, and M. I. Nathan, Tunneling hot-electron transfer amplifier: A hot-electron GaAs device with current gain, *Appl. Phys. Lett.* 47, 1105 (1985).

13. M. S. Shur, A. D. Bykhovski, R. Gaska, M. Asif Khan, and J. W. Yang, AlGaN–GaN–AlInGaN induced base transistor, *Appl. Phys. Lett.* 76, 3298 (2000).

14. A. S. Mayorov, R. V. Gorbachev, S. V. Morozov, L. Britnell, R. Jalil, L. A. Ponomarenko, P. Blake, K. S. Novoselov, K. Watanabe, T. Taniguchi, et al., Micrometer-scale ballistic transport in encapsulated graphene at room temperature, *Nano Lett.* 11, 2396–2399 (2011).

15. S. Sonde, F. Giannazzo, C. Vecchio, R. Yakimova, E. Rimini, and V. Raineri, Role of graphene/substrate interface on the local transport properties of the two-dimensional electron gas, *Appl. Phys. Lett.* 97, 132101 (2010).

16. F. Giannazzo, S. Sonde, R. Lo Nigro, E. Rimini, and V. Raineri, Mapping the density of scattering centers limiting the electron mean free path in graphene, *Nano Lett.* 11, 4612–4618 (2011).

17. B. D. Kong, Z. Jin, and K. W. Kim, Hot-electron transistors for terahertz operation based on two-dimensional crystal heterostructures, *Phys. Rev. Appl.* 2, 054006 (2014).

18. S. Dasgupta, N. A. Raman, J. S. Speck, and U. K. Mishra, Experimental demonstration of III-nitride hot-electron transistor with GaN base, *IEEE Electron. Device Lett.* 32, 1212–1214 (2011).
19. Z. Yang, Y. Zhang, D. N. Nath, J. B. Khurgin, and S. Rajan, Current gain in sub-10nm base GaN tunneling hot electron transistors with AlN emitter barrier, *Appl. Phys. Lett.* 106, 032101 (2015).
20. G. Gupta, E. Ahmadi, K. Hestroffer, E. Acuna, and U. K. Mishra, Common emitter current gain >1 in III-N hot electron transistors with 7-nm GaN/InGaN base, *IEEE Electron. Device Lett.* 36, 439–441 (2015).
21. C. S. Vaziri, G. Lupina, C. Henkel, A. D. Smith, M. Ostling, J. Dabrowski, G. Lippert, W. Mehr, and M. Lemme, A graphene-based hot electron transistor, *Nano Lett.* 13, 1435 (2013).
22. C. Zeng, E. B. Song, M. Wang, S. Lee, C. M. Torres, J. Tang, B. H. Weiller, and K. L. Wang, Vertical graphene-base hot electron transistor, *Nano Lett.* 13, 2370 (2013).
23. S. Vaziri, M. Belete, E. Dentoni Litta, A. D. Smith, G. Lupina, M. C. Lemme and M. Östlinga, Bilayer insulator tunnel barriers for graphene-based vertical hot-electron transistors, *Nanoscale* 7, 13096–13104 (2015).
24. A. Zubair, A. Nourbakhsh, J.-Y. Hong, M. Qi, Y. Song, D. Jena, J. Kong, M. Dresselhaus, and T. Palacios, Hot electron transistor with van der Waals base-collector heterojunction and high-performance GaN emitter, *Nano Lett.* 17, 3089–3096 (2017).
25. A. F. J. Levi and T. H. Chiu, Room-temperature operation of hot-electron transistors, *Appl. Phys. Lett.* 51, 984 (1987).
26. T. S. Moise, Y.-C. Kao, and A. C. Seabaugh, Room-temperature operation of a tunneling hot-electron transfer amplifier, *Appl. Phys. Lett.* 64, 1138–1140 (1994).
27. F. Giannazzo and V. Raineri, Graphene: Synthesis and nanoscale characterization of electronic properties, *Rivista del Nuovo Cimento* 35, 267–304 (2012).
28. F. Giannazzo, G. Fisichella, A. Piazza, S. Agnello, and F. Roccaforte, Nanoscale inhomogeneity of the Schottky barrier and resistivity in MoS_2 multilayers, *Phys. Rev. B* 92, 081307(R) (2015).
29. F. Giannazzo, G. Fisichella G. Greco, S. Di Franco, I. Deretzis, A. La Magna, C. Bongiorno, G. Nicotra, C. Spinella, M. Scopelliti, et al., Ambipolar MoS_2 transistors by nanoscale tailoring of Schottky barrier using oxygen plasma functionalization, *ACS Appl. Mater. Interfaces* 9, 23164–23174 (2017).
30. S. Vaziri, A. D. Smith, M. Östling, G. Lupina, J. Dabrowski, G. Lippert, W. Mehr, F. Driussi, S. Venica, V. Di Lecce, et al., Going ballistic: Graphene hot electron transistors, *Solid State Commun.* 224, 64–75 (2015).
31. C. M. Torres, Y. W. Lan, C. Zeng, J. H. Chen, X. Kou, A. Navabi, J. Tang, M. Montazeri, J. R. Adleman, M. B. Lerner, et al., High-current gain two-dimensional MoS_2-base hot-electron transistors, *Nano Lett.* 15 (12), 7905–7912 (2015).
32. W. Mehr, J. Dabrowski, J. C. Scheytt, G. Lippert, Y.-H. Xie, M. C. Lemme, M. Ostling, and G. Lupina, Vertical graphene base transistor, *IEEE Electron. Device Lett.* 33, 691–693 (2012).
33. F. Giannazzo, S. Sonde, V. Raineri, and E. Rimini, Screening length and quantum capacitance in graphene by scanning probe microscopy, *Nano Lett.* 9, 23 (2009).
34. F. Driussi, P. Palestri, and L. Selmi, Modeling, simulation and design of the vertical graphene base transistor, *Microelectron. Eng.* 109, 338–341 (2013).
35. V. Di Lecce, R. Grassi, A. Gnudi, E. Gnani, S. Reggiani, and G. Baccarani, Graphene base transistors: A simulation study of DC and small-signal operation, *IEEE Trans Electron. Devices* 60, 3584–3591 (2013).
36. V. Di Lecce, R. Grassi, A. Gnudi, E. Gnani, S. Reggiani, and G. Baccarani, Graphene-base heterojunction transistor: An attractive device for terahertz operation, *IEEE Trans. Electron. Devices* 60, 4263–4268 (2013).
37. V. Di Lecce, A. Gnudi, E. Gnani, S. Reggiani, and G. Baccarani, Simulations of graphene base transistors with improved graphene interface model. *IEEE Trans. Electron. Devices* 36, 969–971 (2015).
38. U. K. Mishra and J. Singh, *Semiconductor Device Physics and Design*, Springer, Dordrecht, 2008, ISBN 978-1-4020-6480-7.
39. G. Fisichella, G. Greco, F. Roccaforte, and F. Giannazzo, Current transport in graphene/AlGaN/GaN vertical heterostructures probed at nanoscale, *Nanoscale*, 6, 8671–8680 (2014).

40. F. Giannazzo, G. Fisichella, G. Greco, and F. Roccaforte, Challenges in graphene integration for high-frequency electronics, *AIP Conference Proc.* 1749, 020004 (2016).
41. F. Giannazzo, G. Fisichella, G. Greco, A. La Magna, F. Roccaforte, B. Pecz, R. Yakimova, R. Dagher, A. Michon, and Y. Cordier, Graphene integration with nitride semiconductors for high power and high frequency electronics, *Phys. Status Solidi A* 214, 1600460 (2017).
42. R. H. J. Vervuurt, W. M. M. Kessels, and A. A. Bol, Atomic layer deposition for graphene device integration, *Adv. Mater. Interfaces* 4, 1700232 (2017).
43. G. Fisichella, E. Schilirò, S. Di Franco, P. Fiorenza, R. Lo Nigro, F. Roccaforte, S. Ravesi, and F. Giannazzo, Interface electrical properties of Al_2O_3 thin films on graphene obtained by atomic layer deposition with an in situ seed-like layer, *ACS Appl. Mater. Interfaces* 9, 7761–7771 (2017).
44. T. Araki, S. Uchimura, J. Sakaguchi, Y. Nanishi, T. Fujishima, A. Hsu, K. K. Kim, T. Palacios, A. Pesquera, A. Centeno, et al., Radio-frequency plasma excited molecular beam epitaxy growth of GaN on graphene/Si(100) substrates, *Appl. Phys. Express* 7, 071001 (2014).
45. J. Kim, C. Bayram, H. Park, C.-W. Cheng, C. Dimitrakopoulos, J. A. Ott, K. B. Reuter, S. W. Bedell, and D. K. Sadana, Principle of direct van der Waals epitaxy of single-crystalline films on epitaxial graphene, *Nature Commun.* 5, 4836 (2014).
46. Y. Shi, W. Zhou A.-Y. Lu, W. Fang, Y.-H. Lee, A. L. Hsu, S. M. Kim, K. K. Kim, H. Y. Yang, L.-J. Li, et al., van der Waals epitaxy of MoS_2 layers using graphene as growth templates, *Nano Lett.* 12, 2784–2791 (2012).
47. G. Lupina, J. Kitzmann, I. Costina, M. Lukosius, C. Wenger, A. Wolff, S. Vaziri, M. Östling, I. Pasternak, A. Krajewska, et al., Residual metallic contamination of transferred chemical vapor deposited graphene, *ACS Nano* 9, 4776–4785 (2015).
48. G. Fisichella, S. Di Franco, F. Roccaforte, S. Ravesi, and F. Giannazzo, Microscopic mechanisms of graphene electrolytic delamination from metal substrates, *Appl. Phys. Lett.* 104, 233105 (2014).
49. J.-Y. Hong, Y. C. Shin, A. Zubair, Y. Mao, T. Palacios, M. S. Dresselhaus, S. H. Kim, and J. Kong, A rational strategy for graphene transfer on substrates with rough features, *Adv. Mater.* 28, 2382–2392 (2016).
50. J.-K. Choi, J. Kwak, S.-D. Park, H. D. Yun, S.-Y. Kim, M. Jung, S. Y. Kim, K. Park, S. Kang, S.-D. Kim, et al., Growth of wrinkle-free graphene on texture-controlled platinum films and thermal-assisted transfer of large-scale patterned graphene, *ACS Nano* 9, 679–686 (2015).
51. H. H. Kim, S. K. Lee, S. G. Lee, E. Lee, and K. Cho, Wetting-assisted crack- and wrinkle-free transfer of wafer-scale graphene onto arbitrary substrates over a wide range of surface energies, *Adv. Funct. Mater.* 26, 2070–2077 (2016).
52. C. Berger, Z. Song, X. Li, X. Wu, N. Brown, C. Naud, D. Mayou, T. Li, J. Hass, A. N. Marchenkov, et al., Electronic confinement and coherence in patterned epitaxial graphene, *Science* 312, 1191–1195 (2006).
53. K. V. Emtsev, A. Bostwick, K. Horn, J. Jobst, G. L. Kellogg, L. Ley, J. L. McChesney, T. Ohta, S.A. Reshanov, J. Rohrl, et al. Towards wafer-size graphene layers by atmospheric pressure graphitization of silicon carbide, *Nat. Mater.* 8, 203–207 (2009).
54. C. Virojanadara, M. Syvajarvi, R. Yakimova, L. I. Johansson, A. A. Zakharov, and T. Balasubramanian, Homogeneous large-area graphene layer growth on 6H-SiC(0001), *Phys. Rev. B* 78, 245403 (2008).
55. A. Michon, S. Vézian, E. Roudon, D. Lefebvre, M. Zielinski, T. Chassagne, and M. Portail, Effects of pressure, temperature, and hydrogen during graphene growth on SiC(0001) using propane-hydrogen chemical vapor deposition, *J. Appl. Phys.* 113, 203501 (2013).
56. C. Bouhafs, A. A. Zakharov, I. G. Ivanov, F. Giannazzo, J. Eriksson, V. Stanishev, P. Kühne, T. Iakimov T. Hofmann, M. Schubert, et al., Multi-scale investigation of interface properties, stacking order and decoupling of few layer graphene on C-face 4H-SiC, *Carbon* 116, 722–732 (2017).
57. J.-H. Lee, E. K. Lee, W.-J. Joo, Y. Jang, B.-S. Kim, J. Y. Lim, S.-H. Choi, S. J. Ahn, J. R. Ahn, M.-H. Park, et al., Wafer-scale growth of single-crystal monolayer graphene on reusable hydrogen-terminated germanium, *Science* 344, 286–288 (2014).

58. F. Varchon, R. Feng, J. Hass, X. Li, B. N. Nguyen, C. Naud, P. Mallet, J. Y. Veuillen, C. Berger, E.H. Conrad, et al. Electronic structure of epitaxial graphene layers on SiC: Effect of the substrate, *Phys. Rev. Lett.* 99, 126805 (2007).

59. G. Nicotra, Q. M. Ramasse, I. Deretzis, A. La Magna, C. Spinella, and F. Giannazzo, Delaminated graphene at silicon carbide facets: Atomic scale imaging and spectroscopy, *ACS Nano* 7, 3045–3052 (2013).

60. F. Giannazzo, I. Deretzis, A. La Magna, F. Roccaforte, and R. Yakimova, Electronic transport at monolayer-bilayer junctions in epitaxial graphene on SiC, *Phys. Rev. B* 86, 235422 (2012).

61. S. Sonde, F. Giannazzo, V. Raineri, R. Yakimova, J.-R. Huntzinger, A. Tiberj, J. Camassel, Electrical properties of the graphene/4H-SiC (0001) interface probed by scanning current spectroscopy, *Phys. Rev. B* 80, 241406(R) (2009).

62. F. Speck, J. Jobst, F. Fromm, M. Ostler, D. Waldmann, M. Hundhausen, H. Weber, and T. Seyller, The quasi-freestanding nature of graphene on H-saturated SiC(0001), *Appl. Phys. Lett.* 99, 122106 (2011).

63. A. D. Koehler, N. Nepal, M. J. Tadjer, R. L. Myers-Ward, V. D. Wheeler, T. J. Anderson, M. A. Mastro, J. D. Greenlee, J. K. Hite, K. D. Hobart, et al., Practical challenges of processing III-nitride/graphene/SiC devices, *CS MANTECH Conference*, May 18–21, (2015), Scottsdale, Arizona.

64. N. Nepal, V. D. Wheeler, T. J. Anderson, F. J. Kub, M. A. Mastro, R. L. Myers-Ward, S. B. Qadri, J. A. Freitas, S. C. Hernandez, L. O. Nyakiti, et al., Epitaxial growth of III-nitride/graphene heterostructures for electronic devices, *Appl. Phys. Express* 6, 061003 (2013).

65. A. Michon, A. Tiberj, S. Vezian, E. Roudon, D. Lefebvre, M. Portail, M. Zielinski, T. Chassagne, J. Camassel, and Y. Cordier, Graphene growth on AlN templates on silicon using propane-hydrogen chemical vapor deposition, *Appl. Phys. Lett.* 104, 071912 (2014).

66. R. Dagher, S. Matta, R. Parret, M. Paillet, B. Jouault, L. Nguyen, M. Portail, M. Zielinski, T. Chassagne, S. Tanaka, et al., High temperature annealing and CVD growth of few-layer graphene on bulk AlN and AlN templates, *Phys. Status Solidi A* 214, 1600436 (2017).

6

Tailoring Two-Dimensional Semiconductor Oxides by Atomic Layer Deposition

Mohammad Karbalaei Akbari and Serge Zhuiykov

CONTENTS

6.1 Introduction

The development of the scientific field of two-dimensional (2D) materials was prompted by the discovery of graphene and its extraordinary properties, which opened up new horizons for materials scientists [1–6]. The field evolved with the invention of a number of 2D materials beyond graphene with unexpected and distinctive physical and chemical properties, leading to considerable attention being devoted to understanding the properties of this new class of nanomaterials [7–11]. Ultra-thin 2D semiconductor oxides introduced another class of quasi-2D films whose properties often deviated from their respective microstructural counterparts. The nanoscale dimensions of 2D materials have a much greater level of structural, thermodynamic and kinetic freedom compared with those constraints in microstructural materials, caused by the microscopic extension of three-dimensional structures, making it possible to achieve unprecedented properties in 2D nano-films. The rise of these unexpected 2D semiconductor oxides properties is in close relation with their ultra-thin thickness. However, deviation from regular physical and chemical behavior is not determined by the specific film thickness, but depends on the intrinsic properties of oxide materials, their synthesis methods, structural features and the targeted properties. Accordingly, the addition or reduction of a monolayer can appreciably affect the properties of ultra-thin films and deviations may occur as a function of film thickness [11–15]. The definition of a 2D nanostructure is quite broad since the various synthesis techniques give very different structural and geometrical features to the 2D nanostructures produced. The concept of 2D oxide structures can be assigned to ultra-thin oxide films with a thickness of less than 1 nm (monolayer) up to a few nanometers (several fundamental layers). So far, the attention of researchers has mostly concentrated on the development of ultra-thin oxide films with a thickness of only one atomic layer or a single polyhedron layer of oxide materials, which approaches the behavior of oxide materials at the 2D limit with a high level of approximation [16]. In this context, characterizing and exploiting the properties of 2D oxide materials for practical applications and acquiring knowledge of device fabrication have attracted tremendous attention together with advancing the miniaturization of devices and shrinking the component dimensions [17–19]. Apart from pure scientific exploration of 2D semiconductor oxides, nowadays novel technologies use quasi ultra-thin oxide structures as the main components of novel nano-devices, which can be upgraded and evolved into new configurations. Consequently, promising applications of 2D oxide nanostructures are envisaged. Ultra-thin oxide materials have already found several applications in solid oxide fuel cells [20], catalyst films [21], corrosion protection layers [22], gas sensors [23], spintronic devices [24] and ultraviolet (UV) and visible light sensors [25]. Two-dimensional oxide semiconductors are considered one of the main components of metal-oxide-semiconductor field-effect transistors (MOSFETs) and other novel nanoelectronic devices [26]. Solar energy cells [27], plasmonic devices [28,29], data storage applications [30], biocompatibility and biosensing features [31] and, lately, supercapacitance properties [32,33] and electrochemical sensing [34] are among the recently announced applications of 2D oxide semiconductors.

A large number of ultra-thin layered materials can be synthesized by mechanical and chemical exfoliation, as well as deposition and growth techniques [15,35,36]. Achieving freely suspended, large-area 2D films is extremely challenging since the structural stability and molecular integrity of 2D films can be easily destroyed. Accordingly, one of the main challenges is the versatile and conformal deposition of these quasi-2D oxide nanostructures on convenient substrates including conducting, semiconducting and dielectric sublayers [37–45]. Precise control of thickness, monitoring of conformity and continuity of 2D oxide films over the substrate are among the major technical challenges, especially when the wafer-scale growth of 2D nanostructures is the main scope of discussion [46]. The growth of 2D oxides is achievable by using high-tech deposition techniques mostly exploiting the benefits of physical vapor and chemical vapor deposition (CVD) approaches by exerting ultra-high vacuum conditions during the fabrication process [46–48]. Since thin film growth is inherently a non-equilibrium process, discussion of the stoichiometry of deposited 2D oxide films, the stabilization of meta-stable oxide phases, the structural controllability and phase complexities of oxide–metal interface systems are of vital importance.

Atomic layer deposition (ALD) is a cyclic vapor–phase deposition process that has the advantages of the temporally separated and self-limiting reactions of two or more reactive precursors [49]. The sequential order and restricting nature of the ALD process enable the deposition of ultra-thin films of one atomic layer thickness at a time. Despite the capabilities of the ALD technique, the conformal deposition of 2D semiconductor oxide nanostructures on a desired surface is an extremely challenging process since the complete monolayer film is not deposited even if the surface of the substrate is covered entirely and saturated by precursors. In this case, the reactivity of the substrate surface to precursors and the later sequential ligand exchanges and self-limiting reactions are strongly engaged in the successful deposition of monolayer structures. These are considered as interwoven parameters affecting the successful or unsuccessful outcomes of the ALD process of the 2D nanostructures [50,51]. It turns out to be a more intricate case when the focus of the study shifts from the deposition of simple 2D oxides to ternary or complex oxide structures or heterojunctions.

The aim of this chapter is therefore to provide an overview of the development of 2D nanostructures using the ALD technique. Attention will be concentrated on the deposition of 2D semiconductor oxides (e.g. films with a thickness of a few or several monolayers) on a conducting or semiconducting substrates (e.g. metal or silicon/silicon oxide substrate). The chapter will begin with an explanation of interfacial and structural concepts in deposited 2D oxide films. Then, the principles of ALD will be introduced focusing on the development of 2D nanostructures. In doing so, the ALD development of 2D WO_3, TiO_2 and Al_2O_3 oxides will be discussed as practical cases. The developed recipes can be used as technical examples and adjusted to deal with the other challenges in the development of 2D nanostructured semiconductor oxides. The chapter will end with discussions on the practical applications of ALD-developed 2D oxide semiconductors.

6.2 Interfacial and Structural Concepts in Deposited 2D Oxides

The growth behavior of 2D oxide films is affected by a series of interconnected parameters, including the surface and interfacial energy of the oxide–metal system, the stoichiometry and thickness of 2D oxide films, the lattice mismatch between oxide film and metal substrate and the growth conditions [41,52]. Here, the focus is on interfacial and structural concepts in metal-supported 2D oxide films to provide a deeper understanding for future discussions.

6.2.1 Substrate Effects

The atomic structure is one of the main fundamental characteristics of materials from which many other key properties are determined. From a structural point of view, the crystalline bulk of oxide materials can be considered as the sequential order of alternating crystalline planes or, from another point of view, they can be envisaged as the combination of metal oxygen polyhedral coordination blocks that are connected via shared corners, edges and the plane of polyhedral coordination blocks [14]. The crystalline structures of some simple and complex 2D oxide nanostructures are presented in Figure 6.1 [53]. Taking a typical titanium oxide as an example, the host layer consists of TiO_6 octahedra, which are connected via edge sharing to form a 2D structure. Theoretically, a stoichiometric TiO_2 composition is gained by full occupation of the octahedral site by Ti atoms, while the simple lack of Ti sites or the substitution of Ti atoms by other metal atoms results in a negative charge in the host TiO layer [54]. The complexity of stoichiometry issues will increase as the structure of the 2D oxide nanostructure becomes more complicated.

In 2D structures, the limitations caused by the structural features of bulk materials are removed, alleviated or altered [55]. The epitaxial effects of a substrate on the growth of 2D oxide films alone, the interaction between the oxide ultra-thin films and their substrates experiences both charge distribution and mass transfer [56]. Compared with a bulk oxide,

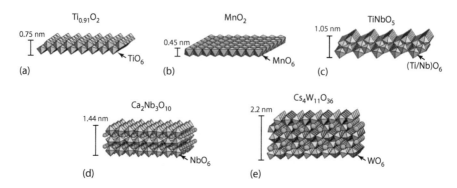

FIGURE 6.1
The structures of selected oxide nanosheets (a) $Ti_{0.91}O_2$, (b) MnO_2, (c) $TiNbO_5$, (d) $Ca_2Nb_3O_{10}$, (e) $Cs_4W_{11}O_{36}$. (Reproduced from Osada, M. and Sasaki, T., *Nanofabrication*, InTech, 2011.)

the chemical bonding, the electronic structure and its levels and interfacial interactions are fundamentally changed in 2D oxide nano-films [57–59], with each individual alteration unpredictably affecting the properties of ultra-thin nanostructures. For practical targets, the deposited monolayer oxides are usually supported by a solid substrate [14,60,61]. The atomically clean surface of inert metals is the most convenient substrate for the deposition of 2D oxide films [14]. The rigid support for a 2D oxide layer can be gained by using noble metals of Group Ib (Cu, Ag, Au) as the substrate. The resulting interface would be abrupt, but not necessarily a chemically inert one, keeping in mind, however, that the interplay and interaction between metal atoms of a substrate and the metal oxide components would be effective in determining the geometrical features of deposited 2D oxide films and overwhelmingly affect the resulting properties of 2D oxide nanostructures [41–43]. The conducting characteristic of metal substrates is beneficial for the charge transfer from an oxide monolayer to a metal substrate for subsequent experimental measurements of the electrical properties of 2D oxide materials [44]. The conductance of metal substrates also gives the opportunity to characterize the structural and chemical features of 2D oxide films by electron-based probe techniques, scanning tunneling microscopy (STM) and electron diffraction techniques [41,45].

Planar mono and few-layer oxides are deposited by various methods on the surface of metal substrates. The stabilization of ultra-thin films can be attained through the interaction between the metal substrate and the deposited oxide films. This new oxide–metal configuration introduces a new class of materials with novel properties that don't exist inherently in nature [60–62]. Depending on the thickness of the 2D oxide film, which is considered a 2D polar structure, the interaction between the oxide and metal films can significantly modify the structural properties, stability and electronic structure of polar oxide films [63]. At the interface of a thick polar oxide film and the metal substrate, the polarity effects are strongly alleviated by the transfer of compensating charges from the oxide film to the metal substrate [64]. For ultra-thin oxide films and especially in the case of uncompensated polarity, the interface of the oxide film and the metal substrate experiences an interfacial charge transfer that no longer originated from the requirements of polarity compensation [63]. In this case, the thin film undergoes a structural disorientation to respond to the electrostatic field created due to the interfacial charge transfer between the ultra-thin oxide film and the metal substrate. The created rumpling causes a separation between the atomic planes of cations and anions of a 2D polar film [65]. The rumpling can be ignored in the case of unsupported oxide films. However, it plays a significant role when the ultra-thin oxide film is grown on a metal substrate [65,66]. The direction and the extent of the charge transfer at the 2D oxide–metal interface are determined by the electronegativity of the metal substrate. When the metal substrate has high electronegativity, the electrons are transferred from the 2D oxide to metal support. Consequently, due to electrostatic forces, the anions of the oxide layer are pushed outward, as presented in Figure 6.2a [67]. For example, in the case of MgO monolayers deposited on transition metals (Ag, Au, Pt, substrates with high electronegativity), dynamic functional theory (DFT) has shown that the electrons flow from the monolayer oxide film into the metal substrate inducing a positive rumpling in which oxygen atoms of 2D films relax outward. Vice versa, in the metal substrate with low electronegativity the electron current will flow from the metal substrate to the 2D metal oxide. The displacement of cations of the oxide layer to the outward face of the 2D films is the feature of these 2D oxide–metal structures, shown in Figure 6.2b [67]. The negative rumpling was predicted by DFT when the 2D MgO film was deposited on the metal surfaces with low electronegativity (Al and Mg) resulting in a closer positioning of the oxygen atoms (anions) to the metal surface [67–69].

The interface of 2D oxide with Au can be considered a practical example of the interaction between 2D oxide films and metal substrates. Once the oxide film has grown, the mass transfer between the oxide film and the metal substrate is usually ignored, since the Au surface is inherently inert. However, the interfacial structure should be considered for the growth of the first monolayer. Thus, the bonding energies between Au and the oxygen (O) or metal (M) component of the 2D oxide film play a significant role in determining the chemical bonding structures at the 2D oxide–Au interface [56,70]. Considering the weak adsorption of oxygen on the Au surface and the facile decomposition of the Au oxide, the preferential bonding between Au and metal atoms of the oxide film is predicted, mostly resulting in the development of Au–M–O interfaces [71–73]. Surprisingly, the work function of the metal substrates can be tuned by the appropriate selection of a 2D oxide–metal interface providing the opportunity to design nano-devices by nano-surface engineering. For instance, while the growth of an MgO thin film decreases the work function of an Au substrate, the deposition of TiO_2 and SiO_2 results in the raised work function of the Au [74–76].

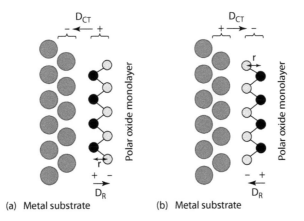

FIGURE 6.2
(a, b) A graphical scheme of the interfacial charge transfer (DCT) and film rumpling (DR) dipole momentum for an oxide monolayer (where the small black and white circles represent cations and anions, respectively) grown on a metal substrate (large gray circle are metal atoms). Regarding the electronegativity of metal substrates, electrons can transfer from or into the 2D oxide film. (Reproduced with permission from Goniakowski, J., et al., *Phys. Rev. B*, 80, 125403, 2009.)

6.2.2 Structural Concepts

The level and quality of the interaction between 2D oxide films and a metal substrate can tangibly affect the structural growth of the developed oxide thin films. A strong interaction between a metal substrate and oxide thin films facilitates the adaptive growth of 2D oxide films from the crystalline structure of a metal substrate [77–79]. A variety of 2D oxide films were grown on Au substrates including TiO_x [71,79], VO_x [80], CoO [81], MoO_3 [82], MgO [74], ZnO [83] and WO_x [84]. The bulk Au has a face-centered-cubic (fcc) crystalline structure, while different crystalline planes of Au can provide different crystalline orientations. As an example, the unconstructed Au(111) consists of hexagonal lattices, while a reconstructed Au(111) plane shows a complex structure [41,52]. It was observed that various deposited oxide films on an Au substrate lifted the herringbone reconstruction, which is caused by the strong interaction between the Au(111) facet and the ultra-thin oxide film. Nevertheless, with the formation of an oxide–Au interface, there is no longer an energetic advantage to adapt from the top layer of Au films. Consequently, the original hexagonal lattice form of the Au(111) facet acts as a template for the growth of ultra-thin oxide films [79]. In the case of TiO_x film grown on the Au(111) substrate, the co-existence of honeycomb and pinwheel structures with different stoichiometries was reported, as displayed in Figure 6.3 [71,79]. When the Ti atoms occupy the three-fold hollow sites of an Au lattice and O atoms position at the bridge sites of Ti atoms, a honeycomb structure with a stoichiometry of Ti_2O_3 is formed, as shown in Figure 6.3a [71,85]. In the other mechanism, the superposing of a metal/O lattice over the Au(111) surface forms moiré patterns resulting in the growth of different appearance pinwheel structures with the stoichiometry of TiO, as demonstrated in Figure 6.3b and c [71,79]. The hexagonal structures are the most commonly observed moiré patterns of ultra-thin 2D oxide films grown on Au(111) substrates. In addition to TiO, the hexagonal moiré patterns have been characterized in FeO [86], CoO [87] and ZnO [83] ultra-thin films developed on the Au(111) substrate.

Generally, the stoichiometry of 2D oxide films is the function of a host number of engaged parameters. Apart from the effects of substrates, the stoichiometry of a film can be altered by changing the growth conditions, oxidation parameters and post treatments. As a rule, the combined effects of various parameters will determine the final stoichiometry of a 2D oxide film. For example, even in the case of a non-metallic substrate, the growth of a distorted hexagonal arrangement of metastable 2D TiO_2 island films was observed on the surface of the single crystal rutile TiO_2, which exhibited completely different symmetry from the rectangular (011) plane of the bulk substrate [88]. The degree of lattice mismatch

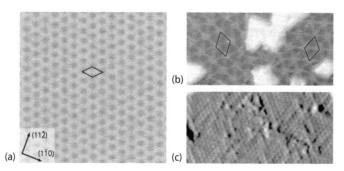

FIGURE 6.3

An STM image of (a) a honeycomb Ti_2O_3 structure, (b) a pinwheel TiO monolayer grown on Au(111) and (c) an atomically resolved STM image of pinwheel structures on Au(111). (Reproduced with permission from Wu, C., et al., *J. Phys. Chem. C*, 115, 8643–8652, 2011.)

is another factor determining the structure of 2D oxide films. The effect of induced stress due to the substrate lattice mismatch can be relaxed and thus becomes weaker as the thickness of the oxide film grows. The $TiO_x/Au(111)$ system can be considered a typical example of the effect of thickness, when the triangular structures start to grow as the second layer on top of the first honeycomb Ti_2O_3 and pinwheel TiO monolayers [71]. While it is difficult to determine the atomic structure of the second monolayer, it can be inferred that the second monolayer is still affected by the lattice structure of the Au substrate. Further discussions on interfacial and structural concepts of a 2D oxide–metal substrate are beyond the scope of this chapter.

6.3 Atomic Layer Deposition of 2D Nanostructures

The utilization of the properties of 2D nanostructures needs the feasibility of fabricating 2D material–based devices. To fulfill this ambition, the atomically thin 2D ultra-thin nanostructures should be tailored to devices. Field-effect transistors (FET) based on 2D materials is a typical example of tailoring 2D nanostructures in practical devices that can be employed in measuring the channel properties of these ultra-thin structures [89]. Advanced fabrication techniques are an essential prerequisite for the fabrication of 2D-based devices. However, existing technologies face serious challenges, especially in the case of conformal deposition of ultra-thin 2D films [46]. The most conventional deposition techniques for the development of an ultra-thin film are sputtering, evaporation and CVD. These techniques are mostly suitable for the fabrication of Si-based devices; however, so far they are not suitable for tailoring 2D materials to device fabrication [46,89]. For example, sputtering uses the harsh environment so achieving flawless 2D structures is impossible. The evaporation and CVD-based techniques cannot achieve high-quality 2D nano-films with precise control of their thickness over a large substrate area. These techniques are developed for thin film deposition in the range of tens to hundreds of nanometers. By employing typical CVD or physical vapor deposition (PVD) techniques it would be rather difficult to achieve continuous films with ultra-thin 2D thickness within a few nanometer range [46,89,90]. At best, the surface of a substrate consists of several island domain structures with thicknesses ranging from a few to several nanometers. These island structures do not theoretically or practically refer to uniformly deposited 2D nanostructures [91]. Furthermore, full coverage of the substrates with a high aspect ratio is not attainable using these methods, which are necessary for practical applications such as various sensors and supercapacitors.

ALD as the ultra-thin film deposition technique is based on controlled and self-limited chemical reactions on the surface of substrates. In practical implementation, it is similar to CVD, except that the precursors are not introduced into the reactor simultaneously, but are typically separated in time by inert gas purges. As distinct from other CVD techniques, in ALD the source vapors are pulsed into the reactor alternately, one at a time, separated by purging, or evacuation periods. Each precursor exposure step saturates the surface with a (sub)monolayer of that precursor [92]. This results in a unique self-limiting film growth mechanism with a number of advantageous features. In fact, the self-saturation mechanism of chemical reactions is the main advantage of ALD, facilitating the atomic-scale control of the thickness of ultra-thin films. As a result, the reaction is limited to the monolayer of a reactant that has been adsorbed on the substrate [93,94]. Through this modification, excellent conformity and uniformity over geometrically complicated substrates and on a

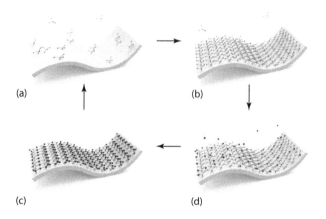

(a)   (b)

(c)   (d)

FIGURE 6.4
A schematic of an ALD cycle and its consecutive cycles: (a) exposure of a precursor, (b) removal of unreacted products molecules by purging of inert gas, (c) exposure of counter reactants and (d) purging with inert gas to remove unreacted counter-reactant molecules. (Reproduced with permission from Kim, H. G. and Lee, H. B. R., *Chem. Mater.*, 29, 3809–3826, 2017.)

large spatial resolution can be attained. As a general point of view, one ALD cycle consists of four individual steps, including exposure of precursors, purging of reacted chemicals and reaction products, counter-reactant exposure and finally purging, as demonstrated in Figure 6.4 [89]. Regarding the mechanism of ALD and its sequential exposure of precursors and counter reactants, the chemical reactions are restricted to the substrate surface properties and strongly depend on the surface characteristics. To be clearer, on an individual substrate, the ALD growth behavior shows different characteristics on two areas with different surface properties. The surface sensitivity of ALD is too strong, providing the opportunity for selective deposition, which is routinely called area-selective ALD [95]. The self-limiting nature of ALD also allows excellent composition control and the fabrication of multilayer 2D nanostructures. Compared with the parallel methods of typical thin film deposition including CVD, PVD and sputtering, the ALD process is performed under much milder conditions, which are necessary for the uniform deposition of 2D structures. Another outstanding feature of ALD is its slow growth rate. This ALD feature makes the deposition of high-quality 2D nano-films possible. All the foregoing ALD advantages turn this fabrication method into the most promising technique for wafer-scale deposition of ultra-thin 2D nano-films for practical applications and places ALD in a unique position to address many challenges in the development of 2D semiconductors [46]. Table 6.1 provides a general overview of the existing thin film fabrication techniques and their capabilities ratings [46]. Considering the information presented in Table 6.1, it is evidently understood that the ALD method has significant advantages over other deposition techniques in terms of the deposition process itself and control over various parameters of fabrication. Although several articles report on the ALD of oxide semiconductor films, they mostly deal with the ALD of oxide films whose thicknesses are far beyond 2D limits, so they do not fall under the category of 2D nanostructures. The following sections will concentrate on the ALD of 2D oxide semiconductor films by focusing on the development of 2D semiconductor oxides.

6.3.1 ALD Window

As previously mentioned, the ALD is a surface-sensitive method and its mechanisms are the function of precursors, reactant agents and substrate surface properties; furthermore, it

TABLE 6.1

Properties of Various Modern Techniques for Thin Film Deposition

	Deposition Technique					
Property	Chemical Vapor Deposition (CVD)	Molecular Beam Epitaxy (MBE)	Atomic Layer Deposition (ALD)	Pulsed Layer Deposition (PLD)	Evaporation	Sputtering
Deposition rate	Good	Fair	Poor	Good	Good	Good
Film density	Good	Good	Good	Good	Fair	Good
Lack of pinholes	Good	Good	Good	Fair	Fair	Fair
Thickness uniformity	Good	Fair	Good	Fair	Fair	Good
Sharp dopant profiles	Fair	Good	Good	Varies	Good	Poor
Step coverage	Varies	Poor	Good	Poor	Poor	Poor
Sharp interfaces	Fair	Good	Good	Varies	Good	Poor
Low substrate temp	Varies	Good	Good	Good	Good	Good
Smooth interfaces	Varies	Good	Good	Varies	Good	Varies
No plasma damage	Varies	Good	Good	Fair	Good	Poor

Source: Reprinted with permission from Zhuiykov, S., et al., *Appl. Surf. Sci.*, 392, 231–243, 2017.

depends on ALD fabrication parameters including process design, chamber pressure and temperature, plasma control and other ALD parameters [96,97]. Therefore, the developed ALD recipe for 2D materials is technically regarded as valuable data. The self-saturation phenomena and their mechanisms are the main parameters determining the conformity of ALD film [98–100]. The grow rate of the ALD process is controlled by a process temperature range, in which the ALD reactions occur. This temperature range is called the "ALD window" as shown in Figure 6.5a. To promote or initiate the reactions and to reduce the amount of unreacted precursors on a substrate, an elevated temperature must be selected for the ALD process. An inefficient and low-temperature ALD process may result in the occurrence of the physical adsorption of precursors on the substrate surface, which would consequently lead to precursor condensation on the substrate. However, the ALD window is restricted by excess condensation or thermal activation at the low-temperature range and is generally bounded by precursor degradation or desorption at high temperatures [101,102]. A typical ALD window for a successful ALD process is in the range of 200°C–400°C, as shown in Figure 6.5b [46]. However, it can be changed for specific ALD recipes. The self-controlled growth in the ALD window is the consequence of chemisorption. In some cases, due to lack of thermal energy, the ALD process is assisted by using plasma. Plasma-enhanced atomic layer deposition (PEALD) makes it possible to expand the number of precursors. PEALD is typically employed in the deposition of oxide films to circumvent the need for H_2O as oxygen precursors or to increase the growth rate [104]. However, the ability of conformal coating of 2D structures by a high aspect ratio is probably limited by PEALD because of ozone generation and the inefficient mean free path of plasma ions [103–105].

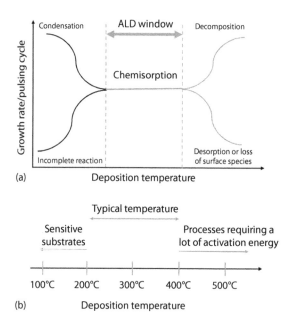

(a)

(b)

FIGURE 6.5

(a) Dependence of the deposition temperature on the growth rate for ALD and (b) a typical temperature range for the most common ALD precursors. (Reproduced with permission from Zhuiykov, S., et al., *Appl. Surf. Sci.*, 392, 231–243, 2017.)

6.3.2 ALD Precursors

The precursors in an ALD process can be classified into three major types, including inorganic precursors that do not contain any carbon; metal-organic precursors that possess organic ligands, but without metal to carbon bonds; and organometallic precursors that contain organic ligands wherein metal to carbon bands exist [46,96]. The successful design of an ALD process is highly dependent on the proper selection of precursors. As chemical products, precursors should be synthesized by easy routes and need to be conveniently available for practical ALD targets. Self-decomposition, volatility, dissolution and etching effects on substrates, sufficient purity and byproducts are the main parameters that should be considered during synthesis and for the selection of precursors. Despite the significant progress in the synthesis of new chemical compounds, fulfilling the requirement of a suitable precursor for ALD of 2D nanostructures needs a great deal of work. The saturation of the growth rate as a function of the precursor pulse duration is directly attributed to the ALD window temperature. To achieve a high-quality ALD-deposited 2D nanostructure, the self-saturation process should be controlled. Each ALD precursor shows a characteristic sticking coefficient that can be changed on various substrates. By changing the ALD conditions, the sticking coefficient directly affects the precursor purge time. For instance, the growth per cycle (GPC) obtained from an insufficient purge time can be twice as high as when obtained at the optimized ALD recipe [106,107]. The same challenges should be encountered when the second reactant precursor comes into action. For example, in the case of metal oxide, sufficient H_2O pulse time is the essential parameter that should be considered to obtain high-quality 2D oxide nano-films. Evidently, insufficient oxidant agents will change the growth regimes or affect the composition of ALD films [106]. Another important parameter is the reactivity of precursors, especially when considering that most of the strategies use in the design of an ALD recipe are based on thermally driven reactions [108–110]. In this case, the reactivity becomes a more critical issue when thermally stable precursors are used or when low-temperature deposition is required owing to the requirements of an ALD process. Regarding the aforementioned ALD variables, a vital role that has been established in the design of an ALD recipe for 2D semiconductor oxides is the correct selection of ALD precursors [46].

6.3.3 Case Study: ALD of 2D WO$_3$

Increasing interest in the employment of ultra-thin semiconductor oxides for advanced applications in functional devices has resulted in more attention being paid to the development of an ALD technique. During the last decade, several scientific reports have confirmed the capability of ALD to fabricate thin film semiconductor oxides such as indium oxide [111,112], tantalum oxide [113], ruthenium oxide [114,115], titanium dioxide [116,117], zinc oxide [118,119], iron oxides [120], tin oxide [121], cobalt oxide [122], tungsten oxide [123] and barium titanate [124]. While the ALD of semiconductor oxide films has been well investigated and the process parameters have been extensively studied, our knowledge of the development of ultra-thin 2D oxide nanostructures is deficient. Accordingly, the number of published articles is low, opening up an uncharted area in nanofabrication topics. Take the WO$_3$ semiconductor oxide as an example. To date, there have only been a few reported ALD processes for the fabrication of thin film WO$_3$. A recipe was developed by considering the various parameters of ALD to enable the deposition of a monolayer WO$_3$ on a large surface [46,84]. The bis(ter-butylimido) bis(dimethylamino) tungsten(VI) (^{t}BuN)$_2$W(NMe$_2$)$_2$ as the tungsten precursor and H$_2$O as the oxygen source were used and the deposition was carried out on an Si/SiO$_2$ substrate using the Savannah S100 (Ultratech/Cambridge Nanotech) ALD instrument, as graphically demonstrated in Figure 6.6. To confirm the ALD growth of an oxide film, in situ ellipsometry was used during the deposition process. In experiments, the dependence of the growth rate on the precursor pulse times, substrate temperature and cycle numbers was investigated. It must be stressed that the deposition of ultra-thin 2D films with a thickness of less than 1.0 nm is an extremely challenging target. Non-uniformity and a significantly less deposition rate than the desired thickness were observed after the first deposition. However, a noticeable improvement was observed when the temperature was increased to 350°C. Ellipsometry analysis of several depositions also indicated significantly less practical deposition than the intended desired thickness. To use ellipsometry for thickness measurements, a sufficient coating is needed to

FIGURE 6.6

(a) A schematic interpretation of precursors used (^{t}BuN)$_2$(Me$_2$N)$_2$W and H$_2$O, (b) an optical image of a wafer-scale developed 2D WO$_3$ film made and diced on Si/SiO$_2$ and (c) an image of MCN Savannah S100 ALD apparatus connected to the glove-box used for WO$_3$ ALD. (Reproduced with permission from Zhuiykov, S., et al., *Appl. Surf. Sci.*, 392, 231–243, 2017 and Zhuiykov, S., et al., *Appl. Mater. Today*, 6, 44–53, 2017.)

build reliable optical constants. As the deposition thickness yielded 10 nm, the comparison between the measured and the literature values of the refractive index showed good agreement. Changing other ALD parameters, especially precursor dosing, and optimizing the device temperature resulted in an improved deposition rate. An example of the ALD recipe for a 2D WO_3 film is presented in Table 6.2 [84].

As shown in Figure 6.7a and b, the growth rate is sensitive to the precursor doses. As expected, the growth rate stabilized to a maximum thickness per cycle when the precursor dosing achieved full monolayer coverage. Growth rate saturation was observed after 50 ms for H_2O and after 2 s for $(^tBuN)_2W(NMe_2)_2$, indicating the self-limiting growth characteristics of the ALD. In addition, it was established that the growth was also highly sensitive to the substrate temperature. The first deposition at an operating temperature of ~300°C confirmed very limited growth of the 2D WO_3 film. Consequently, the optimum growth condition yielded a stable WO_3 growth (see Figure 6.7c and d). The subsequent ellipsometry map analysis of final deposited films confirmed that the obtained thicknesses for 2D WO_3 were ~0.7 nm and ~1.2 nm. It should be considered that the developed recipe is attributed to the ALD of WO_3 on an Si/SiO_2 substrate with its specific growth condition. Due to the natural roughness of an Si/SiO_2 substrate, the development of a more conformal and consistent monolayer was not possible [84]. An investigation of a wafer-scale ALD of WO_3 on an Au substrate showed that the uniformity of the films decreases and their thicknesses gradually deviate from the targeted value, as the films grow thicker. This can be caused by non-uniform spatial distribution of precursors and the temperature discrepancy in the substrate [86].

Figure 6.8a shows a cross-sectional view of a 2D WO_3 film deposited on an Si/SiO_2 using the ALD recipe presented in Table 6.2. A scanning transmission electron microscopy (STEM) image displays the uniform development of the ALD WO_3 film across a large area of an Si/SiO_2 wafer. The film was semi-amorphous without a clearly identified crystalline

TABLE 6.2

Developed ALD Recipe for 2D WO_3 Using $(tBuN)_2W(NMe_2)_2$ and H_2O Precursors

Deposition Parameters	1	2	3	4	5	6	7
Inner heater (°C)	300	350	350	350	350	350	350
Outer heater (°C)	300	300	280	280	280	280	280
W precursor heater (°C)	80	80	80	80	80	80	80
Isolate pump	–	–	*	*	*	*	*
Exposure	–	–	–	–	–	–	–
Initiate pump	–	–	–	*	–	–	–
Pulse H_2O (s)	0.05	0.05	0.06	0.06	0.06	0.06	0.06
Exposure (s)	–	–	5	5	5	5	5
Initiate pump	–	–	*	*	*	*	*
Purge H_2O (s)	4	4	12	12	12	12	12
Isolate pump	–	–	–	*	*	–	*
Pulse W precursor	1	1	1	1	1	2	1
Exposure	–	–	–	6	6	–	6
Initiate pump	–	–	–	*	*	–	*
Purge W precursor (s)	10	6	6	14	14	14	14
Number of cycle	400	300	300	300	250	250	185
Thickness (nm)	8	30.9	30.9	30.9	28.5	28.5	19.1

Source: Reprinted with permission from Zhuiykov, S., et al., *Appl. Mater. Today,* 6, 44–53, 2017.
* Process is carried out; – process is exempted.

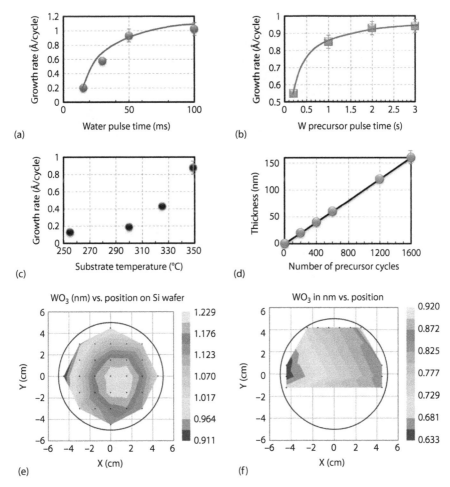

FIGURE 6.7
Experimental data for the growth of a WO_3 film based on the conception of (a) H_2O and (b) $(tBuN)_2W(NMe_2)_2$ precursors. The stabilization of the growth rate to a maximum thickness per cycle after the achievement of a full monolayer coverage. (c) The initial growth rate vs. temperature and (d) the thickness variation of a WO_3 film per ALD cycle number. (e, f) An ellipsometry map for 1.2 nm thick and 0.7 nm thick WO_3 on Si/SiO$_2$ after an optimized ALD process. (Reproduced with permission from Zhuiykov, S., et al., *Appl. Mater. Today*, 6, 44–53, 2017.)

structure, as depicted in Figure 6.8b. Atomic force microscopy (AFM) measurements of the 2D WO_3 film and its corresponding scanning electron microscopy (SEM) observations clearly demonstrated the uniformly developed 2D films with a high level of accuracy in thickness, as shown in Figure 6.8c and d. Due to the low conductivity and semi-amorphous nature of the ALD-developed 2D WO_3 film, its surface appeared to be fuzzy and smooth with the visual absence of clearly distinguished grain boundaries. AFM and focused ion beam SEM measurements, which are not presented here, confirmed a high level of correlation of the results obtained by two independent material characterization techniques for the monolayer WO_3.

The surface chemical characterization of ALD 2D WO_3 films by x-ray photoelectron spectroscopy (XPS) gives valuable information about the nature of the bonding and surface chemical composition of 2D films. The study of the XPS peaks corresponding to O 1s-levels of oxygen atoms O^{2-} in the lattice of SiO_2 and WO_3 revealed that, when the samples

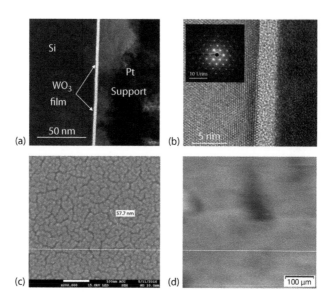

FIGURE 6.8
(a) STEM cross-sectional view of a ~3.3 nm thick 2D WO_3 film deposited on Si/SiO_2 wafer by ALD and (b) a high magnification image of the interface between Si/SiO_2 and WO_3 film. The inset shows the SAED pattern indicating the crystalline structure of SiO_2, (c) SEM and (d) AFM image of the surface morphology of an ALD-developed WO_3 film. (Reproduced with permission from Zhuiykov, S., et al., *Appl. Mater. Today*, 6, 44–53, 2017.)

consisted of a double layer of the oxygen-octahedron structure, the bottom oxygen was shared with SiO_2. Evidence was provided that metal precursors were adsorbed initially on the surface of SiO_2, thus metal–oxygen bonding was formed on the substrate. XPS spectra depicted that the oxygen-octahedron structure in the 10.0 nm thick WO_3 sample was totally free from the substrate effect. The studies also showed that tungsten (W) in both the monolayer and few-layer films was present in the six-valent state (W^{6+}). From the broadening of the core-level spectra of W, it was concluded that the monolayer WO_3 film was not fully oxidized and it was not fully crystalline [125,126]. By increasing the thickness, the crystallinity of thicker 2D films was developed [126]. Further studies of O 1s peaks of XPS analysis confirmed the development of stoichiometric crystalline structures in thicker WO_3 films. The adsorption of OH species on the oxygen vacancies of a monolayer WO_3 and the obtained peaks confirmed the development of a sub-stoichiometric WO_{3-x} compound in the monolayer WO_3 film. Fourier transform infrared spectroscopy (FTIR) further demonstrated the characteristic mode of a WO_3 structure. Modifications in the tungsten oxide framework were suggested by studies of the FTIR spectra of W-O stretching modes, possibly because the ALD-developed surface of a 2D WO_3 was not a fully defect-free surface and this was later confirmed by the absorption of surface-active sites on a 2D film.

6.3.4 Case Study: ALD of 2D TiO_2

The wafer-scale synthesis and conformal growth of a 2D TiO_2 film on an Si/SiO_2 substrate was achieved by ALD using a tetrakis (dimethylamino) titanium (TDMAT) precursor and H_2O as the oxidation agent [127,128]. Regarding the challenging nature of developing 2D oxide nanostructures, the reliability of the fresh TDMAT precursor was initially confirmed. To establish the experimental dependence of the deposition rate on the ALD temperature for the fresh TDMAT precursor, a few blank depositions were done.

Then, the optimum number of cycles, deposition rates and predicted thicknesses were calculated and averaged for each operating temperature. For precise fabrication of extremely thin 2D TiO$_2$ films, where their thickness must be below ~0.5 nm, the optimum operating temperature of 250°C was selected. To achieve complete coverage of a monolayer TiO$_2$, two individual ALD super cycles were designed [127]. The first super cycle consisted of 10 consecutive cycles of pulse/purge TDMAT to ensure reliable coverage of the surface by precursors. Afterward, 10 pulse/purge stages of H$_2$O were performed to ensure the completion of oxidation [127,128].

The reactions that occurred during ALD fabrications of a monolayer TiO$_2$ are schematically presented in Figure 6.9. From the thermodynamic and kinetic points of view, the development of TiO$_2$ by TDMAT and H$_2$O can be separated into two half reactions [128]. The initial growth stage is expected to be influenced by the number of chemisorption sites on the Si/SiO$_2$ substrate. As graphically demonstrated in Figure 6.9a, the first stage is the chemisorption of TDMAT molecules by active sites of the SiO$_2$ surface. The next step is the

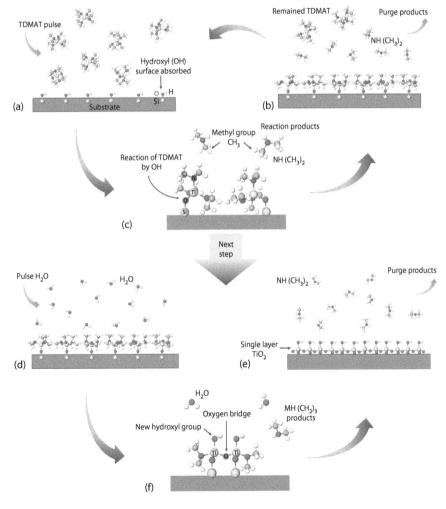

FIGURE 6.9
A graphical scheme of the development of atomic-layered titanium oxide by ALD. (Reproduced with permission from Zhuiykov, S., et al., *Mater. Des.*, 120, 99–108, 2017.)

ligand exchange that usually occurs easily even at very low operating temperatures and it is attributed to the high reactivity of both TDMAT and OH groups [129]. Figure 6.9b shows a scheme of the ligand exchange process. The subsequent step is the purge of the remaining TDMAT and reaction products (NH (CH$_3$)$_2$) from the reaction chamber [130]. During the next step, by introducing H$_2$O into the ALD chamber, the oxidation process continues. Figure 9d and e demonstrates the oxidation process and the development of Ti–O–Ti bridge bonding in the deposited film, respectively. On completion of the oxidation process, the 2D TiO$_2$ film is formed on the surface where all the Ti atoms are connected to each other by oxygen atoms. The interaction between the surface SiO$_2$ and the deposited TiO$_2$ was confirmed by observation of the Si-O-Ti spectra in the FTIR measurement of an ALD TiO$_2$ monolayer film [127]. It confirmed the development of bonding between Ti atoms and Si-O atoms of the SiO$_2$ substrate. Further studies by XPS measurements showed the characteristic spectra of Si–O–Ti binding energy [127]. Comparative XPS studies confirmed the fact that the bottom oxygen in the atomic-layered TiO$_2$ film was shared with the SiO$_2$ substrate [127,131]. The binding energy of Si was also altered after the deposition of the monolayer TiO$_2$. The position of Si^{4+} shifted to lower energy during ALD of the atomic-layered TiO$_2$ film. Changes in the binding energies in both Si 2p and Ti 2p peaks after the deposition of 2D TiO$_2$ could be elucidated by the development of Si–O–Ti bonds at the interface between the native SiO$_2$ and the deposited 2D TiO$_2$ film. As titanium was introduced into the Si–O bonds during the ALD process, the binding energy of Si^{4+} decreased slightly from its bulk SiO$_2$, which was related to the fact that the silicon was more electronegative than titanium. The evidence of oxygen bridge bonding was also demonstrated by characterization of the vibrational mode of Ti–O–Ti bonds in the FTIR spectrum of the monolayer TiO$_2$ film and also by investigation of the XPS spectrum of the monolayer film. The detection of the FITR characteristics of Ti^{4+} confirmed the existence of Ti(IV)O oxide in a crystalline state [127].

In another study, the growth of 2D TiO$_2$ films over Si/SiO$_2$ and Au substrates was investigated by spectroscopic ellipsometry and x-ray reflectivity [132]. A growth rate of 0.75 and 0.87 Å/cycle was measured for the deposition of 2D TiO$_2$ film on Si/SiO$_2$ and Au substrates, respectively [132]. In both substrates, the growth rates were linear, excluding a short delay in the first ALD cycles of TiO$_2$ on Si/SiO$_2$. The growth incubation in this ALD process was attributed to the preparation of hydroxyl groups on the surface. It seems that it took longer time for TDMAT to react with OH groups on the Si/SiO$_2$ surface compared with that on the Au surface [133,134].

Studies using micro-Raman spectroscopy revealed the presence of both crystalline anatase and rutile phases of TiO$_2$ in the ALD-developed 2D TiO$_2$ films. Typical Raman spectra for 2D TiO$_2$ films are demonstrated in Figure 6.10a. Raman studies showed that the final structure was indeed a combination of two crystalline phases of TiO$_2$. To investigate the impact of layer thickness on the Raman characteristic modes of a 2D film, the Raman spectra of 7.0 and 0.7 nm thick TiO$_2$ films were compared, and the results are shown in Figure 6.10b and c. Compared with a 7.0 nm thick TiO$_2$ film, a blue shift was observed in the E_g vibrational mode of a 0.7 nm thick TiO$_2$ film, which was accompanied by a decrease in the peak intensity and width broadening. The impacts of phonon confinement on the blue shift and broadening of the E_g Raman mode of anatase TiO$_2$ were reported previously [135]. The phonon confinement within 2D TiO$_2$ nanostructures resulted in incremental phonon momentum distribution and broadening of phonon momentum scattering accompanied by a shift in Raman characteristic bands [136]. From the stoichiometric point of view, the reduced intensity and width broadening of the E_g vibrational mode of anatase can also be attributed to the deviation from the stoichiometric level by reducing the film thickness down to a single fundamental layer [137,138]. XPS studies of 2D TiO$_2$ films developed on an

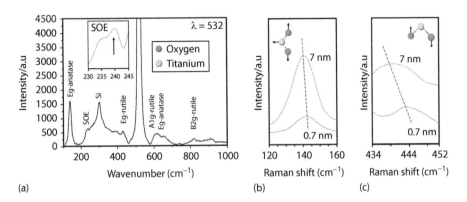

FIGURE 6.10
(a) The Raman spectra of ALD 2D TiO$_2$ films, (b) and (c) the thickness dependence of Raman peaks. (Reproduced with permission from Karbalaei Akbari, M., et al., *Mater. Res. Bull.*, 95. 380–391, 2017.)

Au substrate provide valuable information for the case of a surface chemical structure of 2D films. The XPS results of 7 and 0.7 nm thick TiO$_2$ films confirmed that titanium atoms (Ti) exist in a four-valent state (Ti^{4+}). The XPS results indicated that a thicker film (7 nm) was more oxidized than a thinner film (0.7 nm). The development of a stoichiometric oxide crystalline structure in a thicker film (7 nm) was confirmed by the characterization of an O 1 s peak of XPS measurements [132]. Furthermore, the presence of weakly adsorbed species on the surface of a 0.7 nm thick 2D TiO$_2$ film was shown by characterization of OH groups on the surface of a 0.7 nm thick 2D TiO$_2$ film indicating the development of non-stoichiometric structures with oxygen-deficient sites on the surface of a 0.7 nm thick 2D TiO$_2$ nanostructure.

6.3.5 Case Study: ALD of 2D Aluminum Oxide

ALD of a monolayer alumina film over Cu and Cu$_2$O/Cu substrates was investigated using trimethylaluminum (TMA) as the Al precursor and O$_2$ as the oxidant agent [139]. The interaction of TMA with a Cu surface faces several challenges. Regarding the DFT calculations, the adsorption and dissociation of TMA on the surface of a pure Cu(111) are endothermic, confirming that the interaction of TMA with a pure Cu surface is thermodynamically unfavorable in the absence of hydroxyl groups. XPS measurements and high-resolution electron energy loss spectroscopy did not show any characteristic vibration of Al atoms on a pure Cu surface. On the other hand, first-principle calculations demonstrated that TMA is capable of reacting with a copper oxide surface in the absence of hydroxyl species. The adsorption of Cu by a Cu$_2$O/Cu surface is limited by the initial amount of oxygen in a Cu$_2$O lattice structure. This adsorption results in the reduction of some surface Cu^{1+} to metallic copper (Cu^0) and the formation of copper aluminate compounds. The XPS studies revealed that the Al:O atomic percentage ratio at the first deposited layer was approximately 0.46 while the stoichiometric Al$_2$O$_3$ yielded an Al:O ratio of 0.66. It confirmed the development of non-stoichiometric 2D alumina layers. Further XPS measurements showed the formation of copper aluminate, most likely CuAlO$_2$ on the basis of the Al:O ratio of 0.5. The STM image of the surface of a TMA-exposed Cu$_2$O/Cu film demonstrated the growth of 2D islands of CuAlO$_2$ with an average height of approximately 0.19 nm with flat and uniform surfaces, which was close to Cu-O and Al-O bond length, as demonstrated in Figure 6.11a and b. After the second ALD cycle, TMA continued to react with surface Al–O, forming

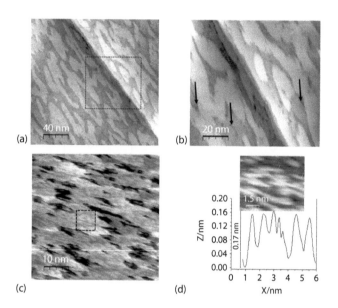

FIGURE 6.11

(a) Low and (b) high magnification STM images of the $Cu_2O/Cu(111)$ surface exposed to first cycle and (c) second cycle of TMA, and (d) zoom-in region of the highlighted section of the image (c) and the line profile along the solid line indicated in the image. (Reproduced from Gharachorlou, A., et al., *ACS Appl. Mater. Interfaces*, 7, 16428–16439, 2015.)

stoichiometric Al_2O_3. The observed morphological changes after consecutive ALD cycles proposed the development of second alumina islands with a thickness of 0.17 nm, as shown by the STM image in Figure 6.11c and d [139]. It was confirmed that TMA readily reacted with oxide surfaces even in the absence of co-adsorbed hydroxyls. For ALD applications on an air-exposed Cu surface, large domains of oxides might still exist that can facilitate the selective ALD of metal surfaces. This is of great importance in thin film applications like microelectronics and catalysis where only a few ALD cycles are desirable.

6.4 The Properties and Applications of ALD 2D Oxide Film

Interest in research relevant to ALD-developed nanostructures is increasing as evident by the steady growth of annual scientific publications [46]. This fact independently acknowledges the great impact of ALD-developed nanomaterials on their various applications. It further confirms the growing interest in the distinguished capabilities of 2D nanostructures that can be obtained in the nano-architecture of the different 2D materials using the ALD technique. Although ALD is currently mainly utilized in the microelectronic industry, its ability for conformal deposition into high aspect ratios offers great potential for a breakthrough in the field of ultra-thin 2D nano-materials [46]. The ALD offers the opportunity to engineer either dielectric or metallic properties of materials by creating nano-laminates or alloy films. This ability of ALD combines the advantageous properties of different materials [140,141]. ALD also has the potential for applications in the field of 2D nano-electrodes of various chemical sensors and biosensors, where high-quality, conformal and ultra-thin films in aggressively shaped nano-electrodes are often used. In fact, the low deposition rate of ALD does not hinder its applications for thin film electrodes [141]. Therefore, in comparison to other ultra-thin film deposition techniques, the ALD technique offers several advantages in the development of 2D nanostructures.

6.4.1 The Catalytic Applications

Metal oxides represent one of the widely employed materials for catalytic applications. Their reducibility is an important property that facilitates the catalysis performance of metal oxides. While the catalytic applications of crystalline thin-oxide films, such as crystalline TiO_2 and MgO, have been the main focus of research activities during the past decades, ultra-thin amorphous oxides have recently confirmed their beneficial role as supports for catalytic applications [142–147]. Despite the modulation of 2D oxide films deposited on metal substrates, the oxide phases can be typically considered in terms of transitional or quasi transitional structures [14,148]. This is due to the incommensurate mismatch between the crystalline structure of the metal supports and 2D oxide films. In actual cases, since the thermodynamic and entropic parameters in the growth process are not precisely controllable, the real structure of ultra-thin oxide films is not fully crystalline [141,148,149]. It especially occurs in the case of 2D oxide films grown using the ALD technique. As a consequence, one of the main drawbacks for the growth of 2D oxides on crystalline metal substrates would be the transition from crystalline-like structures into semi-amorphous or disordered structures. The advantage of the ALD technique is its technological structure, which can be scaled up to meet the requirements of industrial applications. One of the most desired catalyst structures is the embedded catalytic nanostructures in 2D oxide phases [150, 151]. The mentioned catalytic nanostructures have various geometrical shapes, ranging from large nanoparticles to sub-nanometer particles, nano-island structures and nanoclusters. In general, the activity of the catalytic metals can be optimized when they are used in the form of supported small sub-nanoclusters or nanoparticles. To support catalytic nanostructures on the surface, an oxide ultra-thin layer can be used [148,149]. These composite nanostructures, i.e. catalytic nanostructures supported by 2D oxides, offer tremendous advantages. From the theoretical point of view, the charge transfer between catalytic nanostructures and metal supports is facilitated by the tunable conductance of 2D oxide films. From a technological point of view, since the charge transfer is created between a metal substrate and catalytic structures, the deposition of catalytic components is facilitated on 2D structure. Furthermore, the charge states of catalytic particles can be controlled by the precise selection of 2D oxide–metal substrate pairs and the accurate control of 2D oxide film thickness. In doing so, fine-tuning of the catalytic properties of these nanostructures can be gained [149]. Since the trends are inclined toward the application of supported sub-nanometer catalytic structures, the role of 2D oxide films as the support of catalysts has become even more important [152,153]. The direct epoxidation of propylene to propylene oxide by molecular oxygen was confirmed at low temperatures by using sub-nanometer Ag aggregates grown on 2D alumina supports. The 2 nm height silver clusters were grown on three monolayer alumina films developed by ALD [154,155]. Despite the fact that the surface morphology and termination of the ALD-developed oxide films are complex, this 2D oxide structure provides several advantages. The stabilization of catalytic particles against sintering under reaction conditions and the control of structural crystallinity versus amorphicity are just two of them [156]. As an example, high conversion accompanied by high selectivity was gained when 20 nm diameter silver nanoclusters were grown on 7.2 Å thick 2D ALD alumina on a naturally oxidized silicon substrate. The surface of this catalyst structure before and after epoxidation reactions are shown in Figure 6.12a and b, respectively. No noticeable changes were observed in the size of the nanostructured clusters and their distribution before and after the catalytic reactions, while the aspect ratio (height/diameter) of the nanoclusters was changed dramatically [154]. The topographical STM images of the surface, presented in Figure 6.12c, confirmed the amorphous structure of the ALD 2D alumina-support film. X-ray absorption characterization showed that this particular 2D alumina film can be best described as a mixture of tetrahedral and octahedral building units. The same photocatalytic activity was also reported in the case of selective propene epoxidation on immobilized Au clusters by ALD TiO_2 film [153].

Furthermore, the ALD technique provides the opportunity to develop 2D oxide films with a multicomponent structure around the catalyst particles to tune the performance of the system. The ALD of an ultra-thin TiO_2 coat on Pd catalysts grown on an amorphous Al_2O_3 substrate showed the improved resistance of catalyst particles against sintering at high-temperature degradation [155]. A schematic of Pd nanoparticles grown on an Al_2O_3 substrate and a graphical model of the development of a 2D TiO_2 film below and around the same Pd clusters are presented in Figure 6.13a and b. A significant enhancement in the sintering resistance of Pd clusters was observed by grazing-incidence small-angle scattering (GISAXS) (Figure 6.13c). The onset of sintering by Pd clusters supported on Al_2O_3 was around 170°C; an improvement in sintering resistance by 40°C in comparison with the individual Pd clusters. The thin ALD TiO_2 overcoat (TiO_2/Pd/Al_2O_3) improved the sintering resistance of the clusters, without any loss in its catalytic activity. This is the central point for the production of new catalytic materials for long-term applications.

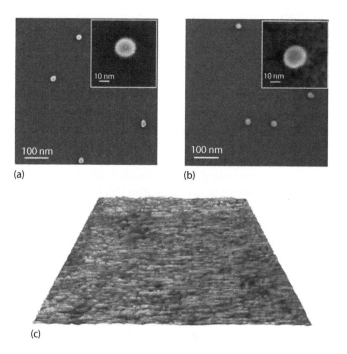

(a) (b) (c)

FIGURE 6.12
Representative HRSEM images of silver nanoclusters (a) before the epoxidation reaction and (b) after reaction cycles. (c) Topographic STM image of the amorphous alumina support layer. (Reproduced with permission from Molina, L.M., et al., *Catal. Today*, 160, 116–130, 2011.)

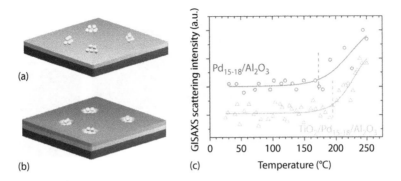

FIGURE 6.13
A graphical image of (a) Pd/Al$_2$O$_3$ and (b) titanium oxide over coated cluster sample TiO$_2$/Pd/Al$_2$O$_3$ and (c) integrated GISAXS scattering intensities as a function of temperature for Pd clusters without and with a titania protective overcoat. (Reproduced with permission from Lee, S., et al., *J. Phys. Chem. C*, 114, 10342–10348, 2010.)

This structure exhibited further enhanced stability during the reaction, with no indication of particle growth up to about 200°C [155].

The catalytic performance of 2D oxide films is not merely restricted to the embedded photocatalyst nanostructures in 2D oxide films. The efficient photocatalytic degradation of palmitic acid (PA) by 2D TiO$_2$ films was attained under both UV and visible light illumination [132]. In this study, 2D TiO$_2$ films with various thicknesses were deposited on Si/SiO$_2$ and Au substrates. The strong interaction between 2D TiO$_2$ films and PA molecules was confirmed by an investigation of the FTIR spectrum of 2D TiO$_2$ films covered by PA molecules. It was shown that a 2D TiO$_2$ film with a thickness of 3.5 nm can efficiently degrade the PA acid molecules under UV light illumination, while the partial degradation of PA molecules was attained for 0.7 nm thick 2D TiO$_2$ films. The optical images and FTIR measurements presented in Figure 6.14 show the gradual degradation of PA on the surface of a 2D TiO$_2$ photocatalyst deposited on Si/SiO$_2$ and Au substrates. At the beginning of the test, the surface of the photocatalyst film was entirely covered by a thick layer of PA film. By increasing the UV radiation time up to 24 h, integrated island structures with irregular shapes and sizes appeared. Regarding the observations, the photocatalysis process sequentially continued up to the formation of isolated islands and then the separation of PA particles. The surface was finally cleaned from PA molecules after 100 h illumination. The most interesting results were attributed to the efficient degradation of PA molecules under visible light, which were attained by the 2D TiO$_2$ films deposited on the Au substrate. Heterogeneous photocatalysis can be considered as the main mechanism engaged in the degradation of PA molecules by a 2D TiO$_2$–Au bilayer. The excitation process, the separation and the transfer of charge carriers under visible light illumination in the Au–TiO$_2$ bilayer were engaged in the photocatalytic performance [157,158]. Considering the structural and interfacial concepts of 2D oxide–metal heterostructures, the TiO$_2$–Au interface experiences a charge transfer from Au into TiO$_2$ semiconductors [149]. This property plays an important role when the plasmonic-assisted visible light electron generation of an Au substrate starts the electron injection into a photocatalyst 2D TiO$_2$ film [159]. In doing so, the electrons needed for photocatalytic reactions will be provided by the Au sublayer under visible light illumination. The important factors here are the geometrical features and the relation between the thickness of ultra-thin 2D oxide films and the roughness of an Au substrate. The deposition of an ultra-thin 2D oxide on the rough surface of an Au film facilitated plasmonic-assisted electron generation under visible light illumination [160,161].

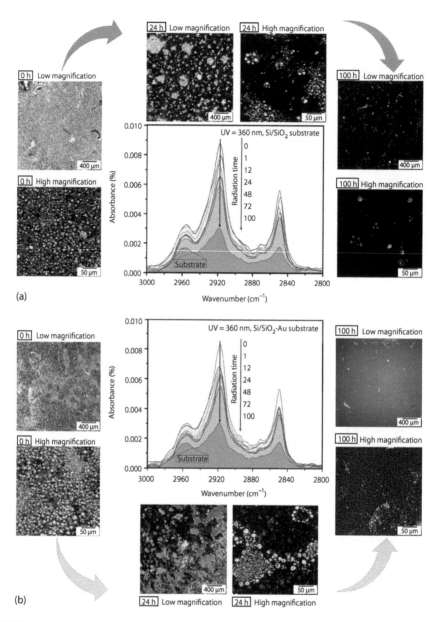

FIGURE 6.14
FT-IR absorbance spectra confirming the degradation of PA coated on TiO_2 film on (a) Si/SiO_2 and (b) Si/SiO_2–Au substrates, under UV illumination, as a function of irradiation time, and its corresponding optical image, representing degradation stages. (Reproduced with permission from Karbalaei Akbari, M., et al., *Mater. Res. Bull.*, 95, 380–391, 2017.)

6.4.2 The Photovoltaic Applications

The idea of employing 2D semiconductor materials in photovoltaic and photodetection applications arises from the fact that 2D films, monolayers or few-layer nanostructures can facilitate the ultra-fast transfer of electrons, and consequently, can pave the way for a new generation of photodetectors with high photoresponding behavior [162,163]. In this regard, although the concept of improving the functional capabilities of 2D nanostructured photodetectors was recently investigated, most of the reported studies only focused on the properties of individual monolayers with micron-sized longitudinal dimensions. A valuable alternative is the ALD technique, which reliably develops 2D nanostructures with precise thickness controllability and with flawless continuity over a large surface area. Furthermore, ALD can be used successfully for device fabrication. Thus, the optimization of an ALD recipe, the geometrical control of 2D oxide films, the stoichiometry and the chemical states of a 2D structure and the electrical contact between the semiconductor 2D oxide film and the metal electrode individually can play a vital role in the photovoltaic and photodetection performance of 2D oxide films.

As a practical example, a UV-A photodetector based on a 0.7 nm thick 2D WO_3 was fabricated by ALD on the surface of 4-inch diameter $Si/SiO_2/Au$ wafers [164]. The Au film was used both as the support for 2D WO_3 and as the electrode for electrical measurements. The fabricated wafer-scale photoelectrode, its graphical scheme and TEM cross-sectional view of a 2D WO_3 film are presented in Figure 6.15a–c, respectively. The UV-A detecting performance of the 2D WO_3 photodetector showed improved photoresponsivity under various bias voltages, as shown in Figure 6.15d. The ultra-fast response time (40 μs, Figure 6.15e and f) of the 2D WO_3-based photodetector was the outstanding feature of this 2D nanostructure.

The measured response time was at least a 400-fold improvement on the previous reports for other WO_3-based UV-photodetectors that have been fabricated [164]. This ultra-fast response was attributed to the 2D nature of the WO_3 film and also to the presence of a metal–semiconductor junction in the fabricated photodetector [164]. The developed 2D photodetectors demonstrated long-term stability and reliable sensitivity to variation in the power density. The results of photoresponsivity measurements proved conclusively that the device has a generally distinguished response to the UV-A. However, the 2D

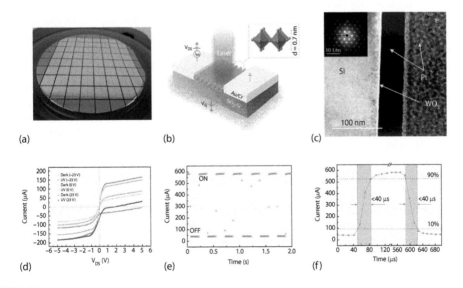

(a) (b) (c)

(d) (e) (f)

FIGURE 6.15
(a) Optical image of 2D WO_3 thin films deposited on the 4-inch diced Au wafer, (b) three-dimensional schematic view of the 2D WO_3 photodetector, (c) the cross-sectional TEM image of 2D WO_3, (d) the I-V characteristics of device under UV light illumination at different back-gate voltages (increasing from top to down of graph), (e) time-resolved photoresponse of the device under on-off illumination operation, (f) measured photoresponse time of the developed device. (Reproduced with permission from Hai, Z., et al., *Appl. Surf. Sci.*, 405, 169–177, 2017 and Hai, Z., et al., *Sensor Actuat. B Chem.*, 245, 954–962, 2017.)

WO$_3$-based photodetectors still provided some responses to the incident light with energy levels lower than the bandgap of WO$_3$. This was attributed to structural defects in the 2D WO$_3$ film, which can be responsible for photoresponsivity at longer wavelengths of light [165]. A similar ultra-fast photoresponse behavior was also reported when a 0.37 nm thick 2D TiO$_2$ semiconductor was used as a photodetector [166]. The performance of a 2D TiO$_2$-based photodetector was investigated under the illumination of various light sources and the measured photocurrent was quick and stable with an ultra-fast response (30 µs) and recovery (63 µs) time. This is a significant improvement on previous results for the response time of other nanostructured TiO$_2$-based photodetectors [166]. The calculated photoresponsivity (R_λ) of this 2D TiO$_2$-based photodetector was about 733 times higher than the reported R_λ for a photodetector with a 20 nm thick ALD TiO$_2$ semiconductor film [166,167]. The fabricated 2D TiO$_2$ photodetector has demonstrated rewardable long-term stability and a sensitive reaction to the variation of light intensity. A rectifying behavior was observed in the measurement of the photoresponsivity of a 2D TiO$_2$-based photodetector deposited on an Au substrate. This rectifying response is very similar to the normal metal–semiconductor junction behavior under various back-gate voltages [168,169]. This evidence confirmed that the possible interaction in the 2D oxide–metal interface can change the electrical response of these devices.

Visible light responsivity is a desired property in semiconductor materials, which can facilitate the employment of visible light sources for energy conversion. Due to the large bandgap of oxide semiconductors, the illumination of UV light sources is needed to stimulate the electrons of the valance band of semiconductor oxides [170,171]. Visible light photoresponsivity was observed when 2D semiconductor oxide films were deposited on an Au substrate and then used as a photosensor. A strong photoresponsivity to the $\lambda = 785$ nm visible light laser was measured when 2D WO$_3$ mono and few-layer films were grown by ALD on an Au substrate [172]. The same behavior was observed when 2D TiO$_2$ oxide films were deposited by ALD on an Au substrate to harvest the visible light energy for catalytic reactions [133]. The I-V (electrical current to voltage of photodetectors) curves of Figure 6.16a and b show the improved visible light photovoltaic performances of a 2D TiO$_2$–Au bilayer film compared with that of 2D TiO$_2$ films developed on an SiO$_2$ bare substrate. This visible light sensitivity of 2D TiO$_2$–Au bilayer films also made it possible to recruit the simulated sunlight for photocatalysis of fatty acid molecules. The improved visible light responsivity of the TiO$_2$–Au bilayer film can be attributed to the effect of TiO$_2$–Au plasmonic coupling, which is in direct relation to the thickness and geometrical features of 2D oxide films and is further affected by the TiO$_2$–Au interface. The excitation process, separation and transfer of charge carriers in the Au–TiO$_2$ bilayer film under visible light illumination can be attributed to various individual mechanisms. The tunneling of high energy carriers or hot electrons into the conduction band of a 2D oxide semiconductor is considered as one of the main mechanisms [173,174]. This mechanism is graphically depicted in Figure 6.16c. In fact, the Au film can be considered as a rough substrate for ultra-thin 2D TiO$_2$ [160]. This process includes photo-plasmon coupling accompanied by the generation of hot electrons by the decay of a surface plasmon. It continues by the injection or tunneling of hot electrons into 2D TiO$_2$. It results in the creation of activated electrons with different levels of energy in a TiO$_2$ conduction band [160,173]. In this mechanism, hot electrons must overcome the metal–semiconductor (MS) Schottky barrier at the interface of the Au–TiO$_2$ bilayer [160,161]. In another mechanism, the excitation of surface plasmons by visible light enables energy transfer into a 2D TiO$_2$ film. As graphically shown in Figure 6.16d, in this process the valance-band electrons of 2D TiO$_2$ are excited resulting in the generation of electrons in TiO$_2$ and holes in an Au film. In detail, the intense electric field called "hot spot" is effectively introduced into

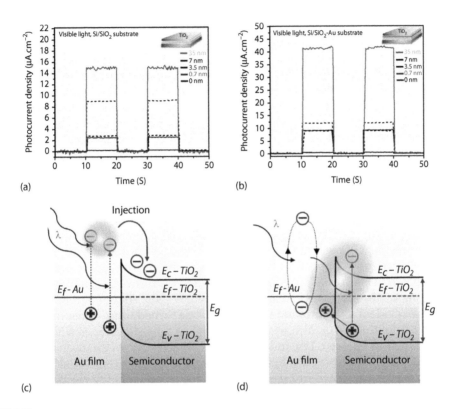

FIGURE 6.16
The I-t plots of photodetectors under visible light illumination. (a) The increasing photocurrent density with increasing the thickness of TiO_2 films deposited on Si/SiO_2 and (b) Au substrates. The mechanisms of photo-electron generation in an $Au–TiO_2$ bilayer under visible light, (c) the injection of hot electrons into a conduction band of TiO_2, whose potential energy is high enough to overcome the potential barrier between $Au–TiO_2$ bilayer and (d) the energy transfer by the generation of an electric field in 2D TiO_2 film. (Reproduced with permission from Karbalaei Akbari, M., et al., *Mater. Res. Bull.*, 95, 380–391, 2017.)

the semiconductor layer, exciting TiO_2 electrons from the valence to the conduction band [172,173]. Since a part of a 2D TiO_2 film can be covered by hot spots, the required energy to overcome the Schottky barrier can be provided by the incident of high energy photons of the visible light [153,161]. These mechanisms can adequately explain the improved visible light photocurrent density and photocatalytic performance of a 2D $TiO_2–Au$ bilayer film. The extraction of plasmonic-derived charge carriers from a multilayer stack, comprising a monolayer of Au nanoparticles deposited on the 2D $TiO_2–Au$ bilayer, was independently confirmed by other researchers [174]. This multilayer structure has also demonstrated the broadband and intense absorption of light, which leads to a significant increase in incident photon-to-electron conversion efficiency [174].

6.4.3 Supercapacitance Performance of 2D Oxide Semiconductors

Supercapacitors, as a very promising type of energy storage device, have attracted much attention in recent years owing to their superior properties such as high power density, fast charge/discharge rate, excellent reversibility and long cyclic life [175–177]. While the effects of electrode thickness on capacitor performance are recognized [178], most of them focus on micro and submicron films and only a few scientific reports have focused comprehensively on nanoscale and sub-nanoscale thicknesses [179]. Therefore, an investigation of the effect of the thickness of 2D electrodes at the few-layered level down to the single fundamental layer would be intriguing and influential. Common fabrication methods do not have the capability for such precise thickness and conformity control for wafer-scale deposition of ultra-thin 2D films, which hinders the development of atomically thin 2D supercapacitor electrodes. As a technological advancement, the employment of the ALD technique with special precursors made this investigation possible [180,181]. Despite the short history of ALD application in the development of supercapacitors, the advantages of this precise deposition technique have been proven in several cases. ALD-developed Co_3O_4 [182], NiO [183] and VO_x thin films are some of the examples of using the ALD technique in the development of supercapacitors [184].

The supercapacitance performance of ALD-developed 2D WO_3 electrodes with thickness variations from 6.0 nm down to 0.7 nm was recently reported [126]. The thickness-dependent super-capacitance behaviors of 2D WO_3 electrodes were studied using the cyclic voltammetry (CV) technique. The obtained results indicated that the capacitance values were apparently changed as the thickness of the electrodes decreased from 6.0 to 0.7 nm, as shown in Figure 6.17a. Besides the impressive improvement of the specific capacitance judging by the integrated areas of the CV curves, the shapes of the CV curves also underwent an evident transformation as the electrode turned thinner. For instance, the CV curves of a 6.0 nm thick WO_3 film were close to the rectangular shapes without any peak, as clearly shown in Figure 6.17b, indicating the characteristic of an electric double-layer capacitor [185]. The CV curves exhibited almost ideal rectangular shapes along with larger integrated areas as presented in Figure 6.17a–c. This means that the electric double-layer capacitance was the dominant capacitance mechanism in these 2D WO_3 supercapacitors. Even when the thickness of the 2D WO_3 film decreased to 0.7 nm, the general pattern of the CV curves was still shaped like a rectangle, as demonstrated in Figure 6.17d. The appearance of distinct redox peaks demonstrated that the mechanism shifted from the electric double-layer capacitance for a few-layered 2D WO_3 to the pseudo-capacitance for a monolayer WO_3. This phenomenon possibly resulted from the full exposure of the surface atoms during the redox reactions in a monolayer WO_3. The mechanism shift can also be contributed by the sub-stoichiometric structure of a monolayer WO_3, while the few-layered 2D film adopted an adsorption/separation process potentially due to its 2D-layered nature with its rough surface characteristics [186]. The galvanostatic charge-discharge (GCD) measurements confirmed the non-faradic capacitance behavior of the few-layered 2D WO_3, which is consistent with the CV measurements. The pseudo-capacitance behavior of a monolayer WO_3 film was further confirmed by the study of CV curves. The specific capacitances calculations for the 2D WO_3 electrodes suggested a negative correlativity between the specific capacitance of the WO_3 thin film and its thickness. A comparative study in terms of the specific capacitance is summarized in Table 6.3 [126]. Data show that the monolayer WO_3 film (0.7 nm thick) had the highest specific capacitance among most of the WO_3 nanostructured electrodes, demonstrating the superior advantages of the atomically thin 2D WO_3 as supercapacitor electrodes. Cycling stability and electrochemical impedance spectroscopy (EIS) measurements revealed that the monolayer WO_3 film has lower charge transfer resistance compared with that of thicker WO_3 films [126].

It was suggested that the development of heterojunction semiconductors could have distinctive modulation effects on electrical conductivity and ion diffusion behavior [187–189]. The supercapacitance properties of an atomically thin 2D WO_3/TiO_2 heterojunction as an electrochemical capacitor have also been investigated [190]. The ALD-enabled approach to the fabrication of a 2D heterojunction is especially advantageous when the film quality and

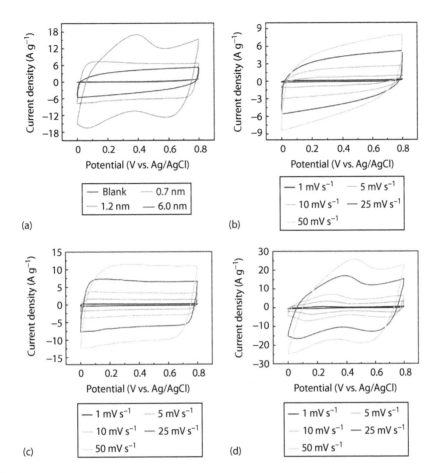

FIGURE 6.17
(a) CV curves of 2D WO$_3$ samples with different thickness. The CV curves of (b) 6.0 nm, (c) 1.2 nm and (d) 0.7 nm thick WO$_3$ film at different scan rates. Higher scan rate resulted in more capacitance or larger CV curve. (Reproduced with permission from Hai, Z., et al., *Electrochim. Acta*, 246, 625–633, 2017.)

thickness are critical for the surface/near-surface reactions in high performance supercapacitors [191]. The CV measurements revealed the ideal capacitance behavior of WO$_3$ 2D films while the 2D TiO$_2$ films showed the pseudo-capacitance behavior [190]. Furthermore, the energy storage mechanism of 2D WO$_3$/TiO$_2$ was shown to primarily be a non-faradaic reaction. The GCD studies revealed a longer discharge time for a 2D WO$_3$/TiO$_2$ heterojunction compared with that of a 2D WO$_3$. A longer discharge time indicated the contribution of a 2D TiO$_2$ film in the improvement of the capacitance of a 2D WO$_3$/TiO$_2$ film. The discharge performance of the 2D nanostructures showed that the specific capacitance of a 2D WO$_3$/TiO$_2$ heterojunction was much higher than the specific capacitance of individual 2D TiO$_2$ and 2D WO$_3$ films. Long-term stability (over thousands of cycles) was another distinguishing characteristic of this 2D supercapacitor electrode. The capacitance retention of a 2D WO$_3$/TiO$_2$ heterojunction was slightly lower than that of 2D TiO$_2$. The long-term cycling stability of 2D WO$_3$/TiO$_2$ was improved compared with that of WO$_3$, which is attributed to the stabilization of the internal interface by the deposition of 2D TiO$_2$ on 2D WO$_3$ [126].

The surface functionalization of 2D materials with other nanostructures is one of the proposed strategies to impact the capacitance properties of ultra-thin films [33,192].

TABLE 6.3

Supercapacitor Performance of Various Nanostructured WO_3 Electrodes

Electrode Materials	Electrolyte	Cs (F g^{-1})	T (a.u.)	Cr (%)
WO_3 film (0.7 nm)	1.0 M H_2SO_4	650.3	2000	65.8
WO_3 film (1.2 nm)	1.0 M H_2SO_4	396.9	2000	75.5
WO_3 film (6.0 nm)	1.0 M H_2SO_4	225.4	2000	91.7
WO_3 nanoflower	1.0 M H_2SO_4	196	5000	85
WO_3 nanoflower	0.5 M H_2SO_4	127	1000	83.7
WO_3 nanorod	1.0 M Na_2SO_4	463	1000	97
WO_3 nanorod	1.0 M H_2SO_4	114	/	/
WO_3 nanoparticle	2.0 M KOH	255	/	/
WO_3 nanoparticle	0.5 M H_2SO_4	54	/	/
h-WO_3 thin film	1.0 M H_2SO_4	694	2000	87
h-WO_3.n.H_2O	0.5 M H_2SO_4	498	/	/
WO_3 thin film	1.0 M Na_2SO_4	530	2000	84
Mesoporous WO_3	2.0 M H_2SO_4	109	/	/
WO_3/WO_3.0.5 H_2O	0.5 M H_2SO_4	290	/	/
GNS/WO_3	1.0 M H_2SO_4	143.6	/	/
rGO/WO_3.H_2O	1.0 M H_2SO_4	244	900	97
Co/WO_3	2.0 M KOH	45	/	/
Ni/WO_3	2.0 M KOH	171.28	/	/

Source: Reprinted with permission from Hai, Z., et al., *Electrochim. Acta*, 246, 625–633, 2017.

In this regard, TiO_2 nanoparticle-functionalized 2D WO_3 films were fabricated by the facile two-step ALD process followed by the post-annealing process at 380°C [33]. The supercapacitance properties of these nanostructured electrode materials were investigated. To fabricate a nanostructured electrode, initially a 6 nm thick 2D WO_3 was deposited on a polycrystalline Au substrate followed by ALD of a 1.5 nm thick TiO_2 film over WO_3. A post-annealing process at air atmosphere resulted in the formation of uniformly distributed TiO_2 nanoparticles on the surface of a 2D WO_3 film. The detailed SEM images of TiO_2 nanoparticle-functionalized 2D WO_3 films are shown in Figure 6.18a and b. XPS and Raman analyses confirmed the development of stoichiometric TiO_2 nanoparticles with an anatase crystalline structure on the surface of a WO_3 film. The TiO_2 functionalized 2D WO_3 electrode demonstrated a distinct improvement in the specific capacitance compared with that of a 2D WO_3 electrode. Studies confirmed that the capacitance in a 2D WO_3 electrode was stored by the accumulation of electrolyte ions between the electrolyte–electrode interfaces, which is known as the electric double-layer capacitance mechanism [193]. The GCD tests also confirmed the dominance of the electric double-layer capacitance mechanism of the galvanostatic charge and the discharge performance of a 2D WO_3 film. After TiO_2 functionalization of a WO_3 2D film, the mechanism shifted to pseudo-capacitance, as demonstrated in Figure 6.18c. The transition of electrochemical mechanisms is ascribed to the following facts. The TiO_2-functionalized WO_3 film contains more active sites. This can result in a higher surface-to-volume ratio and better electron mobility compared with that of pure 2D WO_3 electrodes, facilitating ion adsorption and diffusion on the active material. The improved ion adsorption and diffusion leads to an enhanced electrochemical performance. Furthermore, the improved capacitance performance is also related to the formation of a TiO_2–WO_3 heterojunction with enhanced conductance modulation [194,195]. The characteristics of pseudo-capacitance behavior were also detected by the study of galvanostatic charge and discharge analysis of TiO_2 functionalized 2D WO_3 films. The remarkably 1.5-times enhancement of the specific capacitance and faster charge transfer are the other benefits of the functionalization of 2D WO_3 films with TiO_2 nanoparticles.

6.4.4 Electrochemical Sensors Based on 2D Oxide Semiconductors

The fast and accurate sensing of environmentally hazardous components has been the focus of attention during the last decade [196]. Recently, ALD-developed 2D WO_3 nano-structures were used for the electrochemical detection of hydrazine in aqueous solutions [34]. The main idea behind using 2D nanostructures as sensors is attributed to the employment of the 2D-confined effects of 2D nanostructures to facilitate the efficient sensing of ultra-low concentration chemicals in an aqueous environment [34]. The ALD technique has the capability to develop ultra-thin oxide films with precise thickness controllability. This is important especially when it is known that the electro-catalytic activity of nano-structured materials is significantly dependent on their distribution on the substrate. Traditional methods, such as electrochemical deposition and hydrothermal synthesis, have no reasonable control over the uniform distribution of materials. Furthermore, traditional techniques cannot satisfy the necessity of the long-term stability of sensors, which is due to the gradual detachment and dissolution of the catalyst materials from the substrate [197]. ALD as an alternative technique can fulfill the requirements of a uniformly developed oxide sensor film. The studies of EIS of 2D WO_3 films confirmed that the monolayer WO_3 has the lowest electron transfer resistance compared with that of thicker 2D WO_3 films. The redox peak current of a monolayer WO_3 was higher than that of other thicker 2D WO_3 films. This is due to the largest effective surface area of a monolayer WO_3 resulting in an increase in current and surface charge density. The results of CV tests are shown in Figure 6.19a. The measurements indicate that 2D WO_3 nano-films possess a lower oxidation potential and a higher oxidation peak current for hydrazine electro-oxidation. This can be attributed to faster electron transfer kinetics in 2D WO_3 nano-films owing to a high effective surface area. The efficient electro-catalytic activity of a monolayer WO_3 film without any fouling effects showed the fast electron transfer reactions on the surface of a 2D WO_2 film. From the I-t curves, depicted in Figure 6.19b, fast response and all steady states were achieved within just 2 s for 2D WO_3 films. The monolayer WO_3 film exhibited the largest catalytic response current to changes in the hydrazine concentration, which was considerably higher than that of the other 2D WO_3 nano-films. An evaluation of long-term stability by conducting amperometric experiments indicated that the fabricated monolayer WO_3 films

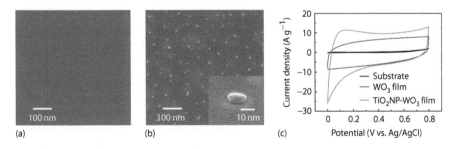

(a) (b) (c)

FIGURE 6.18
SEM images of (a) 2D WO_3 film and (b) 2D WO_3 film functionalized by TiO_2 nanoparticles after post-annealing. Inset in (b) is the HRSEM image of TiO_2 nanoparticle and (c) CV curves of the 2D electrodes. (Reproduced with permission from Hai, Z., et al., *Mater. Today Commun.*, 12, 55–62, 2017.)

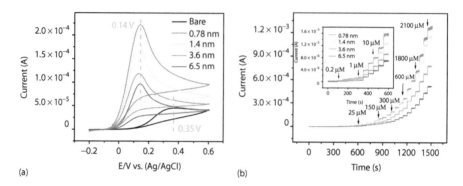

FIGURE 6.19

(a) CV curves of 2D WO_3 films and Au electrodes for hydrazine detection, the lower film thickness demonstrated the higher peak current and the larger catalytic response, (b) the sensitive chronoamperometric response of 2D WO_3 films to the changes of hydrazine concentration. (Reproduced with permission from Wei, Z., et al., *Chem. Electro. Chem.*, 5, 266–272, 2018.)

have excellent long-term stability. As an interesting property, the anti-interfering ability of a monolayer WO_3 was confirmed when the interfering chemicals were added to the aqueous environment containing hydrazine. The amperometric response of monolayer WO_3 nano-films to hydrazine did not change upon the introduction of KCl, $NaNO_3$ and $NaNO_2$, glucose and H_2O_2 into the aqueous solution containing hydrazine [34]. As a practical result, the response of monolayer WO_3 nano-films to hydrazine was not affected by the addition of a 100-fold concentration of inorganic and a 50-fold concentration of biological substances [34]. A sensitivity of 1.23 was recorded for hydrazine sensing achieved by using a monolayer WO_3 in the presence of interfering agents. This value was very close to the value of sensitivity of the monolayer WO_3 (1.24) when there was no interference agent. These observations confirmed the reliability of the 2D WO_3 oxide as electrochemical sensors.

6.5 Summary

The present chapter has highlighted the capabilities of the ALD technique on the development and tailoring of ultra-thin 2D nanostructured semiconductor oxides. Compared with other chemical and PVD techniques, the ALD capabilities for design, characterization and tailoring of these 2D nanostructures have broad appeal as follows:

- The unique self-limiting film growth mechanism originated from the self-saturation nature of chemical reactions in the ALD process. It facilitates the atomic-scale control of the thickness of ultra-thin films.

- The capabilities and advantages of ALD substantiate the employment of a precisely designed ALD recipe for the development of 2D heterostructured oxide films with unique functionalities. These nanostructures offer a combination of 2D oxide films properties that are not available for the single component materials.

- The surface functionalization of 2D oxide films with ALD-assisted complementary processes makes it possible to manipulate the properties of 2D oxide materials. This is considered one of the highly valuable features of ALD for the design and fabrication of the doped and functionalized 2D nanostructures.

Despite the mentioned intrinsic advantage of ALD, the development and monitoring of wafer-scaled 2D oxide films with a high level of uniformity and controlled crystallinity and stoichiometry need the accurate design of an ALD recipe. Herewith, the recent progress in the development of mono and few-layered nanostructured semiconductor oxides was discussed by investigating the practical cases assisted by ALD engineering. The technological ALD parameters were reviewed to provide an overview of the challenges faced during the development of 2D oxide nanostructures. The discussions were then accompanied by complementary material characterization studies to reinforce the scientific aspects of ALD development of 2D nanostructured semiconductor oxides.

Subsequently, the tailoring of ALD-developed 2D semiconductor oxide was introduced by considering the practical applications of these nanostructures. Consequently, the capabilities of ALD for the fabrication of nano-devices based on 2D semiconductor oxides, including photocatalyst nanostructures, photovoltaic and UV and visible light sensors, supercapacitors and electrochemical sensors for hazardous chemicals were presented. In essence, the reliability of the ALD technique for tailoring 2D nanostructured oxide semiconductors for practical applications was confirmed.

References

1. A. K. Geim and K. S. Novoselov, The rise of graphene, *Nat. Mater.* 6 (2007) 183–191.
2. K. S. Novoselov, Z. Jiang, Y. Zhang, S. V. Morozov, H. L. Stormer, U. Zeitler, J. C. Maan, G. S. Boebinger, P. Kim, and A. K. Geim, Room-temperature quantum hall effect in graphene, *Science* 315 (2007) 1379.
3. R. R. Nair, P. Blake, A. N. Grigorenko, K. S. Novoselov, T. J. Booth, T. Stauber, N. M. R. Peres, and A. K. Geim, Fine structure constant defines visual transparency of graphene, *Science* 320 (2008) 1308.
4. F. Schedin, A. K. Geim, S. V. Morozov, E. W. Hill, P. Blake, M. I. Katsnelson, and K. S. Novoselov, Detection of individual gas molecules adsorbed on graphene, *Nat. Mater.* 6 (2007) 652–655.
5. X. Li, W. Cai, J. An, S. Kim, J. Nah, D. Yang, R. Piner, A. Velamakanni, I. Jung, E. Tutuc, et al., Large-area synthesis of high-quality and uniform graphene films on copper foils, *Science* 324 (2009) 1312–1314.
6. S. Bae, H. Kim, Y. Lee, X. Xu, J. S. Park, Y. Zheng, J. Balakrishnan, T. Lei, H. Kim, Y. Song, et al., Roll-to-roll production of 30-inch graphene films for transparent electrodes, *Nat. Nanotechnol.* 5 (2010) 574–578.
7. J. Wrachtrup, 2D materials: Single photons at room temperature, *Nat. Nanotechnol.* 11 (2016) 7–8.
8. M. Chhowalla, D. Jena, and H. Zhang, Two-dimensional semiconductors for transistors, *Nat. Rev. Mater.* 1 (2016) 16052.
9. F. Xia, H. Wang, D. Xiao, M. Dubey, and A. Ramasubramaniam, Two-dimensional material nanophotonics, *Nat. Photonics.* 8 (2014) 899–907.
10. P. L. Cullen, K. M. Cox, M. K. B. Subhan, L. Picco, O. D. Payton, D. J. Buckley, T. S. Miller, S. A. Hodge, N. T. Skipper, V. Tileli, et al., Ionic solutions of two-dimensional materials, *Nat. Chem.* 9 (2017) 244–249.
11. P. Ajayan, P. Kim, and K. Banerjee, Two-dimensional van der Waals materials, *Phys. Today* 69 (2016) 39–44.
12. Y. Liu, N. O. Weiss, X. Duan, H. C. Cheng, Y. Huang, and X. Duan, Van der Waals heterostructures and devices, *Nat. Rev. Mater.* 1 (2016) 16042.
13. D. Jariwala, T. J. Marks, and M. C. Hersam, Mixed-dimensional van der Waals heterostructures, *Nat. Mater.* 16 (2017) 170–181.

14. F. P. Netzer, and S. Surnev, Structure concept in two dimensional oxide materials, In: F. P Netzar and A. Fortunelli (eds.), *Oxide Material at the Two-Dimensional Limit*, (2016) Springer Series in Materials Science, 234, Springer, Cham, Switzerland, pp. 233–250.

15. Z. Lin, A. McCreary, and N. Briggs, 2D materials advances: From large scale synthesis and controlled heterostructures to improved characterization techniques, defects and applications, *2D Mater.* 3 (2016) 042001.

16. S. Surnev, A. Fortunelli, and F. P. Netzer, Structure–property relationship and chemical aspects of oxide-metal hybrid nanostructures, *Chem. Rev.* 113 (2013) 4314–4372.

17. H. Kuhlenbeck, S. Shaikhutdinov, and H. J. Freund, Well-ordered transition metal oxide layers in model catalysis: A series of case studies, *Chem. Rev.* 113 (2013) 3986–4034.

18. J. Shim, H. Park, D. Kang, J. Kim, S. Jo, Y. Park, and J. Park, Electronic and optoelectronic devices based on two-dimensional materials: From fabrication to application, *Adv. Electron. Mater.* 3 (2017) 1600364.

19. F. Wang, Z. Wang, C. Jiang, L. Yin, R. Cheng, X. Zhan, K. Xu, F. Wang, Y. Zhang, and J. He, Progress on electronic and optoelectronic devices of 2D layered semiconducting materials, *Small*, 13 (2017) 1604298.

20. K. Kerman and S. Ramanathan, Performance of solid oxide fuel cells approaching the two-dimensional limit, *J. Appl. Phys.* 115 (2014) 174307.

21. S. Ida and T. Ishihara, Recent progress in two-dimensional oxide photocatalysts for water splitting, *J. Phys. Chem. Lett.* 5 (2014) 2533–2542.

22. G. Pacchioni, Two-dimensional oxides: Multifunctional materials for advanced technologies, *Chem. Eur. J.* 18 (2012) 10144–10158.

23. S. G. Leonardi, Two-dimensional zinc oxide nanostructures for gas sensor applications, *Chemosensors* 5 (2017) 17.

24. E. Kan, M. Li, S. Hu, C. Xiao, H. Xiang, and K. Deng, Two-dimensional hexagonal transition-metal oxide for spintronics, *J. Phys. Chem. Lett.* 4 (2013) 1120–1125.

25. S. P. Ghosh, K. C. Das, N. Tripathy, G. Bose, D. H. Kim, T. I. Lee, J. M. Myoung, and J. P. Kar, Ultraviolet photodetection characteristics of zinc oxide thin films and nanostructures, *IOP Conf. Ser. Mater. Sci. Eng.* 115 (2016) 012035.

26. W. R. Richards and M. Shen, Extraction of two-dimensional metal–oxide–semiconductor field effect transistor structural information from electrical characteristics, *J. Vac. Sci. Technol. B* 18 (2000) 533–539.

27. J. E. T. Elshof, H. Yuan, and P. G. Rodriguez, Two-dimensional metal oxide and metal hydroxide nanosheets: Synthesis, controlled assembly and applications in energy conversion and storage, *Adv. Energy Mater.* 6 (2016) 1600355.

28. M. Alsaif, M. R. Field, T. Daeneke, A. F. Chrimes, W. Zhang, B. J. Carey, K. J. Berean, S. Walia, J. Van Embden, B. Zhang, et al., Exfoliation solvent dependent plasmon resonances in two-dimensional sub-stoichiometric molybdenum oxide nanoflakes, *ACS Appl. Mater. Interfaces* 8 (2016) 3482–3493.

29. Y. Li, Z. Li, C. Chi, H. Shan, L. Zheng, and Z. Fang, Plasmonics of 2D nanomaterials: Properties and applications, *Adv. Sci.* 4 (2017) 1600430.

30. E. Brunet, G. C. Mutinati, S. Steinhaver, and A. Kock, Oxide ultrathin films in sensor applications, In: G. Pacchioni and S. Valeri (eds.), *Oxide Ultrathin Films: Science and Technology*, 1st edn. (2012) Wiley-V-CH, Weinheim, Germany, pp. 239–241.

31. K. Shavanova, Y. Bakakina, I. Burkova, I. Shtepliuk, R. Viter, A. Ubelis, V. Beni, N. Starodub, R. Yakimova, and V. Khranovskyy, Application of 2D non-graphene materials and 2D oxide nanostructures for biosensing technology, *Sensors* 16 (2016) 223.

32. Y. Zhu, C. Cao, S. Tao, W. Chu, Z. Wu, and Y. Li, Ultrathin nickel hydroxide and oxide nanosheets: Synthesis, characterizations and excellent supercapacitor performance, *Sci. Rep.* 4 (2014) 5787.

33. Z. Hai, M. Karbalaei Akbari, Z. Wei, C. Xue, H. Xu, J. Hu, L. Hyde, and S. Zhuiykov, TiO_2 nanoparticles-functionalized two-dimensional WO_3 for high performance supercapacitors developed by facile two-step ALD process, *Mater. Today Commun.* 12 (2017) 55–62.

34. Z. Wei, Z. Hai, M. Karbalaei Akbari, J. Hu, L. Hyde, S. Depuydt, and S. Zhuiykov, Ultra-sensitive, sustainable, and selective electrochemical hydrazine detection by ALD developed two-dimensional WO_3 nano-films, *Chem. Electro. Chem.* 5 (2018) 266–272.

35. R. Ma and T. Sasaki, Two-dimensional oxide and hydroxide nanosheets: Controllable high-quality exfoliation, molecular assembly, and exploration of functionality, *Acc. Chem. Res.* 48 (2015) 136–143.

36. L. Gao, Y. Li, M. Xiao, S. Wang, G. Fu, and L. Wang, Synthesizing new types of ultrathin 2D metal oxide nanosheets via half-successive ion layer adsorption and reaction, *2D Mater.* 4 (2017) 025031.

37. J. Yu, J. Li, W. Zhang and H. Chang, Synthesis of high quality two-dimensional materials via chemical vapor deposition, *Chem. Sci.* 6 (2015) 6705–6716.

38. Z. Yang and J. Hao, Progress in pulsed laser deposited two-dimensional layered materials for device applications, *J. Mater. Chem. C* 4 (2016) 8859–8878.

39. B. Zou, C. Walker, K. Wang, V. Tileli, O. Shaforost, N. M. Harrison, N. Klein, N. M. Alford, and P. K. Petrov, Growth of epitaxial oxide thin films on graphene, *Sci. Rep.* 6 (2016) 31511.

40. T. Tsuchiya, K. Daoudi, I. Yamaguchi, T. Manabe, T. Kumagai, and S. Mizuta, Preparation of tin oxide films on various substrates by excimer laser metal organic deposition, *Appl. Surf. Sci.* 247 (2005) 145–150.

41. C. Wu and M. R. Castel, Ultra-thin oxide film on Au (111) substrate, In: F. P. Netzer and A. Fortunelli (eds.), *Oxide Material at the Two-Dimensional Limit*, (2016) Springer Series in Materials Science 234, Springer, Cham, Switzerland, pp. 149–168.

42. S. Shaikhutdinov and H. J. Freund, Ultrathin oxide films on metal supports: Structure-reactivity relations, *Annu. Rev. Phys. Chem.* 63 (2012) 619–633.

43. S. A. Chambers, Epitaxial growth and properties of thin film oxides, *Surf. Sci. Rep.* 39 (2000) 105–180.

44. N. Nilius, Properties of oxide thin films and their adsorption behavior studied by scanning tunneling microscopy and conductance spectroscopy, *Surf. Sci. Rep.* 64 (2009) 595–659.

45. J. V. Lauritsen and F. Besenbacher, Model catalyst surfaces investigated by scanning tunneling microscopy, In: B. C. Gates and H. Knozinger (eds.), *Advances in Catalysis*, (2006) 50. Elsevier Academic Press Inc, San Diego, CA, pp. 97–147.

46. S. Zhuiykov, T. Kawaguchi, Z. Hai, M. Karbalaei Akbari, and P. M. Heynderickx, Interfacial engineering of two-dimensional nano-structured materials by atomic layer deposition, *Appl. Surf. Sci.* 392 (2017) 231–243.

47. S. Valeri and S. Benedetti, Synthesis and preparation of oxide ultrathin films. In: G. Pacchioni and S. Valeri (eds.), *Oxide Ultrathin Films: Science and Technology*, (2012) Wiley-V-CH, Weinheim, Germany, pp. 1–26.

48. D. C. Grinter and G. Thornton, Characterization tools of ultrathin oxide films. In: G. Pacchioni and S. Valeri (eds.), *Oxide Ultrathin Films: Science and Technology*, (2012) Wiley-V-CH, Weinheim, Germany, pp. 27–46.

49. S. Zhuiykov, Nanostructured two-dimensional materials, In: V. Tewary and Y. Zhang (eds.), *Modelling, Characterization and Production of Nanomaterials*, (2015) Elsevier, Amsterdam, pp. 477–524.

50. D. M. King, X. Liang, and A. W. Weimer, Functionalization of fine particles using atomic and molecular layer deposition, *Powder Technol.* 221 (2012) 13–25.

51. M. Laskela and M. Ritala, Atomic layer deposition chemistry: Recent developments and future challenges, *Angew. Chem. Int. Ed.* 42 (2003) 5548–5554.

52. S. J. Tauster, S. C. Fung, and R. L. Garten, Strong metal-support interactions. Group 8 noble metals supported on titanium dioxide, *J. Am. Chem. Soc.* 100 (1978) 170–175.

53. M. Osada and T. Sasaki, Chemical nanomanipulation of two-dimensional nanosheets and its applications, In: Y. Masuda (ed.), *Nanofabrication*, (2011) InTech.

54. L. Wang and T. Sasaki, Titanium oxide nanosheets: Graphene analogues with versatile functionalities, *Chem. Rev.* 114 (2014) 9455–9486.

55. F. P. Netzer and S. Surnev, Ordered oxide nanostructures on metal surfaces, In: G. Pacchioni and S. Valeri (eds.), *Oxide Ultrathin Films: Science and Technology*, (2012) Wiley-VCH, Weinheim, Germany, pp. 47–73.

56. Q. Fu and T. Wagner, Interaction of nanostructured metal over layers with oxide surfaces, *Surf. Sci. Rep.* 62 (2007) 431–498.
57. F. P. Netzer, F. Allegretti, and S. Surnev, Low-dimensional oxide nanostructures on metals: Hybrid systems with novel properties, *J. Vac. Sci. Technol. B* 28 (2010) 1–16.
58. M. Nagel, I. Biswas, H. Peisert, and T. Chassé, Interface properties and electronic structure of ultrathin manganese oxide films on Ag(001), *Surf. Sci.* 601 (2007) 4484–4487.
59. F. P. Netzer, Small and beautiful: The novel structures and phases of nano-oxides, *Surf. Sci.* 604 (2010) 485–489.
60. A. Dahal and M. Batzill, Growth from behind: Intercalation-growth of two dimensional FeO moiré structure underneath of metal-supported graphene, *Sci. Rep.* 5 (2015) 11378.
61. C. Chang, S. Sankaranarayanan, D. Ruzmetov, M. H. Engelhard, E. Kaxiras, and S. Ramanathan, Compositional tuning of ultrathin surface oxides on metal and alloy substrates using photons: Dynamic simulations and experiments, *Phys. Rev. B* 81 (2010) 085406.
62. G. Pacchioni, Role of structural flexibility on the physical and chemical properties of metal-supported oxide ultrathin films, In: F. P. Netzer and A. Fortunelli (eds.), *Oxide Materials at the Two-Dimensional Limit*, (2016) Springer, Cham, Switzerland, pp. 91–118.
63. C. Noguera and J. Goniakowski, Electrostatics and polarity in 2D oxides, In: F.P. Netzer and A. Fortunelli (eds.), *Oxide Materials at the Two-Dimensional Limit*, (2016) Springer Series in Materials Science 234, Springer, Cham, Switzerland, pp. 201–231.
64. J. Goniakowski and C. Noguera, Microscopic mechanisms of stabilization of polar oxide surfaces: Transition metals on the MgO(111) surface, *Phys. Rev. B.* 66 (2002) 085417.
65. J. Goniakowski and C. Noguera, Polarization and rumpling in oxide monolayers deposited on metallic substrates, *Phys. Rev. B* 79 (2009) 155433.
66. J. Goniakowski, C. Noguera, and L. Giordano, Prediction of uncompensated polarity in ultrathin films, *Phys. Rev. Lett.* 93 (2004) 215702.
67. J. Goniakowski, C. Noguera, L. Giordano, and G. Pacchioni, Adsorption of metal adatoms on FeO(111) and MgO(111) monolayers: Effects of charge state of adsorbate on rumpling of supported oxide film, *Phys. Rev. B* 80 (2009) 125403.
68. L. Giordano, F. Cinquini, and G. Pacchioni, Tuning the surface metal work function by deposition of ultrathin oxide films: Density functional calculations. *Phys. Rev. B* 73 (2006) 045414.
69. S. J. Tauster, S. C. Fung, R. T. K. Baker, and J. A. Horsley, Strong-interactions in supported-metal catalysts, *Science* 211 (1981) 1121–1125.
70. S. Ma, J. Rodriguez, and J. Hrbek, STM study of the growth of cerium oxide nanoparticles on Au(111), *Surf. Sci.* 602 (2008) 3272–3278.
71. C. Wu, M. S. J. Marshall and M. R. Castell, Surface structures of ultrathin TiO$_x$ films on Au(111), *J. Phys. Chem. C* 115 (2011) 8643–8652.
72. S. Sindhu, M. Heiler, K. M. Schindler, and H. Neddermeyer, A photoemission study of CoO-films on Au(111), *Surf. Sci.* 541 (2003) 197–206.
73. G. Barcaro, E. Cavaliere, L. Artiglia, L. Sementa, L. Gavioli, G. Granozzi, and A. Fortunelli, Building principles and structural motifs in TiO$_x$ ultrathin films on a (111) substrate, *J. Phys. Chem. C* 116 (2012) 13302–13306.
74. Y. Pan, S. Benedetti, N. Nilius, and H. J. Freund, Change of the surface electronic structure of Au(111) by a monolayer MgO(001) film, *Phys. Rev. B* 84 (2011) 075456.
75. E. Cavaliere, I. Kholmanov, L. Gavioli, F. Sedona, S. Agnoli, G. Granozzi, G. Barcaro, and A. Fortunelli, Directed assembly of Au and Fe nanoparticles on a TiOx/Pt(111) ultra-thin template: the role of oxygen affinity, *Phys. Chem. Chem. Phys.* 11 (2009) 11305–11309.
76. J. Goniakowski, L. Giordano, and C. Noguera, Polarity of ultrathin MgO(111) films deposited on a metal substrate, *Phys. Rev. B* 81 (2010) 205404.
77. M. Ritter, W. Ranke, and W. Weiss, Growth and structure of ultrathin FeO films on Pt(111) studied by STM and LEED, *Phys. Rev. B* 57 (1998) 7240–7251.
78. T. U. Nahm, Study of high-temperature oxidation of ultrathin Fe films on Pt(100) by using X-ray photoelectron spectroscopy, *J. Korean Phys. Soc.* 68 (2016) 1215–1220.

79. D. Ragazzon, A. Schaefer, M. H. Farstad, L. E. Walle, P. Palmgren, A. Borg, P. Uvdal, and A. Sandell, Chemical vapor deposition of ordered TiO_x nanostructures on Au(111), *Surf. Sci.* 617 (2013) 211–217.

80. J. Seifert, E. Meyer, H. Winter, and H. Kuhlenbeck, Surface termination of an ultrathin V_2O_3-film on Au(111) studied via ion beam triangulation, *Surf. Sci.* 606 (2012) L41–L44.

81. M. Li and E. I. Altman, Cluster-size dependent phase transition of Co oxides on Au(111), *Surf. Sci.* 619 (2014) L6–L10.

82. S. Y. Quek, M. M. Biener, J. Biener, C. M. Friend, and E. Kaxiras, Tuning electronic properties of novel metal oxide nanocrystals using interface interactions: MoO_3 monolayers on Au(111), *Surf. Sci.* 577 (2005) L71–L77.

83. F. Stavale, L. Pascua, N. Nilius, and H. J. Freund, Morphology and luminescence of ZnO films grown on a Au(111) support, *J. Phys. Chem. C* 117 (2013) 10552–10557.

84. S. Zhuiykov, L. Hyde, Z. Hai, M. Karbalaei Akbari, E. Kats, C. Detavernier, C. Xue, and H. Xu, Atomic layer deposition-enabled single layer of tungsten trioxide across a large area, *Appl. Mater. Today* 6 (2017) 44–53.

85. Z. S. Li, D. V. Potapenko, and R. M. Osgood, Using Moiré patterning to map surface reactivity versus atom registration: Chemisorbed trimethyl acetic acid on TiO/Au(111), *J. Phys. Chem. C* 118 (2014) 29999–30005.

86. X. Deng and C. Matranga, Selective growth of Fe_2O_3 nanoparticles and islands on Au(111), *J. Phys. Chem. C* 113 (2009) 11104–11109.

87. M. Li and E. I. Altman, Shape, morphology, and phase transitions during Co oxide growth on Au(111), *J. Phys. Chem. C* 118 (2014) 12706–12716.

88. J. Tao, T. Luttrell, and M. Batzill, A two-dimensional phase of TiO_2 with a reduced bandgap, *Nat. Chem.* 3 (2011) 296–300.

89. H. G. Kim and H. B. R. Lee, Atomic layer deposition on 2D materials, *Chem. Mater.* 29 (2017) 3809–3826.

90. Y. H. Lee, X. Q. Zhang, and W. Zhang, Synthesis of large area MoS_2 atomic layers with chemical vapor deposition, *Adv. Mater.* 24 (2012) 2320–2325.

91. J. Schoiswohl, S. Surnev, and F. P. Netzar, Reaction on inverse model catalyst surface: Atomic view by STM, *Top. Catal.* 36 (2005) 91–105.

92. C. Detavernier, J. Dendooven, S. P. Sree, K. F. Ludwig, and J. A. Martens, Tailoring nanoporous materials by atomic layer deposition, *Chem. Soc. Rev.* 40 (2011) 5242–5253.

93. R. Warren, F. Sammoura, F. Tounsi, M. Sanghadasa, and L. Lin, Highly active ruthenium oxide coating via ALD and electrochemical activation in supercapacitor applications, *J. Mater. Chem. A* 3 (2015) 15568–15575.

94. C. Zhu, P. Yang, D. Chao, X. Wang, X. Zhang, S. Chen, B. Tay, H. Huang, H. Zhang, W. Mai, et al., All metal nitrides solid state asymmetric supercapacitors, *Adv. Mater.* 27 (2015) 4566–4571.

95. N. P. Dasgupta, H. Lee, S. F. Bent, and P. S. Weiss, Recent advances in atomic layer deposition, *Chem. Mater.* 28 (2016) 1943–1947.

96. J. S. Ponraj, G. Attolini, and M. Bosi, Review on atomic layer deposition and application of oxide thin films, *Crit. Rev. Solid State Mater. Sci.* 38 (2013) 203–233.

97. L. Niinisto, J. Paivasaari, J. Niinisto, M. Putkonen, and M. Nieminen, Advanced electronic and optoelectronic materials by atomic layer deposition: An overview with special emphasis on recent progress in processing of high-K dielectrics and other oxide materials, *Phys. Status Solidi A* 201 (2004) 1443–1452.

98. S. M. George, Atomic layer deposition: an overview, *Chem. Rev.* 110 (2010) 111–131.

99. T. Suntola, Atomic layer epitaxy, *Thin Solid Films* 216 (1992) 84–89.

100. I. Jõgi, M. Pärs, J. Aarik, A. Aidla, M. Laan, J. Sundqvist, L. Oberbeck, J. Heitmann, and K. Kukli, Conformity and structure of titanium oxide films grown by atomic layer deposition on silicon substrates, *Thin Solid Films* 516 (2008) 4855–4862.

101. X. Du and S. M. George, Thickness dependence of sensor response for CO gas sensing by tin oxide films grown using atomic layer deposition, *Sens. Actuators B Chem.* 135 (2008) 152–160.

102. M. A. Malik and P. O'Brien, Organometallic and metallo-organic precursors for nanoparticles, In: R. A. Fischer (ed.) *Precursor Chemistry of Advanced Materials: CVD, ALD and Nanoparticles,* (2005) Springer, Berlin, pp. 125–145.

103. S. E. Potts and W. M. M. Kessel, Energy-enhanced atomic layer deposition for more process and precursor versatility, *Coord. Chem. Rev.* 257 (2013) 3254–3270.

104. X. H. Liang, Y. Zhou, J. Li, and A. W. Weimer, Reaction mechanism studies for platinum nanoparticle grown by atomic layer deposition, *J. Nanopart. Res.* 13 (2011) 3781–3788.

105. M. Laskela and M. Ritala, Atomic layer deposition (ALD) from precursors to thin film structures, *Thin Solid Films* 409 (2002) 138–146.

106. A. Philip, S. Thomas, and K. R. Kumar, Calculation of growth per cycle (GPC) of atomic layer deposited aluminum oxide nanolayers and dependence of GPC on surface OH concentration, *Pram. J. Phys.* 82 (2014) 563–569.

107. X. Liang, S. M. George, A. W. Weimer, N. H. Li, J. H. Blackson, and J. D. Harris, Synthesis of a novel porous polymer/ceramic composite material by low-temperature atomic layer deposition, *Chem. Mater.* 19 (2007) 5388–5394.

108. T. Hatanpää, M. Ritala, and M. Leskelä, Precursors as enablers of ALD technology: Contributions from University of Helsinki, *Coord. Chem. Rev.* 257 (2013) 3297–3322.

109. S. W. Lee, B. J. Choi, T. Eom, J. H. Han, S. K. Kim, S. J. Song, W. Lee, and C. S. Hwang, Influences of metal, non-metal precursors, and substrates on atomic layer deposition processes for the growth of selected functional electronic materials, *Coord. Chem. Rev.* 257 (2013) 3154–3176.

110. B. B. Burton, A. R. Lavoie, and S. M. George, Tantalum nitride atomic layer deposition using (tert-butylimido)tris(diethylamido) tantalum and hydrazine, *J. Electrochem. Soc.* 155 (2008) D508–D516.

111. W. J. Maeng, D. Choi, J. Park, and J. S. Park, Indium oxide thin film prepared by low temperature atomic layer deposition using liquid precursors and ozone oxidant, *J. Alloys Comp.* 649 (2015) 216–221.

112. W. J. Maeng, D. Choi, J. Park, and J. S. Park, Atomic layer deposition of highly conductive indium oxide using a liquid precursor and water oxidant, *Ceram. Int.* 41 (2015) 10782–10787.

113. Y. Wan, J. Bullock, and A. Cuevas, Passivation of C-Si surfaces by ALD tantalum oxide capped with PECVD silicon nitride, *Sol. Ener. Mat. Sol. Cells* 142 (2015) 42–46.

114. S. Yeo, J. Park, S. Lee, D. Lee, J. Seo, and S. Kim, Ruthenium and ruthenium dioxide thin films deposited by atomic layer deposition using a novel zero-valent metalorganic precursor (ethyl-benzene)(1,3-butadiene)Ru(0), and molecular oxygen, *Microelectron. Eng.* 137 (2015) 16–22.

115. S. H. Kwon, O. K. Kwon, J. H. Kim, S. J. Jeong, S. W. Kim, and S. W. Kang, Improvement of the morphological stability by stacking RuO_2 on Ru thin films with atomic layer deposition, *J. Electrochem. Soc.* 154 (2007) H773–H777.

116. S. Moitzheim, C. S. Miisha, S. Deng, D. J. Cott, C. Detavernier, and P. M. Vereecken, Nanostructured TiO_2/carbon nanosheet hybrid electrode for high-rate thin-film lithium-ion batteries, *Nanotechnology* 25 (2014) 504008.

117. W. Chiappim, G. E. Testoni, R. S. Moraes, R. S. Pessoa, J. C. Sagás, F. D. Origo, L. Vieira, and H. S. Maciel, Structural, morphological, and optical properties of TiO_2 thin films grown by atomic layer deposition on fluorine doped tin oxide conductive glass, *Vacuum* 123 (2016) 91–102.

118. J. L. Tian, H. Y. Zhang, G. G. Wang, X. Z. Wang, R. Sun, L. Jin, and J. C. Han, Influence of film thickness and annealing temperature on the structural and optical properties of ZnO thin films on Si (100) substrates grown by atomic layer deposition, *Superlatt. Microstr.* 83 (2015) 719–729.

119. O. Bethge, M. Nobile, S. Abermann, M. Glaser, and E. Bertagnolli, ALD grown bilayer junction of ZnO:Al and tunnel oxide barrier for SIS solar cell, *Sol. Energy Mater. Sol. Cells* 117 (2013) 178–182.

120. S. C. Riha, J. M. Racowski, M. P. Lanci, J. A. Klug, A. S. Hock, and A. B. F. Martinson, Phase discrimination through oxidant selection in low-temperature atomic layer deposition of crystalline iron oxides, *Langmuir* 29 (2013) 3439–3445.

121. V. Aravindan, K. B. Jinesh, R. R. Prabhakar, V. S. Kale, and S. Madhavi, Atomic layer deposited (ALD) SnO_2 anodes with exceptional cycleability for Li-ion batteries, *Nano Energy* 2 (2013) 720–725.

122. B. Han, J. M. Park, K. H. Choi, W. K. Lim, T. R. Mayangsari, W. Koh, and W. J. Lee, Atomic layer deposition of stoichiometric Co_3O_4 films using bis(1,4-di-iso-propyl-1,4-diazabutadiene) cobalt, *Thin Solid Films* 589 (2015) 718–722.

123. J. Malm, T. Sajavaara, and M. Karpinen, Atomic layer deposition of WO_3 thin films using $W(CO)_6$ and O_3 precursors, *Chem. Vap. Depos.* 18 (2012) 245–248.

124. M. Tsuchiya, S. Sankaranarayanan, and S. Ramanathan, Photon-assisted oxidation and oxide thin film synthesis: A review, *Prog. Mater. Sci.* 54 (2009) 981–1057.

125. R. Sivakumar, R. Gopalakrishnan, M. Jayachandran, and C. Sanjeeviraja, Investigation of x-ray photoelectron spectroscopic (XPS), cyclic voltammetric analyses of WO_3 films and their electrochromic response in FTO/WO_3/electrolyte/FTO cells, *Smart Mater. Struct.* 15 (2006) 877–888.

126. Z. Hai, M. Karbalaei Akbari, Z. Wei, C. Xue, H. Xu, J. Hu, and S. Zhuiykov, Nano-thickness dependence of supercapacitor performance of the ALD-fabricated two-dimensional WO_3, *Electrochim. Acta* 246 (2017) 625–633.

127. S. Zhuiykov, M. Karbalaei Akbari, Z. Hai, C. Xue, H. Xu, and L. Hyde, Wafer-scale fabrication of conformal atomic-layered TiO_2 by atomic layer deposition using tetrakis (dimethylamino) titanium and H_2O precursors, *Mater. Des.* 120 (2017) 99–108.

128. S. Zhuiykov, M. Karbalaei Akbari, Z. Hai, C. Xue, H. Xu, and L. Hyde, Data set for fabrication of conformal two-dimensional TiO_2 by atomic layer deposition using tetrakis(dimethylamino) titanium (TDMAT) and H_2O precursors, *Data Brief* 13 (2017) 401–407.

129. J. Dendooven, S. P. Sree, K. D. Keyser, D. Deduytsche, J. A. Martens, K. F. Ludwig, and C. Detavernier, In situ X-ray fluorescence measurements during atomic layer deposition: Nucleation and growth of TiO_2 on planar substrates and in nanoporous films, *J. Phys. Chem. C* 115 (2011) 6605–6610.

130. M. Bouman and F. Zaera, Reductive eliminations from amido metal complexes: Implications for metal film deposition, *J. Electrochem. Soc.* 158 (2011) D524–D526.

131. R. Methaapanon and S. F. Bent, Comparative study of titanium dioxide atomic layer deposition on silicon dioxide and hydrogen-terminated silicon, *J. Phys. Chem. C* 114 (2010) 10498–10504.

132. M. Karbalaei Akbari, Z. Hai, Z. Wei, J. Hu and S. Zhuiykov, Wafer-scale two-dimensional Au-TiO_2 bilayer films for photocatalytic degradation of palmitic acid under UV and visible light illumination, *Mater. Res. Bull.* 95 (2017) 380–391.

133. T. Dobbelaere, M. Minjauw, T. Ahmad, P. M. Vereeken and C. Detavernier, Plasma enhanced atomic layer deposition of zinc phosphate, *J. Non. Cryst. Solids* 444 (2016) 43–48.

134. K. Kanomata, P. Pansila, B. Ahmmad, S. Kubota, K. Hirahara and F. Hirose, Infrared study on room-temperature atomic layer deposition of TiO_2 using tetrakis(dimethylamino)titanium and remote-plasma-excited water vapor, *Appl. Surf. Sci.* 308 (2014) 328–332.

135. Y. Zhang, W. Wu, and K. Zhang, Raman studies of 2D anatase TiO_2 nanosheets, *Phys. Chem. Chem. Phys.* 18 (2016) 32178–32184.

136. C. Y. Xu, P. X. Zhang, and L. Yan, Blue shift of Raman peak from coated TiO_2 nanoparticles, *J. Raman Spectrosc.* 32 (2001) 862–865.

137. S. J. Park, J. P. Lee, J. S. Jang, H. Rhu, H. Yu, B. Y. You, C. S. Kim, K. J. Kim, Y. J. Cho, S. Baik, et al., In situ control of oxygen vacancies in TiO_2 by atomic layer deposition for resistive switching devices, *Nanotechnology* 24 (2013) 295202.

138. A. L. Bassi, D. Cattaneo, V. Russo, C. E. Bottani, T. Mazza, P. Piseri, P. Milani, F. O. Ernst, K. Wegner, and S. E. Pratsinis, Raman spectroscopy characterization of titania nanoparticles produced by flame pyrolysis: The influence of size and stoichiometry, *J. Appl. Phys.* 98 (2005) 074305.

139. A. Gharachorlou, M. D. Detwiler, X. K. Gu, L. Mayr, B. Klotzer, J. Greeley, R. G. Reifenberger, W. N. Delgass, F. H. Ribeiro, and D. Y. Zemlyanov, Trimethyl aluminum and oxygen atomic layer deposition on hydroxyl-free Cu(111), *ACS Appl. Mater. Interfaces* 7 (2015) 16428–16439.

140. M. Knez, K. Niesch, and L. Niinistoe, Synthesis and surface engineering of complex nano-structures by atomic layer deposition, *Adv. Mater.* 19 (2007) 3425–3438.
141. B. J. O'Neill, D. H. K. Jackson, J. Lee, C. Canlas, P. C. Stair, C. Marshall, J. W. Elam, T. F. Kuech, J. A. Dumesic, and G. W. Huber, Catalyst design with atomic layer deposition, *ACS Catal.* 5 (2015) 1804–1825.
142. A. Beniya, N. Isomura, H. Hirata, and Y. Watanabe, Morphology and chemical states of size-selected Ptn clusters on an aluminium oxide film on NiAl(110), *Phys. Chem. Chem. Phys.* 16 (2014) 26485–26492.
143. M. S. Chen and D. W. Goodman, Interaction of Au with titania: The role of reduced Ti, *Top. Catal.* 44 (2007) 41–47.
144. L. Gragnaniello, T. Ma, G. Barcaro, L. Sementa, F. R. Negreiros, A. Fortunelli, S. Surnev, and F. P. Netzer, Ordered arrays of size-selected oxide nanoparticles, *Phys. Rev. Lett.* 108 (2012) 195507.
145. D. Stacchiola, S. Kaya, J. Weissenrieder, H. Kuhlenbeck, S. Shaikhutdinov, H. J. Freund, M. Sierka, T. K. Todorova, and J. Sauer, Synthesis and structure of ultrathin aluminosilicate films. *Angew. Chem. Int. Ed.* 45 (2006) 7636–7639.
146. H. Sabbah, Amorphous titanium dioxide ultra-thin films for self-cleaning surfaces, *Mater. Express*, 3 (2013) 171–175.
147. M. Sterrer, T. Risse, U. M. Pozzoni, L. Giordano, M. Heyde, H. Rust, G. Pacchioni, and H. Freund, Control of the charge state of metal atoms on thin MgO films, *Phys. Rev. Lett.* 98 (2007) 096107.
148. M. De Santis, A. Buchsbaum, P. Varga, and M. Schmid, Growth of ultrathin cobalt oxide films on Pt(111), *Phys. Rev. B* 84 (2011) 125430.
149. G. Barcaro, L. Sementa, F. R. Negreiros, I. O. Thomas, S. Vajda, and A. Fortunelli, Atomistic and electronic structure methods for nanostructured oxide interfaces, In: F. P. Netzer and A. Fortunelli (eds.), *Oxide Materials at the Two-Dimensional Limit*, (2016) Springer Series in Materials Science 234, Springer, Cham, Switzerland, pp. 39–90.
150. L. Cheng, C. Yin, F. Mehmood, B. Liu, J. Greeley, S. Lee, B. Lee, S. Seifert, R. E. Winans, D. Teschner, et al. Reaction mechanism for direct propylene epoxidation by alumina-supported silver aggregates: The role of the particle/support interface, *ACS Catal.* 4 (2014) 32–39.
151. Y. Lei, F. Mehmood, S. Lee, J. Greeley, B. Lee, S. Seifert, R. E. Winans, J. W. Elam, R. J. Meyer, P. C. Redfern, et al. Increased silver activity for direct propylene epoxidation via sub nanometer size effects, *Science* 328 (2010) 224–228.
152. J. W. Elam and S. M. George, Growth of ZnO/Al_2O_3 alloy films using atomic layer deposition techniques, *Chem. Mater.* 15 (2003) 1020–1028.
153. S. Lee, L. M. Molina, M. J. López, J. A. Alonso, B. Hammer, B. Lee, S. Seifert, R. E. Winans, J. W. Elam, M. J. Pellin, et al., Selective propene epoxidation on immobilized Au6–10 clusters: The effect of hydrogen and water on activity and selectivity. *Angew. Chem. Int. Ed.* 121 (2009) 1495–1499.
154. L. M. Molina, S. Lee, K. Sell, G. Barcaro, A. Fortunelli, B. Lee, S. Seifert, R. E. Winans, J. W. Elam, M. J. Pellin, et al., Size-dependent selectivity and activity of silver nanoclusters in the partial oxidation of propylene to propylene oxide and acrolein: A joint experimental and theo-retical study, *Catal. Today* 160 (2011) 116–130.
155. S. Lee, B. Lee, F. Mehmood, S. Seifert, J. A. Libera, J. W. Elam, J. Greeley, P. Zapol, L. A. Curtiss, M. J. Pellin, et al., Oxidative decomposition of methanol on subnanometer palladium clusters: The effect of catalyst size and support composition, *J. Phys. Chem. C* 114 (2010) 10342–10348.
156. M. D. Kane, F. S. Roberts, and S. L. Anderson, Effects of alumina thickness on CO oxidation activity over Pd20/alumina/Re(0001): Correlated effects of alumina electronic properties and Pd20 geometry on activity, *J. Phys. Chem. C* 119 (2015) 1359–1375.
157. Y. Shiraishi, N. Yasumoto, J. Imai, H. Sakamoto, S. Tanaka, S. Ichikawa, B. Ohtanie, and T. Hirai, Quantum tunneling injection of hot electrons in Au/TiO_2 plasmonic photocatalysts, *Nanoscale* 9 (2017) 8349–8361.
158. K. Wu, J. Chen, J. R. McBride, and T. Lian, Efficient hot-electron transfer by a plasmon-induced interfacial charge-transfer transition, *Science* 349 (2015) 632–635.

159. X. Zhang, Y. Chen, R. Liu, and D. Tsai, Plasmonic photocatalysis, *Rep. Prog. Phys.* 76 (2013) 046401.
160. F. Tan, T. Li, N. Wang, S. Lai, W. Yu, and X. Zhang, Rough gold films as broad band absorbers for plasmonic enhancement of TiO_2 photocurrent over 400–800 nm, *Sci. Rep.* 6 (2016) 33049.
161. T. Zhang, D. Su, R. Z. Li, S. J. Wang, F. Shan, J. Xu, and X. Zhang, Plasmonic nanostructures for electronic designs of photovoltaic devices: Plasmonic hot carrier photovoltaic architectures and plasmonic electrode structures, *J. Photon. Energy* 6 (2016) 042504.
162. F. Xia, T. Mueller, Y. Lin, A. Valdes-Garcia, and P. Avouris, Ultra-fast graphene photodetector, *Nat. Nanotechnol.* 4 (2009) 839–843.
163. L. Zheng, L. Zhongzhu, and S. Guozhen, Photodetectors based on two dimensional materials, *J. Semicond.* 37 (2016) 091001.
164. Z. Hai, M. Karbalaei Akbari, C. Xue, H. Xu, L. Hyde, and S. Zhuiykov, Wafer-scaled monolayer WO_3 windows ultra-sensitive, extremely-fast and stable UV-A photodetection, *Appl. Surf. Sci.* 405 (2017) 169–177.
165. W. Tian, C. Zhang, T. Zhai, S. L. Li, X. Wang, J. Liu, X. Jie, D. Liu, M. Liao, Y. Koide, et al., Flexible ultra violet photodetectors with broad photoresponse based on branched ZnS-ZnO heterostructure nanofilms, *Adv. Mater.* 26 (2014) 3088–3093.
166. M. Karbalaei Akbari, Z. Hai, S. Depuydt, E. Kats, J. Hu, and S. Zhuiykov, Highly sensitive, fast-responding, and stable photodetector based on ALD-developed monolayer TiO_2, *IEEE Trans. Nanotechnol.* 16 (2017) 880–887.
167. W. J. Wang, C. X. Shan, H. Zhu, F. Y. Ma, D. Z. Shen, X. W. Fan, and K. L. Choy, Metal–insulator–semiconductor–insulator–metal structured titanium dioxide ultraviolet photodetector, *J. Phys. D: Appl. Phys.* 43 (2010) 045102.
168. F. Jing, D. Zhang, F. Li, J. Zhou, D. Sun, and S. Ruan, High performance ultraviolet detector based on $SrTiO_3/TiO_2$ heterostructure fabricated by two steps in situ hydrothermal method, *J. Alloys Compd.* 650 (2015) 97–101.
169. S. Das, J. Kar, and J. Myoung, Junction properties and application of ZnO single nanowire based Schottky diode, In: A. Hashim (ed.), *Nanowires: Fundamental Research*, (2011) In Tech Publishing, Croatia, pp. 174–178.
170. J. Yang, Y. L. Jiang, L. J. Li, E. Muhire, and M. Z. Gao, High-performance photodetectors and enhanced photocatalysts of two-dimensional TiO_2 nanosheets under UV light excitation, *Nanoscale* 8 (2016) 8170–8177.
171. J. Liu, M. Zhong, J. Li, A. Pan, and X. Zhu, Few-layer WO_3 nanosheets for high-performance UV-photodetectors, *Mater. Lett.* 148 (2015) 184–187.
172. Z. Hai, M. Karbalaei Akbari, C. Xue, H. Xu, S. Depuydt, and S. Zhuiykov, Photodetector with superior functional capabilities based on monolayer WO_3 developed by atomic layer deposition, *Sensor Actuat. B Chem.* 245 (2017) 954–962.
173. M. W. Knight, H. Sobhani, P. Nordlander, and N. J. Halas, Photodetection with active optical antennas, *Science* 332 (2011) 698–702.
174. N. G. Charlene, J. J. Cadusch, S. Dligatch, A. Roberts, T. J. Davis, P. Mulvaney, and D. E. Gómez, Hot carrier extraction with plasmonic broadband absorbers, *ACS Nano*, 10 (2016) 4704–4711.
175. Z. Yu, L. Tetard, L. Zhai, and J. Thomas, Supercapacitor electrode materials: Nanostructures from 0 to 3 dimensions, *Energy Environ. Sci.* 8 (2015) 702–730.
176. G. Wang, L. Zhang, and J. Zhang, A review of electrode materials for electro chemical supercapacitors, *Chem. Soc. Rev.* 41 (2012) 797–828.
177. Y. Liu and X. Peng, Recent advances of supercapacitors based on two-dimensional materials, *Appl. Mater. Today* 7 (2017) 1–12.
178. K. Tsay, L. Zhang, and J. Zhang, Effects of electrode layer composition/thickness and electrolyte concentration on both specific capacitance and energy density of supercapacitor, *Electrochim. Acta* 60 (2012) 428–436.
179. A. Yu, I. Roes, A. Davies, and Z. Chen, Ultrathin transparent, and flexible graphene films for supercapacitor application, *Appl. Phys. Lett.* 96 (2010) 1–4.

180. X. Wang and G. Yushin, Chemical vapor deposition and atomic layer deposition for advanced lithium ion batteries and supercapacitors, *Energy Environ. Sci.* 8 (2015) 1889–1904.
181. R. Warren, F. Sammoura, A. Kozinda, and L. Lin, ALD ruthenium oxide-carbon nanotube electrodes for supercapacitor applications, *Proc. IEEE Int. Conf. Micro Electro Mech. Syst.* 16 (2014) 167–170.
182. C. Guan, X. Qian, X. Wang, Y. Cao, Q. Zhang, A. Li, and J. Wang, Atomic layer deposition of Co_3O_4 on carbon nanotubes/carbon cloth for high-capacitance and ultra stable supercapacitor electrode, *Nanotechnology* 26 (2015) 94001.
183. C. Guan, Y. Wang, Y. Hu, J. Liu, K. H. Ho, W. Zhao, Z. Fan, Z. Shen, H. Zhang, and J. Wang, Conformally deposited NiO on a hierarchical carbon support for high power and durable asymmetric supercapacitors, *J. Mater. Chem. A* 3 (2015) 23283–23288.
184. S. Boukhalfa, K. Evanoff, and G. Yushin, Atomic layer deposition of vanadium oxide on carbon nanotubes for high-power supercapacitor electrodes, *Energy Environ. Sci* 5 (2012) 6872–6879.
185. X. Bai, Q. Liu, J. Liu, H. Zhang, Z. Li, and X. Jing, Hierarchical Co_3O_4@Ni(OH)$_2$ core shell nanosheet arrays for isolated all-solid state supercapacitor electrodes with superior electrochemical performance, *Chem. Eng. J.* 315 (2017) 35–45.
186. K. Kalantar-Zadeh, A. Vijayaraghavan, M. H. Ham, H. Zheng, M. Breedon, and M. S. Strano, Synthesis of atomically thin WO_3 sheets from hydrated tungsten trioxide, *Chem. Mater.* 22 (2010) 5660–5666.
187. N. M. Vuong, D. Kim, and H. Kim, Electrochromic properties of porous WO_3-TiO_2 core shell nanowires, *J. Mater. Chem. C* 1 (2013) 3399–3407.
188. Y. Liu, C. Xie, J. Li, T. Zou, and D. Zeng, New insights into the relationship between photocatalytic activity and photocurrent of TiO_2/WO_3 nanocomposite, *Appl. Catal. A: Gen.* 433–434 (2012) 81–87.
189. Y. R. Smith, B. Sarma, S. K. Mohanty, and M. Misra, Formation of TiO_2-WO_3 nanotubular composite via single-step anodization and its application in photoelectrochemical hydrogen generation, *Electrochem. Commun.* 19 (2012) 131–134.
190. Z. Hai, M. Karbalaei Akbari, C. Xue, H. Xu, E. Solano, C. Detavernier, J. Hu, and S. Zhuiykov, Atomically thin WO_3/TiO_2 heterojunction for supercapacitor electrodes developed by atomic layer deposition, *Comp. Comm.* 5 (2017) 31–35.
191. C. Guan and J. Wang, Recent development of advanced electrode materials by atomic layer deposition for electrochemical energy storage, *Adv. Sci.* 3 (2016) 1500405.
192. B. M. Sanchez and Y. Gogotsi, Synthesis of two-dimensional materials for capacitive energy storage, *Adv. Mater.* 28 (2016) 6104–6135.
193. H. Cesiulis, N. Tsyntsaru, A. Ramanavicius, and G. Ragoisha, The study of thin films by electrochemical impedance spectroscopy, In: I. Tiginyanu, P. Topala, and V. Ursaki (eds.), *Nanostructures and Thin Films for Multifunctional Applications, Nanoscience and Technology* (2016) Springer, pp. 3–42.
194. H. Zhang, S. Wang, Y. Wang, J. Yang, X. Gao, and L. Wang, TiO_2(B) nanoparticle functionalized WO_3 nanorods with enhanced gas sensing properties, *Phys. Chem. Chem. Phys.* 16 (2014) 10830–10836.
195. D. Bekermann, A. Gasparotto, D. Barreca, C. Maccato, E. Comini, C. Sada, G. Sberveglieri, A. Devi, and R. A. Fischer, Co_3O_4/ZnO nanocomposites: From plasma synthesis to gas sensing applications, *Appl. Mater. Interfaces* 4 (2012) 928–934.
196. S. Bagheri, New York/New Jersey Nearshore Waters: A case study in NY/NJ, In: S. Bagheri, (ed.), *Hyperspectral Remote Sensing of Nearshore Water Quality a Case Study in New York/New Jersey*, (2017) Springer, pp. 19–27.
197. A. Salimi and M. Roushani, Non-enzymatic glucose detection free of ascorbic acid interference using nickel powder and nafion sol–gel dispersed renewable carbon ceramic electrode, *Electrochem. Commun.* 7 (2005) 879–887.

7

Tunneling Field-Effect Transistors Based on Two-Dimensional Materials

Peng Zhou

CONTENTS

7.1 Introduction

Although advanced complementary metal-oxide-semiconductors (CMOS) technology has achieved many extraordinary improvements in switching speed, density, functionality and cost, metal-oxide-semiconductor field-effect transistors (MOSFETs) are facing the challenges of scaling down the supply voltage with the increase in power density due to the thermionic emission limitation of carrier injection. Rising leakage currents degrade the I_{ON}/I_{OFF} ratio because the leakage currents will increase exponentially with a reduction in the supply voltage. Only the carriers in the exponential tail of the source Fermi distribution above the channel barrier can migrate to the channel. Considering a large enough bias, the drain current $I_d \propto \int dE D(E) v(E) f_s(E)$, where $D(E)$ is the density of states, $v(E)$ is the carrier velocity and $f_s(E)$ is the source Fermi distribution. As $D(E)v(E)$ is a constant for a specific product and $f_s(E) \approx e^{E - E_f^s/kT}$, the SS can be obtained by

$$ SS = \frac{\partial V_g}{\partial (\log_{10} I_d)} = \frac{\partial V_g}{\partial \left(\dfrac{E}{q}\right)} \frac{kT}{q} \ln 10 \tag{7.1} $$

When considering all the gate voltage has been applied to the channel, we get the ideal case,

$$ SS = \frac{kT}{q} \ln 10, $$

the minimal subthreshold swing of 60 mV/decade at room temperature, which can't be overtaken even if scaled down. Although the power dissipation of a MOSFET, $P = I_{OFF} V_{DD}^3$, the power consumption density increases sharply with the increase in I_{OFF} when scaling down the supply voltage. The leakage power increases by 275-fold in commercial bulk CMOS 45 nm technology when lowering the supply voltage from 0.5 V to 0.25 V [1]. Therefore, it is necessary to resort to other physical mechanisms to solve the power consumption problem, like tunneling field-effect transistors (TFETs).

This review will concentrate on TFETs based on two-dimensional (2D) materials. The history of silicon TFET dates back to 1978 when Quinn et al. proposed the gated p-i-n structure [2], and it was experimentally realized in 1995 by Reddick and Amaratunga [3]. Lateral TFETs on silicon-on-insulator (SOI) observed negative conductance at 90 K [4] and TFETs on SOI with the gate overlapping the depletion region only to minimize capacitance [5]. Later, Group III-V materials [6] and nanotubes were used in TFET. Knoch and Appenzeller proposed nanotube TFETs and an experimentally realized SS below 60 mV/dec at room temperature for the first time in 2005 [7]. Vertical silicon TFETs were fabricated by Hansch et al. [8] using molecular beam epitaxy. TFETs achieved an SS below 60 mV/dec for the first time in a carbon nanotube field-effect transistor (FET) in 2004 [9]. Later, an average SS of ~25 mV/dec was achieved in a GaSb/InAs heterojunction [10] and ~7.8 mV/dec in an Si-based, arch-shaped, gate-all-around TFET [11]. Since its discovery in 2004, graphene has been widely studied and applied to chemical sensors [12], optical devices [13], high-frequency circuits [14] and energy generation and storage [15,16] because of its high mobility of 2.5×10^5 cm^2 V^{-1} s^{-1} [17], excellent thermal conductivity and optical properties, wonderful flexibility, high surface area and so on. Subsequently, many other 2D layered materials have attracted unprecedented attention and multitudinous 2D materials have come to the fore, such as transition metal dichalcogenides (TMDCs), hexagonal boron nitride (h-BN), layered Group-IV and Group-III metal chalcogenides, silicene and germanene. They have been widely investigated in recent decades and have been applied to various devices due to their unique properties different from three-dimensional (3D) materials. Many efforts have been made to achieve TFETs based on 2D materials with low power consumption.

7.2 The Principle of TFET

Taking a p-i-n n-type field-effect transistor (NTFET) as an example to demonstrate the physics of a TFET, a schematic cross section of an NTFET is shown in Figure 7.1a. The source is highly n-doped and the drain is highly p-doped so that the Fermi level of the drain is above its conduction band and that of the source is below its valance band, respectively, as shown in Figure 7.1b. In the OFF state (green lines), the electrons in the valance band can't tunnel into the channel as there is no empty state in the bandgap and the barrier for electrons in the conduction band of the drain is too high to inject into the conduction band of the channel by thermal emission. Therefore, the OFF current is ultralow. Increasing the gate voltage to move the conduction band of the channel below the valance band of the source, the electrons in the energy window ($\Delta\Phi$) of the source (green shading) have a probability of tunneling into the conduction band of the channel as many empty states are available in the conduction band of the channel, switched to the ON state. The electrons above the energy window in the tail of the Fermi distribution cannot tunnel into the channel as there are no empty states in the bandgap. It is effective to achieve an SS below 60 mV/dec by cutting off the tail of the Fermi distribution. In reverse, reducing the gate voltage to move the valance band of the channel above the conduction band of the source, the electrons from the conduction band of the source can tunnel into the valance band of the source, as Figure 7.1b and d shows. Therefore, the p+ doped region is defined as the drain for an NTFET while the n+ doped region is defined as the drain for a p-type field-effect transistor (PTFET) and the surface tunneling junction of an NTFET is at the interface between the p+ region and the channel region (red circle in Figure 7.1a) while that of a PTFET is at the interface between the n+ region and the channel region (red circle in Figure 7.1b). The gate voltage should be higher than a specific voltage to switch on for an NTFET and the gate voltage should be lower than a specific voltage to switch on for a PTFET.

In principle, the TFET is an ambipolar device and both a positive and a negative gate voltage can reduce the tunneling barrier, operating as an NTFET with dominant electron conduction and as a PTFET with dominant hole conduction [18]. The transfer characteristics (channel current versus gate voltage) with the n+ region as the drain and the p+ region as the grounded source are shown in Figure 7.2a. The channel current curves obviously shift with the varying of the drain voltage when the gate voltage is negative while they shift a little when the gate voltage is positive. When $V_g < 0$, $V_g - V_{n+}$ should be lower than V_{TP} to switch on the tunneling junction, where V_{n+} is the potential of the n+ region and V_{TP} is the switch-on voltage of the surface tunneling junction, so the corresponding gate voltage when the channel current is over 10^{-10} A moves right with the increasing drain voltage. When $V_g > 0$, the channel current curves move less prominently as the surface tunneling junction moves to the interface of the channel and the p+ region. Therefore, the varying

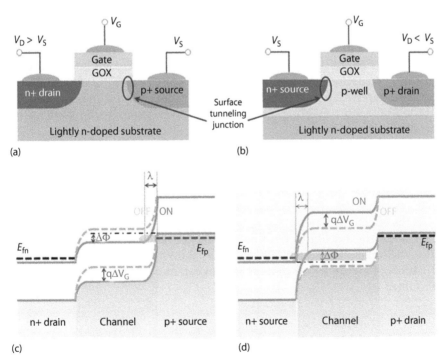

(a)

(b)

(c)

(d)

FIGURE 7.1
Principle of complementary p-i-n TFET. Schematic cross section of NTFET (a) and PTFET (b) with applied gate (V_G), source (V_S) and drain (V_D) voltages. Schematic diagrams of the band structures of NTFET (c) and PTFET (d) in ON and OFF states, respectively. The electrons in the energy window ($\Delta\Phi$) of the p+ source tunnel into the conduction band of the channel while the electrons in the energy band ($\Delta\Phi$) of the channel tunnel into the conduction band of n+ source for PTFET.

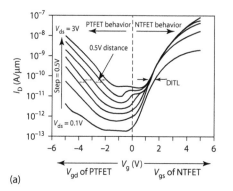
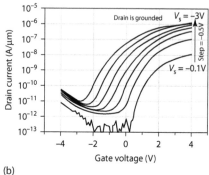

FIGURE 7.2
(a) Transfer characteristics (channel current versus gate voltage) with n+ region as the drain and p+ region as the grounded source. The channel current curves obviously shift with the variety of drain voltage when the gate voltage is negative while they shift a little when the gate voltage is positive due to the changing of the position of the surface tunneling junction. (b) Transfer characteristics with n+ region as the grounded drain and p+ region as the source. The channel current curves obviously shift with the variety of source current when the gate voltage is positive [18].

of the drain voltage cannot influence the switch-on voltage effectively and the V_{TN} (the switch-on voltage of the surface tunneling junction for an NTFET) keeps almost a constant. In converse, if the n+ region is grounded as the drain and the p+ region as the source, the switch-on voltage moves left with the decreasing source voltage, shown in Figure 7.2b. However, the asymmetry of the doping level of the profile and the restriction of the movement of one type of charge carrier with heterostructure (HS) can widen the tunneling barrier at the drain to suppress the ambipolarity and lower the OFF current [18,19].

To describe quantitatively, the tunneling current depends on the transmission probability of the interband tunneling barrier, which can be acquired by Wentzel–Kramers–Brillouin (WKB) approximation,

$$T(E) = \exp\left(-\frac{4\sqrt{2m^* E_g^3}}{3q\hbar\varepsilon}\right),$$

where m^* is the effective carrier mass, E_g is the bandgap, $\hbar$ is Planck's constant divided by 2π and ε is the electrical field in the tunneling junction. Considering an ultrathin body and gate oxides, the screening tunneling length λ (as shown in the schematic diagrams of the band structures of Figure 7.1c and d, the spatial extent of the surface tunneling junction) can be described as

$$\lambda = \sqrt{\frac{\varepsilon_{ch}}{\varepsilon_{ox}} d_{ox} d_{ch}} \,,$$

where d_{ox}, d_{ch}, ε_{ox} and ε_{ch} are the oxide and channel thickness and dielectric constants, respectively [19]. Therefore, the electrical field $\varepsilon = (E_g + \Delta\varnothing)/\lambda$, where E_g is the bandgap of the source and $\Delta\varnothing$ is the energy window as marked in Figure 7.1c and d. The Fermi level of the source is considered several kT lower than its valance band for an NTFET (several kT higher than its conduction band for a PTFET), $f_s(E) \approx 1$ in the energy window [7].

Supposing the states in the energy window of the channel are empty, the tunneling current can be given as

$$I \propto \int dED(E)v(E)T(E)\left[f_s(E) - f_{ch}(E)\right]$$

$$= \int_0^{\Delta\Phi} dED(E)v(E)T(E) \propto \exp\left(-\frac{4\lambda\sqrt{2m^*E_g^{3/2}}}{3q\hbar(E_g + \Delta\varnothing)}\right)\Delta\varnothing \tag{7.2}$$

It suggests that increasing the tunneling probability to approach unity as much as possible signifies a high ON current. The energy filtering cuts off the high-energy part of the source Fermi distribution, which makes the electrons from this part not tunnel to the channel; thus it is possible to achieve an SS below 60 mV/dec [20]. One of the major advantages of TFETs is their weak dependence on the temperature of the OFF currents, which has to be considered seriously in the operation of highly integrated circuits [19,21,22]. Another advantage of TFETs is the decrease in the SS with the reduction in the gate voltage as the SS is approximately inversely related to the gate voltage if the thickness of the gate dielectrics approaches 1 nm. Thus, TFETs are naturally optimized for low-voltage operation [23].

7.3 Performance Optimization

If TFETs aim to replace CMOS transistors, their ON current should be over 100 mA, an SS far below 60 mV/dec for more than five decades of current beyond several microamperes per micrometer, $V_{DD} < 0.5$ V and $I_{ON}/I_{OFF} > 10^5$. The formula for the tunneling current will be the guide to designing TFETs with competitive properties, although the WKB approximation is more suitable for direct bandgap semiconductors than indirect bandgap semiconductors and the case when quantum effects and phonon-assisted tunneling are dominant.

According to the formula derived from the WKB approximation, a high tunneling probability is expected if m^* and E_g of the source material are as small as possible. Replacing the source material with Ge for silicon n-TFETs and with InAs for silicon p-TFETs can improve λ and result in a steep band edge sharpness. However, the OFF currents will increase greatly with a small energy gap as thermal emission becomes pronounced. So, the drain material should be chosen with a large E_g to form a large energy barrier width at the drain side in the off state to keep the OFF current low. Channel materials with an optimized E_g of $(1.1–1.5)qV_{DD}$ show the best performance in TFETs [24], so WTe$_2$ with an E_g of 0.75 eV is expected to show best performance when the supply voltage V_{DD} is about 0.5 V [25,26]. Unfortunately, 2h-WeT2 might be not stable in air [27]. The modulation of the channel bans by the gate is represented by λ, and a small λ results in a high transmission probability from the source to the channel, which depends on gate capacitance, doping profiles, device geometry, dimensions, etc. So, a high-permittivity (high-κ) gate dielectric [19] and a low permittivity (low-κ) channel material should be adopted, especially a high-κ gate dielectric, as a small m^* and a proper E_g are more pivotal for the channel materials. Meanwhile, the gate dielectric and channel material should be as thin as possible to reduce λ [7,19]. The high source doping level must be cut off abruptly to achieve as short a tunneling barrier width as possible, which will result in the bandgap narrowing slightly. The types of band alignment also play a crucial role in HSs TFETs [28,29].

The band alignment was adjusted from staggered to broken by varying the Al context in the InAs/Al$_x$Ga$_{1-x}$ HS TFETs and it was found that a staggered band alignment yielded the best I_{ON}/I_{OFF} and an SS below 60 mV/dec [30,31]. However, due to the nonoptimal device structure with an underlap between the gate and the channel and the phenomenological treatment of scattering, Koswatta considered the proposal of Knoch unreasonable and put forward that TFETs with a broken band alignment delivered superior performance [29].

There is a compromise between a small SS and high ON currents in 3D TFETs [32] as λ of the planar device could not be lower than a certain value because of a limited dopant density in the channel and a limited gate oxide. And relatively high Fermi energies are required to ensure a charge density large enough in the electrodes as λ relies on the charge density in the contacts, which have to yield thin tunneling barriers. One-dimensional (1D) TFETs are expected to achieve high ON currents with steep SS because of their improved electrostatics and the one dimensionality of the density of states [32–36]. The quantum capacitance (C_q) will be far smaller than the oxide capacitance (C_{ox}) even in the case of large gate oxide thickness, which will deteriorate the band edge sharpness [7]. Therefore, 1D TFETs have a better gate modulation than 3D TFETs. Both the channel materials and the gate dielectrics using layered 2D materials can be an atom thick, so a high gate modulation can be expected in 2D TFETs.

Another method to improve the gate modulation of TFETs is to use different device structures. Silicon TFETs were achieved with a planar structure for the first time in 1995 [3]. Later, in 1996, SOI surface TFETs with a negative conductance were fabricated to take advantage of the 2D confinement effect [4]. And it was predicted that the gate overlapping the depletion region only can minimize capacitance instead of overlapping the source and drain region and result in a 10 times increase in I_{ON} and a lower SS [37–39]. Besides TFETs with a planar structure, a vertical TFET with the gate oxide on the vertical side walls was invented in silicon to achieve better gate control [8,40], and incorporating pseudomorphic strained Si$_{1-x}$Ge$_x$ layers between the source and the intrinsic channel results in a significant performance improvement as a direct consequence of tunnel barrier width lowering [41]. Double-gate and gate-all-around structures with high dielectrics can raise ON currents, suppress OFF currents and provide a low SS with a high I_{ON}/I_{OFF} [19,42,43].

7.4 Feasibility of TFETs Based on 2D Materials

Two-dimensional materials are perfect candidates for TFETs mainly due to their rich species, atomic thickness without dangling bonds and high mobility with high ON currents. The discovery of monolayer graphene with unique properties, such as a high mobility and atomic thickness without dangling bonds, gave rise to the prosperity of layered 2D materials in 2004. Layered TMDCs have two hexagonal lattices of MX$_2$ sandwiches, where M is the transition metal and X is chalcogens, and 2H-MX$_2$ is honeycombs with a D$_{3h}$ point group symmetry while 1T-MX$_2$ is center honeycombs with a C$_{3v}$ symmetry [44], as shown in the insets in Figure 7.3. Layered TMDCs can be semiconducting when M=Mo [45,46], W [47], etc., as metallic when M=Nb, Re [48], etc., and even superconducting such as layered CoO$_2$ [49] and NbSe$_2$ [50]. Group IV-VI chalcogenides exist with a cubic NaCl structure, such as SnTe and PbX (X=S, Se, Te), or a rhombohedral structure, such as SnX and GeX (X=S, Se) with bandgaps for both direct and indirect transitions [51,52]. Layered Group III-VI semiconductors generally has a structure similar to layered GaSe [53,54]. Bulk GaSe has great potential for nonlinear optical application with a direct bandgap of approximately 2.1 eV

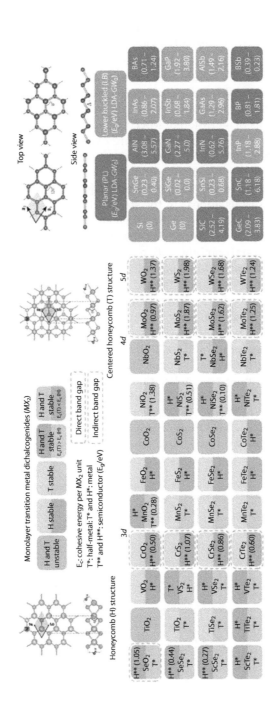

FIGURE 7.3

Summary of the properties of 44 different MX_2 compounds and Group IV elements and Group III-V [60], where M represents transition metal atoms divided into 3d, 4d and 5d groups. +, * and ** represent half-metal, metallic and semiconducting, respectively, and the numbers in the bracket are the corresponding bandgap of these 2D crystals in semiconducting states. The gray shade means that the 2D crystal is unstable. The lower-lying structure (H or T) is the ground states in each box.

and its nanosheets exhibit a higher responsivity of 2.8 A/W and a higher external quantum efficiency of 1367% at 254 nm than 2D nanosheet devices of MoS$_2$ and graphene, although its mobility is lower than 1 cm^2/Vs [55–57]. Except graphene, Group IV element graphene-like 2D sheets, like silicene and germanene formed from Si and Ge, have also attracted wide interest; their existence is predicted with a buckled honeycomb structure [58,59]. Layered binary compounds of Group IV elements, such as SiC, GeC and SnC, are predicted to be planar, similar to graphene and BN, while SnSi, SnGe and SiGe are buckled like silicene and germanene [60]. All of these stable compounds were found to be semiconductors with a variety of bandgaps, but graphene is metallic and BN is an insulator. The stability analysis and semiconducting properties of 44 different MX$_2$ compounds and binary compounds of Group IV elements and Group III-V are shown in Figure 7.3. It is necessary to notice another semiconductor, black phosphorus (BP) with great carrier mobility, outstanding electrostatic modulation, a tunable bandgap of 0.3–2.0 eV and a high ON current [61], which make it widely applied to FET [62,63], battery [64,65], gas sensors [66] and amplitude modulation (AM) demodulator [67]. 2D crystals appear in the stacked layered form in nature and the bandgap of their single layer is usually larger than their stacked layered form [68,69].

Two-dimensional crystal sheets can form a 3D structure by stacking by van der Waals force while the atoms of 3D crystal combine with each other by chemical bonding. Various admixtures of sp^3 bonds result in the electron states at the conduction and valance band edges of 3D semiconductors. The electronic states at the valance band of 3D crystal is more p-like while the conduction band edge is mostly s-like. The directivity of a p-orbital results in the anisotropy of the hole effective mass or the curvature. However, symmetry is desirable for complementary logic devices, like p-type metal-oxide-semiconductor (PMOS) and n-type metal-oxide-semiconductor (NMOS) or PTFET and NTFET, as it could simplify the geometry and layout of complementary circuits. The conduction and valence bands in TMDCs single layer is much more symmetric than 3D semiconductor crystal, although less symmetric than graphene and BN [70,71]. The energy band alignment and relative energy band offset are shown in Figure 7.4 [72]. The relative bandgaps and electron affinity are marked in Figure 7.4 according to the corresponding references [73–75]. A perfect 2D crystal has no dangling bonds on its surface while a 3D crystal necessarily has broken

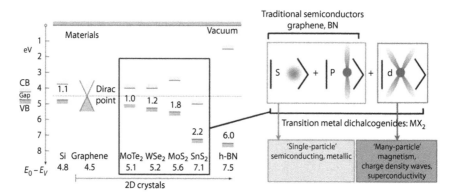

FIGURE 7.4

Energy band alignment of silicon and 2D crystal [72]. The bandgap and electron affinity of various 2D crystals are marked according to the corresponding reports or calculations. The conduction and valence band edge states of TMDCs include s>, |p> and |d> orbitals, while those of traditional semiconductors, graphene and BN only involves the linear combination of |s> and |p> orbitals. Some electronic phenomena of TMDCs require a many-particle effect, such as magnetism, charge density waves and superconductivity.

bonds on its surface, which will be passivated by dielectrics, lattice-matched or -strained HSs. Therefore, the energy gap windows of 2D crystals are not populated by surface states, which are necessary in 3D crystals [74]. It is calculated that the effective masses (m^*) range from $0.34m_0$ to $0.76m_0$ [76], where m_0 is the mass of a free electron, and m^* of 2D BN is ~$0.6m_0$ [77], while graphene is zero-effective mass due to its unique band structure of Dirac cones [78], which have been refined by experimental measurements.

FETs with 2D crystals as the channels, such as graphene, MoS_2 and BP, have shown stable and outstanding electrostatics and transport characteristics. Monolayer MoS_2 transistors with hafnium oxide (HfO_2) as the gate dielectrics have achieved ON current over 10 uA, electron mobility >200 cm^2/Vs and an I_{ON}/I_{OFF} ratio of 10^8 [45]. BP transistors with a thickness of several nanometers have obtained charge-carrier mobility of ~1000 cm^2/Vs with an I_{ON}/I_{OFF} ratio larger than five orders, as shown in Figure 7.5a [79]. Monolayer WSe_2 transistors can work as a p-n junction by electrostatic modulation of the gate voltage [80]. Using high-κ dielectrics as surroundings can damp scattering and improve charge mobility (Figure 7.5b) in transistors with 2D semiconductors as the channel materials because 2D crystals have direct access to the electrons, their spins and atomic vibrations [45,81,82].

To achieve an effective gate electrostatic control over mobile electrons and holes in the channel, the channel thickness has to be reduced when scaling down the source/drain separations in FETs based on 3D crystal semiconductors. However, the surface roughness of a 3D crystal channel will greatly deteriorate when the channels are thinned down. TFETs require a stronger gate control to reduce the screening tunneling length to achieve

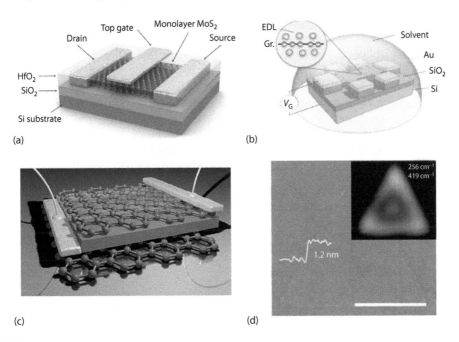

(a) (b) (c) (d)

FIGURE 7.5
(a) Three-dimensional schematic view of an MoS_2 transistor with HfO_2 as top-gate dielectrics on a silicon substrate with an SiO_2 layer [45]. (b) Device schematics of a graphene transistor with an ionic electric double layer (EDL) as its dielectric environment, which can greatly improve the charge mobility [81]. (c) Structure of a graphite/BN/graphite device, where the BN layer works as the tunneling layer [85]. (d) AFM image of a triangular domain of a WS_2-WSe_2 lateral HS with a thickness of 1.2 nm. The inset shows the composite image including Raman mapping at 256 cm^{-1} and 419 cm^{-1}. There is no apparent gap or overlap between the WS_2 and WSe_2 signals, meaning that the successive lateral epitaxy of the HS [90].

a high tunneling probability. Therefore, 2D crystals are a wonderful solution to solve these problems as layered 2D can be atomic thick, thinner than 1 nm. And using 2D dielectrics, like h-BN as the gate oxide [83] or tunneling layer (Figure 7.5c) [84,85], can enhance the control of the gate voltage as 2D h-BN can be thinned to a monolayer.

Chemical doping in 2D crystal is still a challenge and not effective as expected, so TFETs of 2D crystal HSs are not experimentally realized as they depend on electrostatic doping to form an energy band offset instead of the differences in chemical composition. So, taking HSs as the channels is a soothing choice for TFETs. There should be few lattice mismatches and defects in the interface as there are no dangling bonds on the surface of layered 2D crystal and the different layers attach to each other by van der Waals force. Various 2D materials can be applied to the permutation and combination of 2D HSs [86–89]. Besides vertical or out-of-plane HSs, in-plane HSs of unpassivated edge growth have been realized by the successive lateral epitaxial growth of a thermal CVD process (Figure 7.5d) [90]. But intrinsic strain will be induced in the lateral epitaxial growth of in-plane HS, which will result in a lattice mismatch in the lateral interface, reduce the coupling strength between p- and d-orbitals and change the band structure [91].

TFETs based on 2D crystals have an electrostatic advantage and simplicity for device structures. The tunneling current flows laterally from the source to the drain in lateral TFETs based on 3D materials (Figure 7.6a). To enhance the gate control over the channel thickness, the channel should be thinned down (Figure 7.6b) as the gate field is vertical, which would result in increasing the bandgap due to quantum confinement and thus decreasing the interband tunneling [72]. The quantum confinement will be suppressed if the channels are replaced by 2D crystal layers (Figure 7.6c) with little consideration of the surface states–related trap states. As tunneling only happens at the interface of the source and the channel in lateral TFETs, the tunneling area is a line when the channel thickness is atomic thin and the tunneling current will be reduced significantly. Therefore, vertical TFETs are taken into consideration. 3D vertical TFETs with side gates are shown in Figure 7.6d, which can be designed with a gate-all-round structure to enhance the gate control over the channels. To further enlarge the tunneling area, the vertical top-gate TFETs can be fabricated (Figure 7.6e). If the HS channels are replaced by 2D-2D stacks, vertical double-gate TFETs are available (Figure 7.6f). Either lateral TFETs or vertical TFETs with 2D crystals as the channels take advantage of the structure's simplicity.

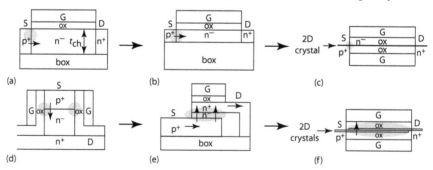

FIGURE 7.6

Device schematics of lateral TFETs based on 3D materials (a), with thinned channel (b) to enhance the gate control over the entire junction thickness. (c) Device schematic of lateral TFETs based on 2D crystal. (d) 3D vertical TFETs with side gates, where the tunneling current flows vertically. (e) 3D vertical TFETs with top gate, where the tunneling current flows in a zigzag way, as indicated by the arrows, and over an area instead of a line, as the shading shows. (f) Vertical double-gate TFETs with p+ and n+ 2D crystal layers. The tunneling junction is marked by the red shading and the arrows indicate the direction of the tunneling current flows [72].

7.5 Current Status of TFETs Based on 2D Materials

The current situation of TFETs based on 2D crystals will be discussed theoretically and experimentally, and can be divided into in-plane tunneling for lateral TFETs and interlayer tunneling for vertical TFETs.

The published simulation works about lateral TFETs based on 2D materials are demonstrated in Figure 7.7. In the case of chemically doped TFET, the source and the drain region is highly doped, especially the source region, as the tunneling junction is formed between the interface of the source region and the electrostatically gated channel. A double-gate structure (Figure 7.8a) is available for TFETs with 2D materials as the channel because of their atomically thin thickness, which makes the gate control over the channel more effective. Currently, most simulation works on TFETs are based on the model structure.

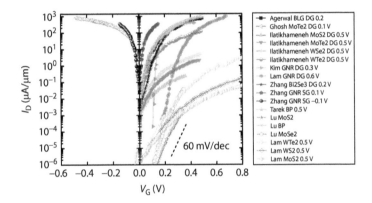

FIGURE 7.7
Simulated transfer characteristics of lateral TFETs based on 2D crystals, such as graphene [100], BP [96,107], MoTe$_2$ [93,98], MoS$_2$ [93,94,96], WSe$_2$ [93], WTe$_2$ [93,94], graphene nanoribbon (GNR) [101,102], Bi$_2$Se$_3$ [104], MoSe$_2$ [96] and WS$_2$ [94].

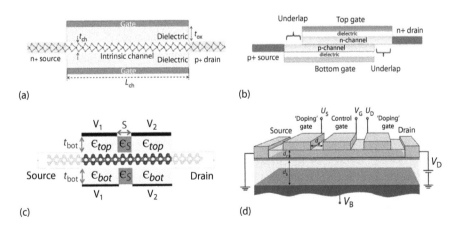

FIGURE 7.8
Two different doping methods for TFETs based on 2D crystals. Schematic cross section of a monolayer 2D crystal-based aligned double gate (a) [98] and a staggered double gate (b) [96] TFET with a chemically doped n+ source, p+ drain and intrinsic channel. (c) Layout of a lateral TFET with electrically doped source and drain region [108]. (d) Electrically doped lateral TFET with control gate [110].

As discussed previously, the source material with a bandgap of about $(1.1–1.5)qV_{DD}$ provides the best performance; therefore, a bandgap ranging from 0.55 to 0.75 eV is the optimal choice for a supply voltage of 0.5 V. The WTe$_2$ monolayer has great potential for high performance TFET applications because of its extremely small effective mass and a direct bandgap of 0.75 eV [92]. A high ON current over 350 µA/µm and an SS below 60 mV/dec even when the channel current is up to 1 µA/µm in WTe$_2$ TFET with a source doping level of 10^{20} cm^{-3} and a channel voltage of 0.5 V [93]. The ON current of WTe$_2$ TFET calculated in another simulation work is of the same order [94], but the SS is larger than that of the former mostly because the bandgap used in the simulation (0.99 eV) is larger than the former one (0.75 eV). Therefore, a small bandgap really counts. By dielectric engineering, lateral WTe$_2$ TFET can achieve a record ON current of ~1000 µA/µm and an SS below 20 mV/dec, where the gates and the channel material are separated by high-κ dielectrics while the spacing part between the gates is filled with low-κ dielectrics [95]. Simulation work has been done on MoS$_2$ lateral TFET as MoS$_2$ has been widely investigated and can be synthetized in large scale. Lateral MoS$_2$ TFETs cannot achieve a high ON current and an SS below 60 mV/dec with the same doping level of drain and applied voltages due to its large bandgap and its heavy effective mass, which could suppress the source-to-drain tunneling and make it unsuitable for TFETs but ideal for ultra-scaled MOSFETs [93,94,96]. The ON current of a lateral MoSe$_2$ TFET is only a little larger than that of lateral MoS$_2$ TFETs under the same simulation conditions probably due to its lighter effective mass and smaller bandgap [96]. Although WSe$_2$ has a larger bandgap (1.56 eV) than MoTe$_2$ (1.08 eV), its smaller reduced effective mass partly compensates for the inferiority resulted from the larger bandgap. Therefore, the ON current of a lateral homogeneous WSe$_2$ TFET is still larger than that of an MoTe$_2$ TFET [93]. Similarly, a lateral homogeneous WS$_2$ TFET has a larger ON current than an MoS$_2$ TFET though a WS$_2$ has a larger bandgap (1.758 eV) than an MoS$_2$ (1.66 eV) [94].

Generally, to compete with advanced CMOS technology, $I_{ON} > 1$ mA/µm, an SS far below 60 mV/dec for five orders, $I_{ON}/I_{OFF} > 10^5$ and $V_{DD} < 0.5$ V for TFETs [20]. The performances of lateral homogeneous TFETs based on TMDCs calculated in these works are dissatisfying as their ON currents are not high enough and the SS is below 60 mV/dec only when the channel currents are small. More specifically, the SS should be far below 60 mV/dec for current levels extending to 1–10 µA/µm when switching on the TFET [97]. However, lateral homogeneous TFETs with five monolayer TMDCs (MoS$_2$, MoSe$_2$, MoTe$_2$, WS$_2$ or WSe$_2$) offer an average I_{ON} of 150 µA/µm and an SS of 4 mV/dec at $V_{DD} = 0.1$ V in Ghosh's simulation work [98], which are more excellent than other simulation works about lateral TFETs based on TMDCs.

Due to the unique band structure of graphene whose Brillouin zone boundary is linearly dispersed, its effective mass approaches zero at the point where the valence band and the conduction band meet. Therefore, the carrier mobility of graphene FETs is ultrahigh and a great number of unusual electronic transport properties follow, such as an anomalous quantum Hall effect [99]. TFETs with graphene as their channel material are expected to achieve a high ON current as its ultralight effective mass benefits the tunneling probability significantly. The characteristics of lateral TFETs based on bilayer graphene were calculated in Agarwal's work [100]. A high ON current over 1 mA/µm, an SS as low as 35 mV/dec and $I_{ON}/I_{OFF} > 2910$ with $V_{DD} = 0.2$ V were acquired using contact-induced doping. The unsatisfactory I_{ON}/I_{OFF} is unsurprising due to its extremely small bandgap. Therefore, opening its bandgap controllably is necessary to increase the I_{ON}/I_{OFF} ratio as well as taking advantage of its high mobility and high ON currents. With the width-dependent energy bandgap of graphene, graphene nanoribbon (GNR) widths between 3 and 10 nm correspond to the energy bandgap in the range of 0.46–0.14 eV. The tunability of its bandgap and the light effective mass of its carriers make it a wonderful candidate for TFETs. A lateral TFET based on GNR 5 nm wide was predicted to have a high I_{ON}/I_{OFF} over 10^7 with an ON current of 800 µA/µm and an effective SS of 0.19 mV/dec [101]. Adding a region with a small bandgap between the source and the channel, the ON and OFF currents can be tuned [102], which improves the tunneling efficiency by choosing source material with a smaller bandgap homojunction compared with Ge-Si HS TFETs. It can avoid the lattice mismatch and also the difficulty of forming HSs. However, GNR TFETs suffer from line edge roughness [103], which would enhance the OFF current significantly and reduce the conductance of GNRs. Therefore, it is necessary to optimize the width of GNR to minimize the OFF current with decreasing the ON current as little as possible, while widening the bandgap will suppress the thermal emission but reduce tunneling probability when narrowing down GNR.

Bi_2Se_3 is a topological insulator with a rhombohedral crystal structure, whose energy bandgap ranges from 41 meV to ~0.5 eV for Bi_2Se_3 thin films. A lateral TFET based on two-quintuple-layer Bi_2Se_3 with a thickness of 1.4 nm and a bandgap of 0.252 eV has been simulated [104]. The Bi_2Se_3 TFET with high source/drain doping level and a drain underlap can achieve an I_{ON}/I_{OFF} current ratio of 10^4, an I_{OFF} of 5 nA/µm and an SS of 50 mV/dec with a supply voltage of 0.2 V, which makes it a dynamic power indicator 10 times lower than MOSFET. The topological insulator has a high static dielectric constant ($\varepsilon_r = 100$), not favorable for short-channel FETs because the lateral electric field penetrates into the channel and broadens the source–channel junction [105]. If considering $\varepsilon_r = 20$, the ON current can be enhanced more than twice as the lateral electric field broadens the source–channel junction, reduces the source carrier injection efficiency and degrades the effective SS.

Another unique layered 2D materials is BP, which is especially attractive as it has electronic properties that lie in between graphene and TMDCs. It has a direct bandgap in both bulk (0.3 eV) and monolayer (2 eV), high mobility, light effective mass of carriers ($0.146m_0$), anisotropic effective mass that increases the density of states near the band edges and a low dielectric constant [106]. All these advantages of BP benefit the performance of BP TFETs as discussed previously. Simulation work predicted that multilayered BP TFETs can achieve an ON current as high as graphene TFETs with a small SS of 24.6 mV/dec [96]. Even scaling down the channel length of BP TFETs to 6 nm, the performance is acceptable with a supply voltage of 0.2 V [107]. In conclusion, compared with lateral TFETs based on graphene Bi_2Se_3 and TMDCs, BP is the most promising candidate for lateral TFETs as a BP lateral TFET can achieve a high ON current of ~1 mA/µm with a high I_{ON}/I_{OFF} ratio of ~10^6 and an ultralow SS due to its merits, though lateral TFETs based on other 2D materials are also predicted to be able to achieve an SS below 60 mV/dec.

Most of the simulation work on lateral TFETs on 2D materials are based on the double-gate structure as shown in Figure 7.8a, where the top gate and bottom gate are aligned and the source and drain are formed by chemical doping. A lateral tunneling window along the channel can be suppressed if lateral offset, like underlap double gates, is adopted (Figure 7.8b), which suppresses the parasitic leakage current [96]. It was observed that lateral tunneling was more pronounced in a lateral BP TFET without underlapping compared with its counterpart with underlapping. As chemical doping in 2D materials is currently a great challenge, the n- and p-type potentials can be defined by two gates with opposite polarities at the two sides of the tunneling junction (Figure 7.8c). Electrical doping avoids fluctuations, threshold voltage shifts and bandgap reduction resulting from doping, which reduces the OFF state performance [108]. Given the channel with ultrathin 2D materials, the pacing between the two gates, the thickness of the gate dielectrics, the dielectric constant of the spacing region and the strain greatly affect the performances of electric doping TFETs, especially the first two factors. A thinner dielectric, smaller spacing and a smaller spacer dielectric constant can reduce the SS and enhance the I_{ON} remarkably. Different from chemically doped TFET, the electric field at the tunneling junction is inversely proportional to the total thickness of the top and bottom dielectrics. What really counts in an electrically doped TFET are the physical dielectrics thickness and distance between the gates [109]. As the bottom gate often uses a thick back oxide, the electric field of the tunneling junction will decrease, which can be avoided by using a back oxide with a low dielectric constant compared to the top oxide ($\varepsilon_{bottom} << \varepsilon_{top}$) [108]. If adding another control gate to the channel and the n- or p-type potentials are fixed at a specific level, the electrically doped lateral TFET will be controlled by the control gate similarly to the chemically doped TFET and can be operated as NTFET or PTFET by positive gate or negative voltage, respectively [110].

The strain on the 2D channel of lateral TFETs influences the device performance. For example, applying a biaxial strain of 3% to the WSe_2 channel would increase the ON currents as 3% biaxial strain reduces the effective mass and bandgap by about 10%–20% [108]. At present, there is no experimental realization of lateral TFETs based on layered 2D materials reported mainly due to the challenge of doping in 2D materials and forming multi-gates within a distance of several nanometers. Fortunately, some progress has been made in doping 2D materials [111–116]. At the same time, many measures have been applied to improve the contact between 2D materials and the electrodes, such as annealing, using graphene as electrodes [117], phase engineering [118] and so on. Moreover, dielectric engineers have investigated solutions to enhance the gate control over the channel, like combining high-κ and low-κ oxides by atomic layer deposition and atomic h-BN.

Due to the challenge of chemical doping in 2D materials and the advantage of free dangling bonds on the surface layer of 2D materials, TFETs based on 2D-2D HSs have been investigated at the same time. The structure of vertical double-gate TFETs with p-type and n-type 2D crystal layers is shown in Figure 7.6f. The simulation work on vertical HS TFETs is demonstrated in Figure 7.9. It is observed that a HS TFET enhances the ON current by an order compared to the homogeneous device, both considered at the same I_{ON}/I_{OFF} ratio in the simulation work, where WTe_2, WS_2 and MoS_2 lateral TFETs and vertical WS_2/

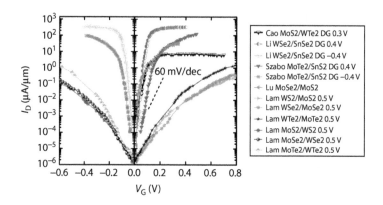

FIGURE 7.9
Simulated transfer characteristics of vertical TFETs based on 2D-2D crystals HSs, such as MoS_2/WTe_2 [119], WSe_2/$SnSe_2$ [121], $MoTe_2$/SnS_2 [120], $MoSe_2$/MoS_2 [96], WS_2/MoS_2, WSe_2/$MoSe_2$ and WTe_2/$MoTe_2$ [94].

MoS_2, WSe_2/$MoSe_2$ and WTe_2/$MoTe_2$ HS TFETs have been simulated, respectively, while $MoTe_2$/WTe_2 HS TFET do not enhance the current significantly compared with a WTe_2 lateral TFET [94]. But the simulation results in this work are not optimistic as the ON current, I_{ON}/I_{OFF} ratio and SS don't meet requirements for current-integrated circuits. The ON current of a vertical MoS_2/$MoSe_2$ HS TFET is also found to be two orders higher than an MoS_2 TFET [96]. However, it is predicted that the vertical TFET based on a monolayer MoS_2 and a WTe_2 heterojunction with a 1 nm thick h-BN interlayer can achieve a low SS of ~40 mV/dec, where the MoS_2 is doped and the h-BN functions as a tunneling barrier [119]. The doping concentration mainly affects the threshold top-gate voltage as mentioned previously [108]. It has been discovered that extending the top gate for a certain length beyond each side of the overlap region of MoS_2 and WTe_2 can improve the SS remarkably because the extending part suppresses the OFF current significantly. For example, the device can attain an ultralow SS of 7 mV/dec with a gate extension length of 20 nm. The ON current of ~10 μA/μm cannot meet the requirement but the TFET is competitive for low power applications. The vertical TFET made of a monolayer $MoTe_2$ and SnS_2 is reported to achieve an ON current >75 μA/μm and an I_{ON}/I_{OFF} ratio of about 10^8 with a supply voltage of 0.4 V [120]. A steep SS of ~14 mV/dec and a high ON current of ~300 μA/μm are estimated theoretically in a vertical TFET based on WSe_2/$SnSe_2$ stacked monolayer HSs [121].

There is a large difference in the work function between VIB- and IVB-TMDCs, providing a solution to form steep HSs with a broken bandgap alignment. It is recommended that VIB-MeX2 (Me=W and Mo; X=Te and Se) as the n-type source and IVB-MeX2 (Me=Zr and Hf; X=S and Se) as the p-type drain can form vertically stacked TFETs due to their unique band edge characters, which results in intervalley scattering during tunneling [92].

Many efforts have been made to experimentally realize vertical TFETs based on 2D-2D HSs. Vertical TFETs based on graphene HSs with h-BN or MoS_2 as the transport barrier exhibit switching ratios of ~50 and ~10,000 at room temperature, respectively [122]. Transistors with a graphene/h-BN/graphene sandwich structure have achieved resonant tunneling and negative differential resistance (NDR), where the tunneling barriers are atomic layers of h-BN [123]. Different from the NDR in the Esaki diode, NDR in these transistors stems from resonant tunneling and appears at both forward and reverse channel bias while NDR in the Esaki diode only appears at the forward bias. Graphene/h-BN/graphene symmetric FETs fabricated by chemical vapor deposition (CVD) at a larger scale are also observed with similar tunneling characteristics, which makes them possible for large-scale applications. Similarly, resonant tunneling is observed in MoS_2/WSe_2/graphene and WSe_2/MoS_2/graphene HSs with NDR characteristics at room temperature, which are synthesized by metal-organic chemical vapor deposition (MOCVD) [124]. Vertical TFETs based on graphene/WS_2/graphene HSs realized an unprecedented I_{ON}/I_{OFF} ratio over 10^6 at room temperature with a very high ON current, where WS_2 functions as an atomically thin barrier between two mechanically exfoliated or CVD-grown graphene layers [125]. More interestingly, NDR can be achieved by a simple three-terminal graphene device due to the competition between electron and hole conduction with the increase in the drain bias [126]. These resonant tunneling devices have potential for applications in high-frequency and logic devices.

A large number of 2D-2D HSs have been investigated, stacked with 2D layers mechanically exfoliated [86,127–130], grown by CVD [131–133] or fabricated by lateral epitaxial growth [90,134,135]. There is no doubt that carriers transfer across the interface very rapidly as the hole transfer from the MoS_2 layer to the WS_2 layer takes place within 50 femtosecond after optical excitation in MoS_2/WS_2 HSs even though there is an air gap of several nanometers in the interface [87] and electrons or holes in the excitons transfer across the interface in subpicosecond time [136]. Some of this work focuses on the application of optoelectronic, photovoltaic or memory devices, while some have developed methods of synthetizing vertical 2D-2D HSs or lateral 2D-2D HSs. Some of these HSs are formed as a p-n diode naturally [127,137] or can be tuned to a p-n junction by the gate voltage [138], which is indispensable for the application of TFETs for ultralow power consumption. As discussed previously, vertical 2D-2D HSs with suitable band alignment can realize tunneling. Dual-gate MoS_2/WSe_2 van der Waals HSs can be gate modulated as an Esaki diode with NDR at 77 K [139]. A type-II alignment between MoS_2 and WSe_2 is determined by micro-beam x-ray photoelectron spectroscopy and scanning tunneling spectroscopy with a valance band offset of 0.83 eV and a conduction band offset of 0.76 eV [140]. However, back-gate MoS_2/WSe_2 van der Waals HSs have also realized NDR even at room temperature [141]. $SnSe_2/BP$ van der Waals heterojunctions with broken-gap band alignment behave as an Esaki diode with evident NDR and a peak-to-valley ratio of 1.8 at room temperature and 2.8 at 80 K [142]. Compared with the simulation results of monolayer $WSe_2/SnSe_2$ HS vertical TFETs [121], a high I_{ON}/I_{OFF} ratio of ~10^7 has been realized in 2D-2D TFETs using $WSe_2/SnSe_2$ HSs in the absence of graphene but the ON current and an SS of 100 mV/dec are not as good as expected [143] (Figure 7.10).

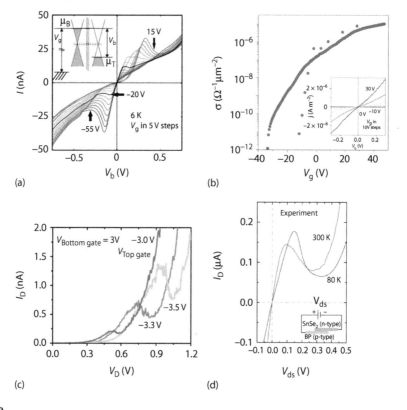

(a)   (b)

(c)   (d)

FIGURE 7.10

Tunneling characteristics of vertical TFETs based on different HSs. (a) Current-voltage characteristics of a specific transistor with a graphene/h-BN/graphene sandwich structure at 6 K with gate voltage ranging from 15 V to −55 V in 5 V steps [123]. (b) Conductivity (slope of the current-voltage curve) of vertical TFETs based on graphene/WS_2/graphene HSs as a function of the gate voltage at zero bias (red circles) and 0.02 V bias (blue circles) [125]. (c) Current-voltage characteristics of MoS_2/WSe_2 HS TFET at $V_{gate-MoS2}=3$ V and $V_{gate-WSe2}$ varied [142]. (d) Current-voltage characteristics of the BP/$SnSe_2$ Esaki tunnel diode at 80 and 300 K. The peak-to-valley ratio increases as the temperature decreases [143].

Besides 2D-2D HS TFETs, the combination of 2D-3D HSs has also been investigated for solar cell and photodetector applications [144–148], as well as TFETs for ultralow power consumption devices to take advantage of 2D and 3D materials. It is surprising that vertical TFETs based on highly doped p-type germanium (p-Ge) and double-layer MoS_2 HSs have realized promising results with a minimum SS of 3.9 mV/dec and an average SS of 31.1 mV/dec for four decades of drain current from 10^{-13} A to 10^{-9} A at room temperature with a supply voltage of 0.1 V [149]. This is the most significant results of TFETs based on 2D materials. Due to the difficulty of doping in 2D materials and the lack of high p-type 2D materials, it is not easy to form TFETs based on 2D-2D HSs for ultralow power consumption. Therefore, highly p-doped Ge with a relatively low electron affinity and a small bandgap is adopted as the source material and n-type MoS_2 with atomic thickness is used as the channel material modulated by the gate to switch on/off the transistor, which results in a small tunneling barrier width. But the promising gate control ability might derive from its polymer complex gate, resulting in a high gate capacitance [150]. It is seen that the steep SS below 60 mV/dec only appears when the channel current is below 10^{-9} A, which can be improved further to meet the requirement for ultralow power consumption applications. 2D/3D MoS_2/GaN Esaki tunnel diodes are also realized experimentally with a high current density and repeatable NDR at room temperature [151]. The 2D/3D HS TFETs take advantage of both layered 2D materials and 3D materials, which have great potential for high performance device applications.

7.6 Conclusion

TFETs based on 2D materials are currently under intense scrutiny. They have great potential to solve the power consumption problem of MOSFETs with the supply voltage scaled down further. The rich species with different band structures, semimetals, metals and even superconductors, and atomic thickness make them unrivalled candidates for TFETs and the gate dielectrics can be atomically thin 2D materials, like BN, which enhances the gate control remarkably. The simplicity of TFETs based on 2D materials takes advantage of the two-dimensionality of layered 2D materials. WTe_2, GNR and BP are the most suitable materials for lateral homojunction TFETs if the doping technology for 2D materials can be overcome. The combination of 2D materials to form HSs opens a door to a brand new world for electronic device technology, including TFETs. Although the results on TFETs based on 2D HSs acquired are not exciting, the theoretical work inspires us to make more significant progress than the current state of the art.

References

1. D. Kim, Y. Lee, J. Cai, I. Lauer, L. Chang, S. J. Koester, et al., Low power circuit design based on heterojunction tunneling transistors (HETTs), in *IEEE International Symposium on Low Power Electronics and Design*, pp. 219–224, 2009.
2. J. J. Quinn, G. Kawamoto, and B. D. Mccombe, Subband spectroscopy by surface channel tunneling, *Surface Science*, vol. 73, pp. 190–196, 1978.

3. W. M. Reddick and G. Amaratunga, Silicon surface tunnel transistor, *Applied Physics Letters*, vol. 67, pp. 494–496, 1995.

4. Y. Omura, Negative conductance properties in extremely thin silicon-on-insulator (SOI) insulated-gate pn-junction devices (SOI surface tunnel transistors), *Japanese Journal of Applied Physics*, vol. 35, pp. L1401-L1403, 1996.

5. C. Aydin, A. Zaslavsky, S. Luryi, S. Cristoloveanu, D. Mariolle, D. Fraboulet, et al., Lateral interband tunneling transistor in silicon-on-insulator, *Applied Physics Letters*, vol. 84, pp. 1780–1782, 2004.

6. T. Baba, Proposal for surface tunnel transistors, *Japanese Journal of Applied Physics*, vol. 31, pp. L455–L457, 1992.

7. J. Knoch and J. Appenzeller, A novel concept for field-effect transistors: The tunneling carbon nanotube FET, *Device Research Conference Digest, 2005, DRC '05. 63rd*, vol. 1, pp. 153–156, 2005.

8. W. Hansch, C. Fink, J. Schulze, and I. Eisele, A vertical MOS-gated Esaki tunneling transistor in silicon, *Thin Solid Films*, vol. 369, pp. 387–389, 2000.

9. S. O. Koswatta, M. S. Lundstrom, M. P. Anantram, and D. E. Nikonov, Simulation of phonon-assisted band-to-band tunneling in carbon nanotube field-effect transistors, *Applied Physics Letters*, vol. 87, pp. 253107-253107-3, 2005.

10. U. E. Avci and I. A. Young, Heterojunction TFET scaling and resonant-TFET for steep sub-threshold slope at sub-9nm gate-length, in *IEEE International Electron Devices Meeting*, pp. 4.3.1–4.3.4, 2013.

11. J. H. Seo, Y. J. Yoon, S. Lee, J. H. Lee, S. Cho, and I. M. Kang, Design and analysis of Si-based arch-shaped gate-all-around (GAA) tunneling field-effect transistor (TFET), *Current Applied Physics*, vol. 15, pp. 208–212, 2015.

12. Y. Liu, X. Dong, and P. Chen, ChemInform abstract: Biological and chemical sensors based on graphene materials, *ChemInform*, vol. 43, pp. 2283–2307, 2012.

13. M. Liu, X. Yin, E. Ulinavila, B. Geng, T. Zentgraf, L. Ju, et al., A graphene-based broadband optical modulator, *Nature*, vol. 474, pp. 64–67, 2011.

14. R. Cheng and X. Duan, High-frequency self-aligned graphene transistors with transferred gate stacks, *Proceedings of the National Academy of Sciences of the United States of America*, vol. 109, pp. 11588–11592, 2012.

15. K. S. Kim, Y. Zhao, H. Jang, S. Y. Lee, J. M. Kim, K. S. Kim, et al., Large-scale pattern growth of graphene films for stretchable transparent electrodes, *Nature*, vol. 457, pp. 706–710, 2009.

16. Y. Zhu, S. Murali, M. D. Stoller, K. J. Ganesh, W. Cai, P. J. Ferreira, et al., Carbon-based supercapacitors produced by activation of graphene, *Science*, vol. 332, pp. 1537–1541, 2011.

17. A. S. Mayorov, R. V. Gorbachev, S. V. Morozov, L. Britnell, R. Jalil, L. A. Ponomarenko, et al., Micrometer-scale ballistic transport in encapsulated graphene at room temperature, *Nano Letters*, vol. 11, pp. 2396–2399, 2011.

18. P. F. Wang, K. Hilsenbeck, T. Nirschl, M. Oswald, C. Stepper, M. Weis, et al., Complementary tunneling transistor for low power application, *Solid-State Electronics*, vol. 48, pp. 2281–2286, 2004.

19. K. Boucart and A. M. Ionescu, Double-gate tunnel FET with high-κ gate dielectric, *Electron Devices IEEE Transactions*, vol. 54, pp. 1725–1733, 2007.

20. A. M. Ionescu and H. Riel, Tunnel field-effect transistors as energy-efficient electronic switches, *Nature*, vol. 479, pp. 329–337, 2011.

21. Y. Liu, R. P. Dick, L. Shang, and H. Yang, Accurate temperature-dependent integrated circuit leakage power estimation is easy, in *Design, Automation and Test in Europe Conference* and *Exposition*, 2007, pp. 1526–1531.

22. K. K. Bhuwalka, M. Born, M. Schindler, M. Schmidt, T. Sulima, and I. Eisele, P-Channel tunnel field-effect transistors down to sub-50 nm channel lengths, *Japanese Journal of Applied Physics*, vol. 45, pp. 3106–3109, 2006.

23. Q. Zhang, W. Zhao, and A. Seabaugh, Low-subthreshold-swing tunnel transistors, *IEEE Electron Device Letters*, vol. 27, pp. 297–300, 2006.

24. H. Ilatikhameneh, G. Klimeck, and R. Rahman, Can homojunction tunnel FETs scale below 10 nm?, *IEEE Electron Device Letters*, vol. 37, pp. 115–118, 2016.

25. H. Ilatikhameneh, T. A. Ameen, G. Klimeck, and J. Appenzeller, Dielectric engineered tunnel field-effect transistor, *IEEE Electron Device Letters*, vol. 36, pp. 1097–1100, 2015.

26. H. Ilatikhameneh, G. Klimeck, and R. Rahman, 2D tunnel transistors for ultra-low power applications: Promises and challenges, in *Energy Efficient Electronic Systems*, 2015, pp. 1–3.

27. C. H. Lee, E. C. Silva, L. Calderin, M. A. T. Nguyen, M. J. Hollander, B. Bersch, et al., Tungsten ditelluride: A layered semimetal, *Scientific Reports*, vol. 5, p. 10013, 2015.

28. A. S. Verhulst, W. G. Vandenberghe, K. Maex, and S. D. Gendt, Complementary silicon-based heterostructure tunnel-FETs with high tunnel rates, *Electron Device Letters IEEE*, vol. 29, pp. 1398–1401, 2009.

29. S. O. Koswatta, S. J. Koester, and W. Haensch, On the possibility of obtaining MOSFET-like performance and sub-60-mV/dec swing in 1-D broken-gap tunnel transistors, *IEEE Transactions on Electron Devices*, vol. 57, pp. 3222–3230, 2010.

30. J. Knoch and J. Appenzeller, Modeling of high-performance p-type III–V heterojunction tunnel FETs, *IEEE Electron Device Letters*, vol. 31, pp. 305–307, 2010.

31. J. Knoch, Optimizing tunnel FET performance: Impact of device structure, transistor dimensions and choice of material, in *International Symposium on VLSI Technology, Systems, and Applications, 2009. VLSI-TSA*, pp. 45–46, 2009.

32. J. Knoch, S. Mantl, and J. Appenzeller, Impact of the dimensionality on the performance of tunneling FETs: Bulk versus one-dimensional devices, *Solid-State Electronics*, vol. 51, pp. 572–578, 2007.

33. J. Appenzeller, Y. M. Lin, J. Knoch, and P. Avouris, Band-to-band tunneling in carbon nanotube field-effect transistors, *Physical Review Letters*, vol. 93, p. 196805, 2004.

34. Y. Yoon and S. Salahuddin, Barrier-free tunneling in a carbon heterojunction transistor, *Applied Physics Letters*, vol. 97, pp. 033102-033102-3, 2010.

35. R. Gandhi, Z. Chen, N. Singh, and K. Banerjee, Vertical Si-nanowire-type tunneling FETs with low subthreshold swing ( ) at room temperature, *IEEE Electron Device Letters*, vol. 32, pp. 437–439, 2011.

36. R. Gandhi, Z. Chen, N. Singh, and K. Banerjee, CMOS-compatible vertical-silicon-nanowire gate-all-around p-type tunneling FETs with ≤50-mV/decade subthreshold swing, *IEEE Electron Device Letters*, vol. 32, pp. 1504–1506, 2011.

37. C. Aydin, A. Zaslavsky, S. Luryi, and S. Cristoloveanu, Lateral interband tunneling transistor in silicon-on-insulator, *Applied Physics Letters*, vol. 84, pp. 1780–1782, 2004.

38. C. Hu, P. Patel, A. Bowonder, and K. Jeon, Prospect of tunneling green transistor for 0.1V CMOS, in *Electron Devices Meeting*, 2010, pp. 16.1.1–16.1.4.

39. R. Asra, M. Shrivastava, K. V. R. M. Murali, R. K. Pandey, H. Gossner, and V. R. Rao, Tunnel FET for V_{DD} scaling below 0.6V with CMOS comparable performance, *IEEE Transactions on Electron Devices*, vol. 58, pp. 1855–1863. 2011.

40. K. K. Bhuwalka, S. Sedlmaier, A. K. Ludsteck, and C. Tolksdorf, Vertical tunnel field-effect transistor, *IEEE Transactions on Electron Devices*, vol. 51, pp. 279–282, 2004.

41. K. K. Bhuwalka, J. Schulze, and I. Eisele, A simulation approach to optimize the electrical parameters of a vertical tunnel FET, *IEEE Transactions on Electron Devices*, vol. 52, pp. 1541–1547, 2005.

42. A. S. Verhulst, B. Sorée, D. Leonelli, W. G. Vandenberghe, and G. Groeseneken, Modeling the single-gate, double-gate, and gate-all-around tunnel field-effect transistor, *Journal of Applied Physics*, vol. 107, pp. 024518-024518-8, 2010.

43. L. D. Seup, H. S. Yang, K. C. Kang, L. Joung-Eob, L. J. Han, S. Cho, et al., Simulation of gate-all-around tunnel field-effect transistor with an n-doped layer, *IEICE Transactions on Electronics*, vol. 93-C, pp. 540–545, 2010.

44. N. V. Podberezskaya, S. A. Magarill, N. V. Pervukhina, and S. V. Borisov, Crystal chemistry of dichalcogenides MX_2, *Journal of Structural Chemistry*, vol. 42, pp. 654–681, 2001.

45. B. Radisavljevic, A. Radenovic, J. Brivio, V. Giacometti, and A. Kis, Single-layer MoS_2 transistors, *Nature Nanotechnology*, vol. 6, pp. 147–150, 2011.

46. W. S. Yun, S. W. Han, S. C. Hong, I. G. Kim, and J. D. Lee, Thickness and strain effects on electronic structures of transition metal dichalcogenides: 2H-MX2 semiconductors (M = Mo, W; X = S, Se, Te)[J]. *Physical Review B Condensed Matter*, vol. 85, pp. 033305-1-033305-5, 2012.

47. S. Das and J. Appenzeller, WSe_2 field effect transistors with enhanced ambipolar characteristics, *Applied Physics Letters*, vol. 103, pp. 103501-103501-5, 2013.

48. K. K. Tiong, C. H. Ho, and Y. S. Huang, The electrical transport properties of ReS_2 and $ReSe_2$ layered crystals, *Solid State Communications*, vol. 111, pp. 635–640, 1999.

49. K. Takada, H. Sakurai, E. Takayama-Muromachi, F. Izumi, R. A. Dilanian, and T. Sasaki, Superconductivity in two-dimensional CoO_2 layers, *ChemInform*, vol. 422, pp. 53–55, 2003.

50. M. Yoshida, J. Ye, T. Nishizaki, N. Kobayashi, and Y. Iwasa, Electrostatic and electrochemical tuning of superconductivity in two-dimensional NbSe2 crystals, *Applied Physics Letters*, vol. 108, p. 202602, 2016.

51. D. J. Xue, J. Tan, J. S. Hu, W. Hu, Y. G. Guo, and L. J. Wan, Anisotropic photoresponse properties of single micrometer-sized GeSe nanosheet, *Advanced Materials*, vol. 24, pp. 4528–4533, 2012.

52. I. Lefebvre, M. A. Szymanski, J. Olivierfourcade, and J. C. Jumas, Electronic structure of tin monochalcogenides from SnO to SnTe, *Physical Review B*, vol. 58, pp. 1896–1906, 1998.

53. J. Robertson, Electronic structure of GaSe, GaS, InSe and GaTe, *Journal of Physics C Solid State Physics*, vol. 12, p. 4777, 2001.

54. L. Plucinski, R. L. Johnson, B. J. Kowalski, K. Kopalko, B. A. Orlowski, Z. D. Kovalyuk, et al., Electronic band structure of GaSe(0001): Angle-resolved photoemission and ab initio theory, *Physical Review B*, vol. 68, 125304, 2003.

55. P. Hu, Z. Wen, L. Wang, P. Tan, and K. Xiao, Synthesis of few-layer GaSe nanosheets for high performance photodetectors, *ACS Nano*, vol. 6, p. 5988, 2012.

56. K. R. Allakhverdiev, M. Ö. Yetis, S. Özbek, T. K. Baykara, and E. Y. Salaev, Effective nonlinear GaSe crystal. Optical properties and applications, *Laser Physics*, vol. 19, pp. 1092–1104, 2009.

57. M. Schlüter, J. Camassel, S. Kohn, J. P. Voitchovsky, Y. R. Shen, and M. L. Cohen, Optical properties of GaSe and mixed crystals, *Physical Review B*, vol. 13, pp. 3534–3547, 1976.

58. K. Takeda and K. Shiraishi, Theoretical possibility of stage corrugation in Si and Ge analogs of graphite, *Physical Review B Condensed Matter*, vol. 50, pp. 14916–14922, 1994.

59. E. Durgun, S. Tongay, and S. Ciraci, Silicon and III-V compound nanotubes: Structural and electronic properties, *Physical Review B*, vol. 72, p. 075420, 2005.

60. M. Xu, T. Liang, M. Shi, and H. Chen, Graphene-like two-dimensional materials, *Chemical Reviews*, vol. 113, p. 3766, 2013.

61. H. Du, X. Lin, Z. Xu, and D. Chu, Recent developments in black phosphorus transistors, *Journal of Materials Chemistry C*, vol. 3, pp. 8760–8775, 2015.

62. S. Das, M. Demarteau, and A. Roelofs, Ambipolar phosphorene field effect transistor, *ACS Nano*, vol. 8, p. 11730, 2014.

63. F. Xia, H. Wang, and Y. Jia, Rediscovering black phosphorus as an anisotropic layered material for optoelectronics and electronics, *Nature Communications*, vol. 5, p. 4458, 2014.

64. M. Buscema, D. J. Groenendijk, G. A. Steele, V. D. Z. Hs, and A. Castellanos-Gomez, Photovoltaic effect in few-layer black phosphorus PN junctions defined by local electrostatic gating, *Nature Communications*, vol. 5, p. 4651, 2014.

65. T. Hong, B. Chamlagain, W. Lin, H. J. Chuang, M. Pan, Z. Zhou, et al., Polarized photocurrent response in black phosphorus field-effect transistors, *Nanoscale*, vol. 6, p. 8978, 2014.

66. L. Kou, T. Frauenheim, and C. Chen, Phosphorene as a superior gas sensor: Selective adsorption and distinct I-V response, *Journal of Physical Chemistry Letters*, vol. 5, p. 2675, 2014.

67. W. Zhu, M. N. Yogeesh, S. Yang, S. H. Aldave, J. S. Kim, S. Sonde, et al., Flexible black phosphorus ambipolar transistors, circuits and AM demodulator, *Nano Letters*, vol. 15, p. 1883, 2015.

68. K. F. Mak, C. Lee, J. Hone, J. Shan, and T. F. Heinz, Atomically thin MoS_2: A new direct-gap semiconductor, *Physical Review Letters*, vol. 105, p. 136805, 2010.

69. A. Splendiani, L. Sun, Y. Zhang, T. Li, J. Kim, C. Y. Chim, et al., Emerging photoluminescence in monolayer MoS_2, *Nano Letters*, vol. 10, pp. 1271–1275, 2010.

70. K. Wood and J. B. Pendry, Layer method for band structure of layer compounds, *Physical Review Letters*, vol. 31, pp. 1400–1403, 1973.

71. K. Kaasbjerg, K. S. Thygesen, and K. W. Jacobsen, Phonon-limited mobility in n-type single-layer MoS_2 from first principles, *Physical Review B*, vol. 85, p. 115317, 2012.

72. D. Jena, Tunneling transistors based on graphene and 2-D crystals, *Proceedings of the IEEE*, vol. 101, pp. 1585–1602, 2013.

73. R. Yan, Q. Zhang, W. Li, I. Calizo, T. Shen, C. A. Richter, et al., Determination of graphene work function and graphene-insulator-semiconductor band alignment by internal photoemission spectroscopy, *Applied Physics Letters*, vol. 101, 022105, pp. 666–35, 2012.

74. W. Mönch, Valence-band offsets and Schottky barrier heights of layered semiconductors explained by interface-induced gap states, *Applied Physics Letters*, vol. 72, pp. 1899–1901, 1998.

75. R. Schlaf, O. Lang, C. Pettenkofer, and W. Jaegermann, Band lineup of layered semiconductor heterointerfaces prepared by van der Waals epitaxy: Charge transfer correction term for the electron affinity rule, *Journal of Applied Physics*, vol. 85, pp. 2732–2753, 1999.

76. L. Liu, S. B. Kumar, Y. Ouyang, and J. Guo, Performance limits of monolayer transition metal dichalcogenide transistors, *IEEE Transactions on Electron Devices*, vol. 58, pp. 3042–3047, 2011.

77. K. Watanabe, T. Taniguchi, and H. Kanda, Direct-bandgap properties and evidence for ultra-violet lasing of hexagonal boron nitride single crystal, *Nature Materials*, vol. 3, pp. 404–409, 2004.

78. L. A. Ponomarenko, R. V. Gorbachev, G. L. Yu, D. C. Elias, R. Jalil, A. A. Patel, et al., Cloning of Dirac fermions in graphene superlattices, *Nature*, vol. 497, pp. 594–597, 2013.

79. L. Li, Black phosphorus field-effect transistors, *Nature Nanotechnology*, vol. 9, pp. 372–377, 2014.

80. J. Ross, P. Klement, A. Jones, N. Ghimire, J. Yan, D. Mandrus, et al., Electrically tunable excitonic light emitting diodes based on monolayer WSe_2 p-n junctions, *Nature Nanotechnology*, vol. 9, p. 268, 2014.

81. A. K. Newaz, Y. S. Puzyrev, B. Wang, S. T. Pantelides, and K. I. Bolotin, Probing charge scattering mechanisms in suspended graphene by varying its dielectric environment, *Nature Communications*, vol. 3, p. 734, 2012.

82. C. Jang, S. Adam, J. H. Chen, E. D. Williams, S. S. Das, and M. S. Fuhrer, Tuning the effective fine structure constant in graphene: Opposing effects of dielectric screening on short- and long-range potential scattering, *Physical Review Letters*, vol. 101, p. 146805, 2008.

83. G. Lu, T. Wu, Q. Yuan, H. Wang, H. Wang, F. Ding, et al., Synthesis of large single-crystal hexagonal boron nitride grains on Cu-Ni alloy, *Nature Communications*, vol. 6, p. 6160, 2015.

84. L. Britnell, R. V. Gorbachev, R. Jalil, B. D. Belle, F. Schedin, A. Mishchenko, et al., Field-effect tunneling transistor based on vertical graphene heterostructures, *Science (New York, N.Y.)*, vol. 335, pp. 947–950, 2012.

85. L. Britnell, R. V. Gorbachev, R. Jalil, B. D. Belle, F. Schedin, M. I. Katsnelson, et al., Electron tunneling through ultrathin boron nitride crystalline barriers, *Nano Letters*, vol. 12, pp. 1707–1710, 2012.

86. F. Wang, Z. Wang, K. Xu, F. Wang, Q. Wang, Y. Huang, et al., Tunable GaTe-MoS_2 van der Waals p-n junctions with novel optoelectronic performance, *Nano Letters*, vol. 15, 7558–7566, 2015.

87. X. Hong, J. Kim, S. F. Shi, Y. Zhang, C. Jin, Y. Sun, et al., Ultrafast charge transfer in atomically thin MoS_2/WS_2 heterostructures, *Nature Nanotechnology*, vol. 9, pp. 682–686, 2014.

88. N. Flöry, A. Jain, P. Bharadwaj, M. Parzefall, T. Taniguchi, K. Watanabe, et al., A $WSe_2/MoSe_2$ heterostructure photovoltaic device, *Applied Physics Letters*, vol. 107, p. 123106, 2015.

89. Y. Gong, S. Lei, G. Ye, B. Li, Y. He, K. Keyshar, et al., Two-step growth of two-dimensional $WSe_2/MoSe_2$ heterostructures, *Nano Letters*, vol. 15, p. 6135, 2015.

90. X. Duan, C. Wang, J. C. Shaw, R. Cheng, Y. Chen, H. Li, et al., Lateral epitaxial growth of two-dimensional layered semiconductor heterojunctions, *Nature Nanotechnology*, vol. 9, p. 1024, 2014.

91. W. Wei, Y. Dai, and B. Huang, Straintronics in two-dimensional in-plane heterostructures of transition-metal dichalcogenides, *Physical Chemistry Chemical Physics PCCP*, vol. 19, pp. 663–672, 2016.
92. C. Gong, H. Zhang, W. Wang, L. Colombo, R. M. Wallace, and K. Cho, Band alignment of two-dimensional transition metal dichalcogenides: Application in tunnel field effect transistors, *Applied Physics Letters*, vol. 103, p. 053513, 2013.
93. H. Ilatikhameneh, Y. Tan, B. Novakovic, G. Klimeck, R. Rahman, and J. Appenzeller, Tunnel field-effect transistors in 2-D transition metal dichalcogenide materials, *IEEE Exploratory Solid-State Computational Devices and Circuits*, vol. 1, pp. 12–18, 2015.
94. K. T. Lam, X. Cao, and J. Guo, Device performance of heterojunction tunneling field-effect transistors based on transition metal dichalcogenide monolayer, *IEEE Electron Device Letters*, vol. 34, pp. 1331–1333, 2013.
95. H. Ilatikhameneh, T. A. Ameen, G. Klimeck, and J. Appenzeller, Dielectric engineered tunnel field-effect transistor, *IEE Electron Device Letters*, vol. 36, pp. 1097–1100, 2015.
96. S. C. Lu, M. Mohamed, and W. Zhu, Novel vertical hetero- and homo-junction tunnel field-effect transistors based on multi-layer 2D crystals, *2d Materials*, vol. 3, p. 011010, 2016.
97. W. G. Vandenberghe, A. S. Verhulst, B. Sorée, W. Magnus, G. Groeseneken, Q. Smets, et al., Figure of merit for and identification of sub-60 mV/decade devices, *Journal of Applied Physics*, vol. 102, pp. 013510–48, 2013.
98. R. K. Ghosh and S. Mahapatra, Monolayer transition metal dichalcogenide channel-based tunnel transistor, *IEEE Journal of the Electron Devices Society*, vol. 1, pp. 175–180, 2013.
99. T. Ohta, A. Bostwick, T. Seyller, K. Horn, and E. Rotenberg, Controlling the electronic structure of bilayer graphene, *Science*, vol. 313, p. 951, 2006.
100. T. K. Agarwal, A. Nourbakhsh, P. Raghavan, and I. Radu, Bilayer graphene tunneling FET for Sub-0.2 V digital CMOS logic applications, *Electron Device Letters IEEE*, vol. 35, pp. 1308–1310, 2014.
101. Q. Zhang, T. Fang, H. Xing, and A. Seabaugh, Graphene nanoribbon tunnel transistors, *IEEE Electron Device Letters*, vol. 29, pp. 1344–1346, 2009.
102. K. T. Lam, D. Seah, S. K. Chin, and S. B. Kumar, A simulation study of graphene-nanoribbon tunneling FET with heterojunction channel, *IEEE Electron Device Letters*, vol. 31, pp. 555–557, 2010.
103. S. G. Kim, M. Luisier, T. B. Boykin, and G. Klimeck, Computational study of heterojunction graphene nanoribbon tunneling transistors with p-d orbital tight-binding method, *Applied Physics Letters*, vol. 104, pp. 329–337, 2014.
104. Q. Zhang, G. Iannaccone, and G. Fiori, Two-dimensional tunnel transistors based on thin film, *IEEE Electron Device Letters*, vol. 35, pp. 129–131, 2014.
105. J. Chang, L. F. Register, and S. K. Banerjee, Topological insulator Bi2Se3 thin films as an alternative channel material in metal-oxide-semiconductor field-effect transistors, *Journal of Applied Physics*, vol. 112, pp. 3045–3067, 2012.
106. Y. Takao and A. Morita, Electronic structure of black phosphorus: Tight binding approach, *Journal of the Physical Society of Japan*, vol. 105, pp. 93–98, 1981.
107. T. A. Ameen, H. Ilatikhameneh, G. Klimeck, and R. Rahman, Few-layer phosphorene: An ideal 2D material for tunnel transistors, *Scientific Reports*, vol. 6, p. 28515, 2015.
108. H. Ilatikhameneh, G. Klimeck, J. Appenzeller, and R. Rahman, Design rules for high performance tunnel transistors from 2D materials, *IEEE Journal of the Electron Devices Society*, 2016.
109. H. Ilatikhameneh, G. Klimeck, J. Appenzeller, and R. Rahman, Scaling theory of electrically doped 2D transistors, *IEEE Electron Device Letters*, vol. 36, pp. 726–728, 2015.
110. G. Alymov, V. Vyurkov, V. Ryzhii, and D. Svintsov, Abrupt current switching in graphene bilayer tunnel transistors enabled by van Hove singularities, *Scientific Reports*, vol. 6, p. 24654, 2016.
111. M. Tosun, L. Chan, M. Amani, T. Roy, G. H. Ahn, P. Taheri, et al., Air-stable n-doping of WSe2 by anion vacancy formation with mild plasma treatment, *ACS Nano*, vol. 10, pp. 6853–6860, Jul 26 2016.

112. H. Fang, S. Chuang, T. C. Chang, K. Takei, T. Takahashi, and A. Javey, High-performance single layered WSe(2) p-FETs with chemically doped contacts, *Nano Letters*, vol. 12, pp. 3788–3792, Jul 11 2012.

113. H. M. Li, D. Lee, D. Qu, X. Liu, J. Ryu, A. Seabaugh, et al., Ultimate thin vertical p-n junction composed of two-dimensional layered molybdenum disulfide, *Nature Communications*, vol. 6, p. 6564, Mar 24 2015.

114. W. Liu, J. Kang, D. Sarkar, Y. Khatami, D. Jena, and K. Banerjee, Role of metal contacts in designing high-performance monolayer n-type WSe_2 field effect transistors, *Nano Letters*, vol. 13, pp. 1983–1990, 2013.

115. S. Chuang, C. Battaglia, A. Azcatl, S. McDonnell, J. S. Kang, X. Yin, et al., MoS_2 P-type transistors and diodes enabled by high work function MoOx contacts, *Nano Letters*, vol. 14, pp. 1337–1342, 2014.

116. D. Mao, X. She, B. Du, D. Yang, W. Zhang, K. Song, et al., Erbium-doped fiber laser passively mode locked with few-layer $WSe_2/MoSe_2$ nanosheets, *Scientific Reports*, vol. 6, p. 23583, 2016.

117. H.-J. Chuang, X. Tan, N. J. Ghimire, M. M. Perera, B. Chamlagain, M. M.-C. Cheng, et al., High mobility WSe_2 p-and n-type field-effect transistors contacted by highly doped graphene for low-resistance contacts, *Nano Letters*, vol. 14, pp. 3594–3601, 2014.

118. R. Kappera, D. Voiry, S. E. Yalcin, B. Branch, G. Gupta, A. D. Mohite, et al., Phase-engineered low-resistance contacts for ultrathin MoS_2 transistors, *Nature Materials*, vol. 13, pp. 1128–1134, 2014.

119. J. Cao, A. Cresti, D. Esseni, and M. Pala, Quantum simulation of a heterojunction vertical tunnel FET based on 2D transition metal dichalcogenides, *Solid-State Electronics*, vol. 116, pp. 1–7, 2016.

120. A. Szabó, S. J. Koester, and M. Luisier, Ab-Initio simulation of van der Waals MoTe 2 –SnS 2 heterotunneling FETs for low-power electronics, *IEEE Electron Device Letters*, vol. 36, pp. 514–516, 2015.

121. M. O. Li, D. Esseni, J. J. Nahas, and D. Jena, Two-dimensional heterojunction interlayer tunneling field effect transistors (thin-TFETs), *Electron Devices Society IEEE Journal of the*, vol. 3, pp. 200–207, 2015.

122. L. Britnell, R. V. Gorbachev, R. Jalil, B. D. Belle, F. Schedin, A. Mishchenko, et al., Field-effect tunneling transistor based on vertical graphene heterostructures, *Science*, vol. 335, pp. 947–950, 2012.

123. L. Britnell, R. V. Gorbachev, A. K. Geim, L. A. Ponomarenko, A. Mishchenko, M. T. Greenaway, et al., Resonant tunnelling and negative differential conductance in graphene transistors, *Nature Communications*, vol. 4, p. 1794, 2013.

124. Y. C. Lin, R. K. Ghosh, R. Addou, N. Lu, S. M. Eichfeld, H. Zhu, et al., Atomically thin resonant tunnel diodes built from synthetic van der Waals heterostructures, *Nature Communications*, vol. 6, pp. 1–10, 2015.

125. T. Georgiou, R. Jalil, B. D. Belle, L. Britnell, R. V. Gorbachev, S. V. Morozov, et al., Vertical field-effect transistor based on graphene-WS_2 heterostructures for flexible and transparent electronics, *Nature Nanotechnology*, vol. 8, pp. 100–103, 2012.

126. Y. Wu, D. B. Farmer, W. Zhu, S. J. Han, C. D. Dimitrakopoulos, A. A. Bol, et al., Three-terminal graphene negative differential resistance devices, *ACS Nano*, vol. 6, pp. 2610–2616, 2012.

127. Y. Deng, Z. Luo, N. J. Conrad, H. Liu, Y. Gong, S. Najmaei, et al., Black phosphorus-monolayer MoS_2 van der Waals heterojunction p-n diode, *ACS Nano*, vol. 8, p. 8292, 2014.

128. P. Rivera, K. L. Seyler, H. Yu, J. R. Schaibley, J. Yan, D. G. Mandrus, et al., Valley-polarized exciton dynamics in a 2D semiconductor heterostructure, *Science*, vol. 351, p. 688, 2016.

129. P. Rivera, Observation of long-lived interlayer excitons in monolayer $MoSe_2$–WSe_2 heterostructures, *Nature Communications*, vol. 6, p. 6242, 2015.

130. S. Bertolazzi, D. Krasnozhon, and A. Kis, Nonvolatile memory cells based on MoS_2/graphene heterostructures, *ACS Nano*, vol. 7, pp. 3246–3252, 2013.

131. K. Wang, B. Huang, M. Tian, F. Ceballos, M. W. Lin, M. Mahjouri-Samani, et al., Interlayer coupling in twisted WSe_2/WS_2 bilayer heterostructures revealed by optical spectroscopy, *ACS Nano*, vol. 10, 2016.

132. B. Li, L. Huang, M. Zhong, Y. Li, Y. Wang, J. Li, et al., Direct vapor phase growth and optoelectronic application of large band offset SnS_2/MoS_2 vertical bilayer heterostructures with high lattice mismatch, *Advanced Electronic Materials*, vol. 2, 2016.

133. Y. Gong, S. Lei, G. Ye, B. Li, Y. He, K. Keyshar, et al., Two-step growth of two-dimensional WSe_2/$MoSe_2$ heterostructures, *Nano Letters*, vol. 15, pp. 6135–6141, 2015.

134. Y. Gong, J. Lin, X. Wang, G. Shi, S. Lei, Z. Lin, et al., Vertical and in-plane heterostructures from WS_2/MoS_2 monolayers, *Nature Materials*, vol. 13, p. 1135, 2014.

135. C. Huang, S. Wu, A. M. Sanchez, J. J. Peters, R. Beanland, J. S. Ross, et al., Lateral heterojunctions within monolayer $MoSe_2$-WSe_2 semiconductors, *Nature Materials*, vol. 13, pp. 1096–1101, 2014.

136. F. Ceballos, M. Z. Bellus, H. Y. Chiu, and H. Zhao, Ultrafast charge separation and indirect exciton formation in a MoS_2–$MoSe_2$ van der Waals heterostructure, *ACS Nano*, vol. 8, pp. 12717–12724, 2014.

137. C. H. Lee, G. H. Lee, V. D. Z. Am, W. Chen, Y. Li, M. Han, et al., Atomically thin p-n junctions with van der Waals heterointerfaces, *Nature Nanotechnology*, vol. 9, pp. 676–681, 2014.

138. P. Chen, J. Xiang, H. Yu, J. Zhang, G. Xie, S. Wu, et al., Gate tunable MoS_2-black phosphorus heterojunction devices, *2d Materials*, vol. 2, p. 034009, 2015.

139. R. Tania, T. Mahmut, C. Xi, F. Hui, L. Der-Hsien, Z. Peida, et al., Dual-gated MoS_2/WSe_2 van der Waals tunnel diodes and transistors, *ACS NANO*, vol. 9, pp. 2017–2019, Jan 2015.

140. M. H. Chiu, C. Zhang, H. W. Shiu, C. P. Chuu, C. H. Chen, C. Y. Chang, et al., Determination of band alignment in the single-layer MoS_2/WSe_2 heterojunction, *Nature Communications*, vol. 6, p. 7666, 2015.

141. A. Nourbakhsh, A. Zubair, M. S. Dresselhaus, and T. Palacios, Transport properties of a MoS_2/WSe_2 heterojunction transistor and its potential for application, *Nano Letters*, vol. 16, 2016.

142. R. Yan, S. Fathipour, Y. Han, B. Song, S. Xiao, M. Li, et al., Esaki diodes in van der Waals heterojunctions with broken-gap energy band alignment, *Nano Letters*, vol. 15, pp. 5791–5798, Sep 9 2015.

143. T. Roy, M. Tosun, M. Hettick, G. H. Ahn, C. Hu, and A. Javey, 2D-2D tunneling field-effect transistors using WSe_2/$SnSe_2$ heterostructures, *Applied Physics Letters*, vol. 108, 083111, pp. 437–5, 2016.

144. M. R. Esmaeili-Rad and S. Salahuddin, High performance molybdenum disulfide amorphous silicon heterojunction photodetector, *Scientific Reports*, vol. 3, p. 2345, 2013.

145. S. Lin, X. Li, P. Wang, Z. Xu, S. Zhang, H. Zhong, et al., Interface designed MoS_2/GaAs heterostructure solar cell with sandwich stacked hexagonal boron nitride, *Scientific Reports*, vol. 5, p. 15103, 2015.

146. M. K. Joo, B. H. Moon, H. Ji, H. H. Gang, H. Kim, G. M. Lee, et al., Electron excess doping and effective Schottky barrier reduction on MoS_2/h-BN heterostructure, *Nano Letters*, vol. 16, 2016.

147. D. Ruzmetov, K. Zhang, G. Stan, B. Kalanyan, G. R. Bhimanapati, S. M. Eichfeld, et al., Vertical 2D/3D semiconductor heterostructures based on epitaxial molybdenum disulfide and gallium nitride, *ACS Nano*, vol. 10, p. 3580, 2016.

148. B. Li, G. Shi, S. Lei, Y. He, W. Gao, Y. Gong, et al., 3D band diagram and photoexcitation of 2D–3D semiconductor heterojunctions, *Nano Letters*, vol. 15, pp. 5919–5925, 2015.

149. D. Sarkar, X. Xie, W. Liu, W. Cao, J. Kang, Y. Gong, et al., A subthermionic tunnel field-effect transistor with an atomically thin channel, *Nature*, vol. 526, pp. 91–95, 2015.

150. M. W. Lin, L. Liu, Q. Lan, X. Tan, K. Dhindsa, P. Zeng, et al., Mobility enhancement and highly efficient gating of monolayer MoS_2 transistors with polymer electrolyte, *Journal of Physics D Applied Physics*, vol. 45, pp. 597–619, 2012.

151. S. Krishnamoorthy, E. W. L. Ii, C. H. Lee, Y. Zhang, W. D. Mcculloch, J. M. Johnson, et al., High current density 2D/3D MoS_2/GaN Esaki tunnel diodes, *Applied Physics Letters*, vol. 109, pp. 147–150, 2016.

8

Surface Functionalization of Silicon Carbide Quantum Dots

Marzaini Rashid, Ben R. Horrocks, Noel Healy, Jonathan P. Goss,
Hua-Khee Chan, and AltonB. Horsfall

CONTENTS

Silicon carbide (SiC) nanostructures are appealing as non-toxic, water-stable and oxidation-resistant nanomaterials. Owing to these unique properties, three-dimensionally confined SiC nanostructures, namely SiC quantum dots (QDs), have found applications in the bioimaging of living cells. Photoluminescence (PL) investigations, however, have revealed that across the polytypes: 3C-, 4H- and 6H-SiC, excitation wavelength–dependent PL is observed for larger sizes but deviates for sizes smaller than approximately 3 nm, thus exhibiting a dual feature in the PL spectra. Additionally, the nanostructures of varying polytypes and bandgaps exhibit strikingly similar PL emission centered at approximately 450 nm. At this wavelength, 3C-SiC emission is above the bulk bandgap as expected of quantum size effects, but for 4H-SiC and 6H-SiC the emissions are below bandgap. 4H-SiC is a suitable polytype to study these effects. Density functional theory (DFT) calculations within the ab initio formalism were performed on OH-, F- and H-terminated 4H-SiC-QDs with diameters in the range of 1–2 nm. The chosen surface terminations relate to the HF/ethanol electrolyte used in the preparation of SiC-QDs and the choice of size coincides with where deviation was observed in experiments. It was found that the absorption onset energies deviate from quantum confinement with -OH and -F terminations, but conform to the prediction when terminated with -H. The weak size-dependent absorption onsets for -OH and -F are due to surface states arising from lone-pair orbitals that are spatially localized to the QD surface where these terminations reside. On the other hand, -H termination shows strong size-dependent absorption onsets due to the delocalization of the electron wave function toward the QD core assisting quantum confinement. It is predicted that the surface-related states dominate up to sizes 2.5 and 2.7 nm for -F and -OH terminations, respectively. As a result, we show that the recombination mechanism involves the interplay between quantum confinement and surface states affecting the resultant energy gap and the resulting PL. Hence, the experimental PL spectra exhibit a dual feature: excitation wavelength independence for small sizes and excitation wavelength dependence for diameters larger than 3 nm, as observed in the experiments reported in the literature.

8.1 The Dual-Feature and Below Bandgap Photoluminescence Spectra in SiC Nanostructures

SiC is a promising material that is both non-toxic and oxidation resistant. 3C-SiC-QDs with an average size of 3.9 nm [1] showed robust PL emission centered at 450 nm with a quantum yield of 17%, comparable to that observed in Si nanoparticles. SiC-QDs were shown to be water stable and oxidation resistant for months without any surface passivation [2]. The properties of SiC such as biocompatibility, chemical inertness, photostability, stability in aqueous environment and resistance to oxidation are highly attractive for a range of applications.

Nonetheless, researchers have observed that the optical properties of SiC-QDs only follow the prediction of quantum confinement for diameters larger than approximately 3 nm, but deviate for smaller sizes. Thus, the expected and required size-tunable optical properties below 3 nm were not achieved. In theory, with the exciton Bohr radii of 2.0, 0.7 and 1.2 nm for 3C-, 6H- and 4H-SiC [2], respectively, strong quantum size effects are expected to occur for diameters smaller than 4 nm in 3C- and 2.4 nm in 4H-SiC-based QDs. The deviation is observed as a dual feature in the PL spectra for the different polytypes of SiC despite the differences in bandgap, along with an unexpected similar emission peak at approximately 450 nm. This 450 nm emission peak is above bandgap for 3C-SiC but below bandgap for 4H-SiC and 6H-SiC, the latter being counterintuitive to quantum confinement. These observations have been attributed to phase transformation into 3C-SiC during ultrasonication or due to surface states, but to date the origin of the dual-feature PL has not been unequivocally determined or explained and is still under investigation. Generally, in order to verify the existence of quantum confinement effects, the red-shifting in PL peak position with increasing excitation wavelengths can be used [2–4]. As the excitation energy is systematically decreased with longer excitation wavelengths, only larger QDs are excited and smaller ones are excluded. Therefore, the PL energy is expected to decrease (red-shift toward longer wavelengths). The majority of studies on the PL spectra of SiC nanoparticles suspended in solvents reported that quantum size effects are observed for larger particles (diameter greater than ~3 nm) but not for smaller diameters [5–7]. In these studies, the PL peak position generally red-shifted with increasing excitation wavelengths for large-sized SiC-QDs, as expected from quantum size effects. In contrast, the PL peak position remained constant (not dependent on excitation wavelength) for small-sized QDs (smaller than 3 nm), which contradicts the predictions from quantum confinement. Mixed results are reported in the literature, but in general the PL emission peak is shown to be insensitive to excitation wavelengths up to around 320 nm. The 'dual-feature' observation is schematically illustrated in Figure 8.1a, where region A depicts the constant PL peak position for excitation wavelengths shorter than 320 nm and region B shows that the PL

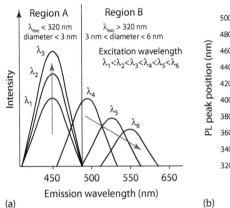
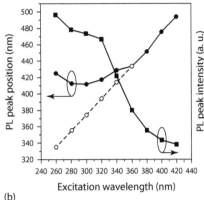

(a)   (b)

FIGURE 8.1

(a) Schematic illustrating the dual-feature PL spectra of SiC-QDs characterized by the constant PL emission peak in region A and the red-shifting PL peak in region B for excitation wavelengths shorter than or longer than 320 nm. (b) The experimentally observed deviation from quantum confinement (deviate from dotted line) in the PL peak position for excitation wavelengths shorter than 320 nm. (From Zhu et al., *Materials Letters*, 2014, 132, 210–213.)

peak position red-shifts for excitation wavelengths longer than 320 nm. For brevity, the associated regions A and B are referred to as 'constant PL' and 'red-shifting PL' and the combination of both trends is referred to as 'dual feature'.

The dual-feature trend in the PL peak position with respect to the excitation wavelength in the experiment is clearly evident in Figure 8.1b, where deviation from the expected size-dependent PL emission (straight dotted line) occurs at approximately 320 nm [8]. The QD's size distribution in [8] was 2–5.5 nm and has been associated with the excitation of QDs smaller than 3 nm.

8.2 DFT Study on the Optoelectronic Properties of OH-, F- and H-Terminated 4H-SiC Quantum Dots

This section presents the results from DFT calculations within the Ab Initio Modeling Program (AIMPRO) on the optical properties of OH-, F- and H-terminated 4H-SiC-QDs in the diameter range of 10–20 Å. The 4H-SiC polytype serves as a good representative to study both the dual feature and below bandgap PL emissions observed in experiments. The QD diameter range from 10 to 20 Å was chosen in order to investigate the discrepancy in quantum confinement effects between the theoretical prediction and the experimentally observed deviation for SiC-QDs smaller than 30 Å. Practically, biomarkers smaller than 50 Å are required for effective renal excretion and so retaining quantum size effects within this size range is important and warrants further understanding. The deviation in the optical properties observed in 4H-SiC-QDs may be contributed by surface effects as opposed to being primarily due to the bulk (polytype). In order to determine the influence of surface effects, different surface termination groups that are likely to be found in the commonly used HF/ethanol electrolyte are simulated, namely -OH, -F and -H. The effect of and contributions from these surface terminations on the onset energies of absorption cross section (ACS) and joint density of states (JDoS) are analyzed. A model is constructed to explain the experimentally observed dual-feature PL.

8.2.1 Computational Method

Calculations based on the DFT are performed within the AIMPRO code [9,10] for quasi-spherical SiC-QDs having core diameters varying from 10 to 20 Å. The modeling of atoms utilizes norm-conserving pseudo potentials [11]. Kohn–Sham orbitals are expanded using sets of independent real-space, atom-centered Gaussian basis sets [12]. In the case of C, O, F and Si, basis sets with four different widths are used, whereas for H three widths are used. To account for polarization in O, C, Si and F, d-Gaussians of one, two, two and four widths are included for the respective atoms.

In constructing the SiC-QDs, a group of atoms within some chosen distance of either an Si or a C atom (Si or C centered) in bulk SiC are selected. Then, atoms at the surface are saturated by either a H atom, an F atom or an OH group.

A plane-wave expansion (Fourier transform) is used to represent the charge density in reciprocal space and the SiC-QDs are modeled using periodic boundary conditions [10], with the cutoff in energy is sufficient to converge the energies to within 10 meV. In this approach, the SiC-QDs are placed in a periodic boundary condition. Following preparation of the SiC core and surface groups, the SiC-QDs are geometrically optimized until all atomic forces are smaller than 0.06 eV/Å.

8.2.2 Results

Firstly, simulations of OH-, F- and H-terminated, 10 Å diameter 4H-SiC-QDs are constructed and subsequently optimized. The corresponding DoS, ACS, JDoS, electron wave function iso-surfaces and electron probability densities for this QD size are calculated and compared for the different terminations. Similar analyses for core diameters of up to 20 Å are performed. A model comprising the core and surface highest occupied/lowest unoccupied (HOMO/LUMO) is constructed to explain the dual-feature PL observed in experiments. Finally, a projection of the QD size range influenced by surface states is made based on the calculations.

8.2.2.1 Optical Properties of 10 Å Diameter 4H-SiC-QD Structures

Relaxed geometries of 10 Å diameter QDs terminated with -OH ($Si_{19}C_{20}O_{40}H_{40}$), -F ($Si_{19}C_{20}F_{40}$) and -H ($Si_{19}C_{20}H_{40}$) are illustrated in Figure 8.2. The average Si-C, O-H, Si-X, C-X (X=H, O and F in each case) bond lengths in angstroms (Å) are calculated to be 1.87 Å,

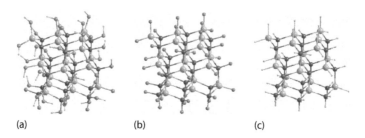

(a)   (b)   (c)

FIGURE 8.2
Geometrically relaxed 10 Å diameter 4H-SiC quantum dots with (a) -OH, (b) -F and (c) -H terminations. Color coding: gray=C, yellow=Si, red=O, green=F and white=H.

1.00 Å, (1.50 Å, 1.65 Å, 1.59 Å), (1.10 Å, 1.46 Å and 1.41 Å) as summarized in Table 8.1. The bond lengths lie within the anticipated range for Si- and C-based single bonds [13,14]. The standard deviation in the Si-C bond lengths across QDs with different surface terminations are within 0.01 Å.

Different surface functionalization (-OH, -F and -H) on the same Si and C core atoms configuration resulted in variation in the HOMO-LUMO energy gap, as shown in Figure 8.3a. The HOMO-LUMO energy gap is clearly distinguishable between different surface terminations, with -H termination showing the largest energy gap (4.50 eV), followed by -F (2.27 eV) and -OH (2.14 eV), respectively.

It is instructive to examine the effects of the surface terminations upon the QDs' ACS, as shown in Figure 8.3b. The absorption onsets correlate with the HOMO-LUMO energy gaps where H termination shows a higher energy optical absorption onset (4.50 eV) compared to -F and -OH in agreement with its larger calculated energy gap. The optical absorption onset of -OH is red-shifted by (2.3 eV) while -F is red-shifted by (1.7 eV) when compared to -H. For optical properties, the JDoS is important as it shows all the possible optical transitions from the valence (HOMO) to the conduction (LUMO) energy levels with energy

TABLE 8.1

10 Å Diameter SiC-QD Bond Lengths

Bond	Si-C	O-H	Si-H	Si-O	Si-F	C-H	C-O	C-F
Average bond length (Å)	1.87	1.00	1.50	1.65	1.59	1.10	1.46	1.41

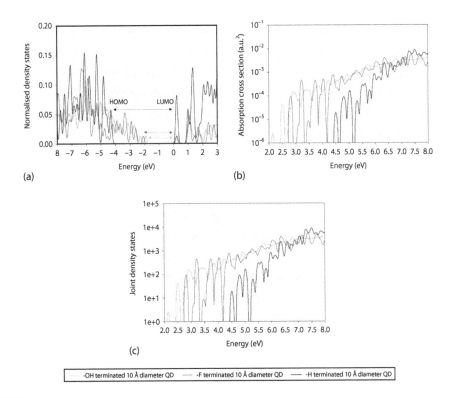

FIGURE 8.3
The respective (a) normalized density of states (DoS), (b) absorption cross section (ACS) and (c) joint density of states (JDoS) for 10 Å diameter -OH, -F and -H functionalized 4H-SiC-QDs.

separation equal to the energy of the absorbed photon. The resemblance in spectral shape and energies of the ACS in comparison to the JDoS indicates that the calculated optical transitions concur with the calculated allowed energies of the electronic DoS.

The difference in the HOMO–LUMO energy gap is further elaborated by examining the spatial distribution of the frontier orbitals (HOMO and LUMO). For -OH termination, Figure 8.4I-a shows HOMO being primarily localized on the surface C and O atoms in the form of C p-orbitals and O lone-pairs, respectively. LUMO orbitals are seen primarily on O atoms (Figure 8.4I-b). As shown in Figure 8.4I-c, a two-dimensional slice confirms the electron probability density (within 1 eV of the HOMO) residing mostly on the surface of the QD. The electron probability density examined is the sum of the squared modulus of the wave function:

$$\sum_{n=i} |\psi_i|^2 (r)$$

where $(\text{HOMO} - E_i) \leq (1 \text{ eV})$.

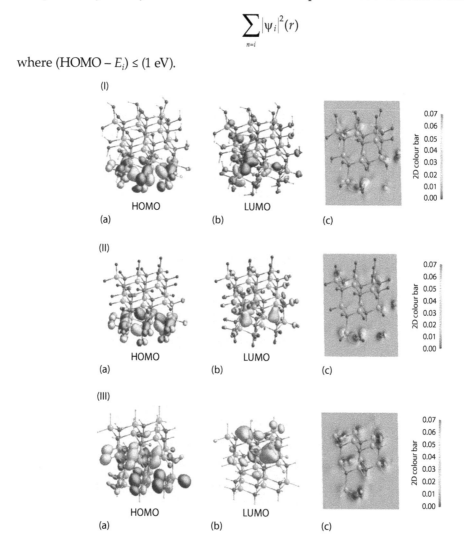

FIGURE 8.4

Wave functions for 10 Å diameter (I) -OH terminated, (II) -F terminated and (III) -H terminated 4H-SiC-QDs with respective 3-D wave function isosurface (0.07 a.u.) for (a) HOMO, (b) LUMO and (c) 2-D slice of electron probability density within 1 eV of the HOMO. The delocalized wave function spatial distribution toward the QD core suggests that the large HOMO-LUMO energy gap for H-termination results from the confining potential of the core of the QD.

For F-termination, lone-pairs on F atoms and p-like wave functions of the nearest neighbors (Figure 8.4II-a and II-c) mainly contribute to the HOMO, while for the LUMO (Figure 8.4II-b), similar to OH-termination, mainly the F atoms contribute. In contrast to -OH and -F terminations, the HOMO orbitals for -H termination (Figure 8.4III-a) are dispersed through the whole cluster, with the electron probability density reaching toward the core of the QD. The LUMO (Figure 8.4III-b) wave function is relatively delocalized to the QD core with minimal contribution from functional H atoms. Relating to the particle in a sphere model, the delocalized wave function spatial distribution toward the QD core suggests that the large HOMO-LUMO energy gap for -H termination results from the confining potential of the core of the QD (quantum confinement) while surface states are dominant and influence the electronic and optical properties of OH- and F-terminated QDs.

While it is shown that the H-terminated 10 Å diameter QD exhibits an optical absorption onset well above that of -F and -OH for a similar cluster size, it would be instructive to investigate the size-dependent quantum confinement effects for a range of QD diameters. In the next sections, the experimentally relevant 10–20 Å diameter range is presented. Full passivation of larger QD surfaces (>10 Å in diameter) with -OH, -F and -H would result in the surface termination species coming closer to each other through which steric repulsion becomes significant. By surface reconstructions with Si and C dimers, these surface species would be better accommodated for larger QD diameters. The effect of surface reconstruction and surface composition on the QDs' electronic and optical properties is presented in the next section.

8.2.2.2 Effect of Surface Composition and Surface Reconstruction

Data relating to optimized SiC-QDs in the diameter range between 10 and 20 Å is presented. The chosen diameter range is within the range where quantum confinement effects are observed in experiments [15], since the exciton Bohr radius in 4H-SiC is 12–18 Å [3,5]. Table 8.2 lists the QD compositions as a function of the QD diameter: N_{Si} (X) and N_C (X) being the number of Si and C atoms for X-centered QD (X for Si or C at the core center of QD), $N_{surface}$ represents the number of surface sites on unreconstructed clusters and N_{recon} (X) is the number of surface sites for reconstructed clusters with an X center (X for Si or C). The choice of atom center of either Si or C does not affect the total number of core atoms but the number of Si or C atoms are interchanged as shown in Table 8.2.

TABLE 8.2

SiC-QD Cluster Composition as a Function of QD Diameter in the Range of 10–20 Å

Diameter (Å)	N_{Si} (Si), N_C(C)	N_{Si} (C), N_C(Si)	$N_{surface}$	N_{recon} (Si)	N_{recon} (C)
10	19	20	40	40	40
11	29	29	56	44	44
12	51	41	82	70	70
13	57	62	94	76	76
14	69	74	100	88	88
15	81	83	112	100	100
16	99	96	126	114	114
17	132	118	160	130	136
18	147	151	172	130	142
19	159	175	190	160	148
20	207	199	208	166	172

The number of surface sites on the unreconstructed clusters is independent of whether the QD is Si or C centered. For the reconstructed clusters, the number of surface sites is different for Si- and C-centered cases.

In the case of unreconstructed clusters, only the positions of Si and C atoms are interchanged without changes to bondings with surface sites. In contrast, for the reconstructed clusters, the formation of dimers (Si-Si or C-C) on the surface depends on the choice of C or Si as the QD center. Additionally, Si-Si with a longer bond length may be formed at certain locations where C-C could not, due to its shorter bond length, thus the C atoms are terminated with more surface sites in comparison to Si for this case.

Figure 8.5 shows the 20 Å SiC-QDs with different centers and surface treatments. The average Si-C, O-H, Si-X, C-X (X=H, O and F in each case), Si-Si and C-C bond lengths in

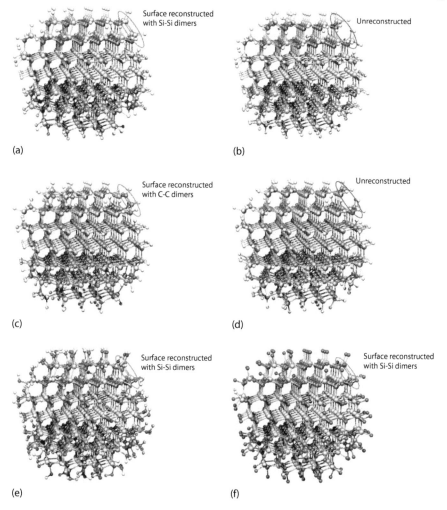

FIGURE 8.5
Geometrically relaxed 20 Å SiC-QDs. (a, b) H-terminated, Si-centered SiC-QDs with and without surface reconstructions, respectively. (c, d) H-terminated, C-centered SiC-QDs with and without surface reconstructions, respectively. (e, f) Si-centered SiC-QDs (with surface reconstructions) terminated with OH- and F-groups, respectively. White, red, yellow, gray and green atoms are H, O, Si, C and F, respectively. The regions indicated by the red and blue ellipses highlight surface sites where there is reconstruction with surface dimers or an unreconstructed surface, respectively.

TABLE 8.3

20 Å Diameter SiC-QD Bond Lengths

Bond	Si-C	O-H	Si-H	Si-O	Si-F	C-H	C-O	C-F	Si-Si	C-C
Average bond length (Å)	1.87	1.00	1.50	1.64	1.59	1.10	1.46	1.42	2.39	1.60

angstroms (Å) are calculated to be 1.87, 1.00, (1.50, 1.64 and 1.59), (1.10, 1.46 and 1.42), 2.39 and 1.60, respectively, which are summarized in Table 8.3. This provides evidence that the optimized structures have reasonable geometries to represent the spherical SiC-QDs observed in experiments. The calculated bond lengths vary only very slightly in comparison to the 10 Å diameter QDs.

To distinguish the impact of surface reconstructions from surface termination, two groups of Si-centered H-terminated QDs within the 10–20 Å size range are compared, one of which includes surface reconstruction.

8.2.2.3 Effect of Surface Termination Groups on Optical Absorption

Figure 8.6a and b shows the ACS for Si-centered, reconstructed, OH- and F-terminated QDs, and can be compared directly with Figure 8.6c. The effect upon the absorption onset for OH- and F-terminations (size independent and flat along ~2.0 eV) in comparison to H-termination (size dependent from 4.5 to ~3.0 eV) is immediately evident. For example, the differences between Figure 8.6a and c (due to different surface termination) outweigh the effects of reconstruction or whether the cluster is centered on the Si or C (Figure 8.6c,d or e,f).

Notably, the quantum confinement effect tends to increase from hydroxyl to fluorine to hydrogen termination, with H-termination showing the clearest size dependence. The absorption onset for H-termination increases from 2.5 to 4.4 eV with reducing size. This is a net increase of ~2 eV in the 10–20 Å diameter range. On the other hand, for F- and OH-terminations, almost no size-dependent change is observed over the same diameter range. As a result, for low core diameters there is a large difference in the onset energy as a function of termination. For example, in the case of 10 Å SiC-QDs, the difference in onset for the ACS for OH- and H-terminations is nearly 2.5 eV.

8.2.2.4 Effect of Surface Termination on Density of States, HOMO/ LUMO Wave Functions and Electron Probability Density

Figure 8.7 shows the DoS for Si-centered, surface-reconstructed, 20 Å diameter SiC-QDs terminated by OH, F and H, with zero energy aligned to the Fermi energy. The significant effect of terminating species is the variation in the HOMO whereby OH, F and H have energies ranging from −0.51 eV to −1.23 eV. In contrast, the LUMO energies are relatively constant, lying in the 1.0–1.1 eV range.

The difference in the energies of the HOMO is related to the degree of wave function localization at the surface. Figure 8.8a shows that the OH-termination HOMO is strongly associated with surface C p-orbitals and oxygen lone-pairs. Likewise, HOMOs of F-terminated clusters are associated with F lone-pairs (Figure 8.8c). The LUMOs of OH-termination (Figure 8.8b) and F-termination (Figure 8.8d) consist of mixtures of surface O (F) lone-pairs and core C p-orbitals for OH- (F-) terminations, respectively. The lone-pair energies are not significantly affected by the core diameter because they are localized on the surface, leading to the relatively weak size dependence in the ACS (Figure 8.6). It is concluded that

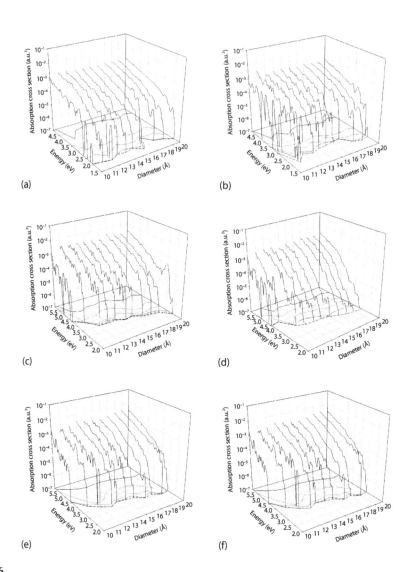

FIGURE 8.6
ACSs for (a) OH-terminated and (b) F-terminated 4H-SiC-QDs. ACSs for Si-centered, H-terminated (c) reconstructed and (d) unreconstructed QD. ACSs for C-centered, H-terminated (e) reconstructed and (f) unreconstructed QDs. The contours indicate 10^{-7} a.u.^{2} to 10^{-3} a.u.^{2}, in factors of 10. The effect of termination species (-OH, -F, -H) outweighs the effects of composition and surface reconstruction.

the seemingly size-independent optical absorption for OH- and F-terminated SiC-QDs is a reflection of the surface states associated with the terminating species.

The influence of surface terminations on the electronic structure is further elucidated by an examination of the energy levels deeper than the frontier electronic levels. Figure 8.9 shows that for -H-terminated Si-centered 20 Å SiC-QDs, the electron probability density of the HOMO (Figure 8.9a) is dominated by electrons within the core of the QD while the LUMO (Figure 8.9b) is surface related arising from the Si dimers. For deeper energy levels, within 0.5 eV of the HOMO (Figure 8.9c), the probability density increases with more electrons populating the QD core while for the LUMO (Figure 8.9d), in addition to the surface, electrons from the core start to contribute. Examining deeper still the energy (within 1 eV) from the frontier electronic levels, it is found that the electron probability density

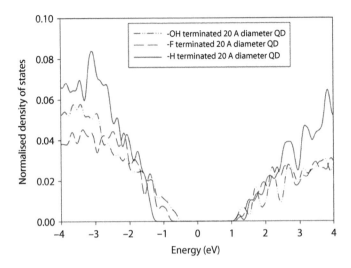

FIGURE 8.7
DoS of 20 Å, Si-centered QDs in the vicinity of the optical gap.

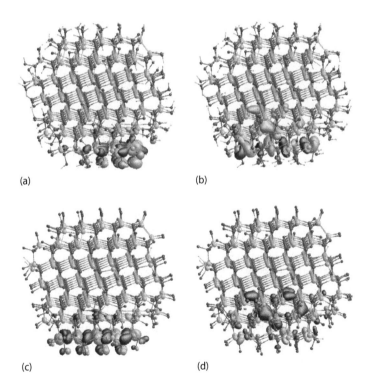

FIGURE 8.8
Wave function iso-surfaces for Si-centered 20 Å SiC-QDs. (a, b) HOMO and LUMO for -OH termination. (c, d) HOMO and LUMO for F-termination. Yellow, gray and white atoms are Si, C and H, respectively. The green and blue iso-surfaces show the molecular orbitals of the positive and negative phase, with an amplitude of 0.04 a.u.$^{-3/2}$. The clusters are shown with the c-axis of the underlying 4H-SiC vertical. Lone-pair energies are not significantly affected by the core diameter because they are localized on the surface, leading to the relatively weak size dependence in the ACS for F- and OH-terminations.

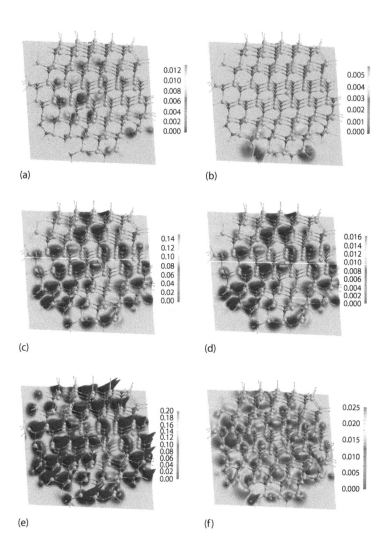

FIGURE 8.9
Two-dimensional slice through the center of an Si-centered 20 Å SiC-QD exhibiting -H electron probability density for (a) HOMO, (b) LUMO, (c) within 0.5 eV of the HOMO, (d) within 0.5 eV of the LUMO, (e) within 1 eV of the HOMO and (f) within 1 eV of the LUMO. The delocalized and localized states of the respective HOMO and LUMO are represented in Figure 8.13a.

is well delocalized throughout the QD core, as can be seen in Figure 8.9e,f. The delocalized electron probability density at higher energy levels (core HOMO and core LUMO) within the QD core would be influenced by the confining potential that supports quantum confinement effects. In contrast, the surface states near the frontal orbital of the LUMO (surface LUMO) are weakly influenced by the confining potential and would obscure quantum size effects. It is clearly illustrated that the energies in the vicinity of the HOMO for H-termination are highly dominated by electrons from the core. On the other hand, for the LUMO, the surface states dominate at lower energies but at higher energies, electrons from orbitals within the core of the QD contribute.

For F-termination, at the frontier orbitals, the HOMO (Figure 8.10a) constitutes electrons that are highly localized to the surface with no contribution from the core, while the LUMO (Figure 8.10b) shows a mix of electrons occupying the surface and core of the QD. Within

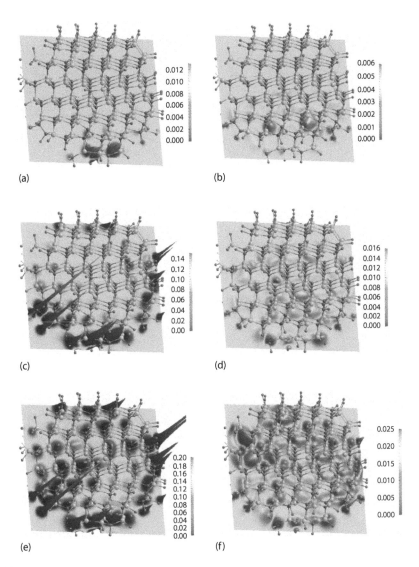

FIGURE 8.10
Two-dimensional slice through the center of an Si-centered 20 Å SiC-QD exhibiting -F electron probability density for (a) HOMO, (b) LUMO, (c) within 0.5 eV of the HOMO, (d) within 0.5 eV of the LUMO, (e) within 1 eV of the HOMO and (f) within 1 eV of the LUMO. The localized states of the respective HOMO and LUMO are represented in Figure 8.13b.

0.5 eV of the HOMO (Figure 8.10c), the electron occupation of surface atoms around the perimeter of the QD is further elaborated and an additional contribution from core-related electrons is observed. The LUMO (Figure 8.10d) in this higher energy interval involves more highly localized electrons at the surface and added contributions from electrons in the core. Higher up in energy to within 1 eV interval from the HOMO and LUMO, more electrons from the QD core participate, such that there is a mix of highly localized surface electrons and core-related electrons in the vicinity of the HOMO (Figure 8.10e), while in the LUMO (Figure 8.10f) the core-related electrons start to dominate.

The trends for -OH termination show that surface localization is greater for -OH, as can be seen for the HOMO (Figure 8.11a), due to surface C p-orbitals and oxygen lone-pairs.

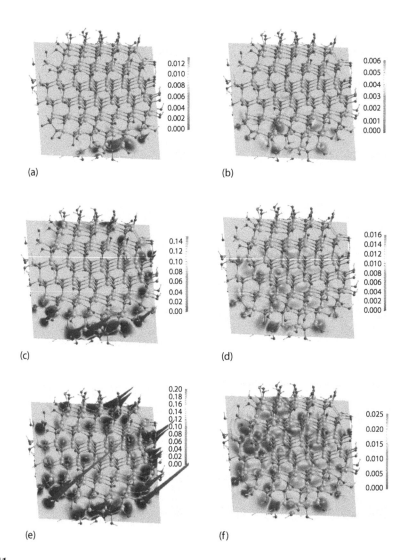

FIGURE 8.11
Two-dimensional slice through the center of an Si-centered 20 Å SiC-QD exhibiting -OH electron probability density for (a) HOMO, (b) LUMO, (c) within 0.5 eV of the HOMO, (d) within 0.5 eV of the LUMO, (e) within 1 eV of the HOMO and (f) within 1 eV of the LUMO. The localized states are represented in Figure 8.13b.

The LUMO (Figure 8.11b) is a mix of surface and core electrons. Toward higher energies of within 0.5 eV of the respective HOMO and LUMO, more electrons occupying the surface of the QD can be observed with still minimal contribution from the QD core in the vicinity of the HOMO (Figure 8.11c), while the electron probability density for energies near the LUMO involves electronic states within the core (Figure 8.11d). It is observed that going higher in energies, the localization of electrons at the surface is balanced out with electrons from within the core, as shown in (Figure 8.11e,f).

The examined electron probability density for -H and -OH terminations spanning the frontier orbital energies toward higher energies indicates that the electrons at the surface and within the QD core are well coupled throughout the investigated energy range for -H. Conversely, for -F and -OH terminations, these are decoupled: the core-related electrons predominantly occupy higher energies whereas surface-related electrons dominate the frontal electronic levels, particularly the HOMO, which causes narrowing of the optical gap. Having inspected the core and the surface electron probability density for 20 Å SiC-QD, further examination of the trends of the JDoS at high energy (core related) versus the JDoS at onset (surface related) for the diameter range 10–20 Å was undertaken. In the next section, the effect of the surface termination groups -H-, -OH and -F upon the optical properties of 4H-SiC-QDs is discussed.

8.2.2.5 Surface-State-Dependent Optical Properties of 4H-SiC-QDs

Figure 8.12a summarizes the energy gap of the QDs as a function of the size and terminating species. H-termination results in a large size dependence, whereas F- and OH-terminations are nearly size independent. For H-termination, the HOMO-LUMO gap increases following the quantum mechanical 'particle in a sphere' picture [16], whereby it is discrete and scales with the QD diameter, d, as d^{-2}. For F- and OH-terminations, additional surface states are present resulting in their energy being governed by the immediate bonding to the SiC surface, and not by the size of the QD. Figure 8.12b allows for a comparison of H-terminated clusters with different compositions and surface reconstructions, which asserts that reconstruction and choice of cluster center have no significant impact upon the energy gap in comparison to the effect of the termination. H-termination consistently exhibits stronger quantum confinement compared to -F and -OH regardless of surface reconstruction. This indicates that the surface termination groups have greater influence than size and surface reconstruction.

To understand the size dependence of SiC-QDs, a simple model, taking into account the different configurations of the surface states, is considered. The states associated with the SiC core exhibiting quantum confinement will result in a narrowing of the optical gap as the core diameter increases. This is represented schematically by the solid lines in Figure 8.13a,b. The second component arises from two surface effects. The first is the introduction of states due to homo-nuclear bonds arising from reconstruction. In the case of H-termination, this introduces an unoccupied state below that of the core, as represented by the higher dashed lines in Figure 8.13a. For a large enough cluster, this surface state is expected to lie outside the core gap, i.e. reconstructions impact the optical properties of relatively small SiC-QDs. However, as previously noted, the impact of the reconstruction is not significant in comparison to surface termination. For F- and OH-terminations (Figure 8.13b), the lone-pair orbitals mean that there are both occupied and empty surface states within the SiC core–related energy gap. Therefore, the predicted optical properties of F- and OH-terminated SiC-QDs can be divided into two regions. In region A, the HOMO and LUMO are associated with the surface states and the energy gap is approximately constant. In region B, the HOMO and LUMO originate from core states, so that quantum confinement effects become significant.

The dual-feature PL spectra of SiC-QDs observed experimentally are represented schematically in Figure 8.13c. In a QD ensemble containing a range of core sizes, short wavelengths excite all QDs [4,5,7], whereas longer wavelengths excite only large QDs. However, 10–60 Å SiC-QDs yield PL peaks that are largely insensitive to excitation wavelength up to around 320 nm. The dual feature of the PL spectra has been discussed in terms of quantum confinement effects only being responsible for the observed size dependence of emission in larger SiC-QDs [4,5,17], but a clear explanation regarding the size-independent features was not established. The observation can be explained if the emission (and excitation) is not purely a core property, but rather has a contribution from the surface states (Figure 8.13a,b). Red-shifts for excitation wavelengths above around 360 nm show

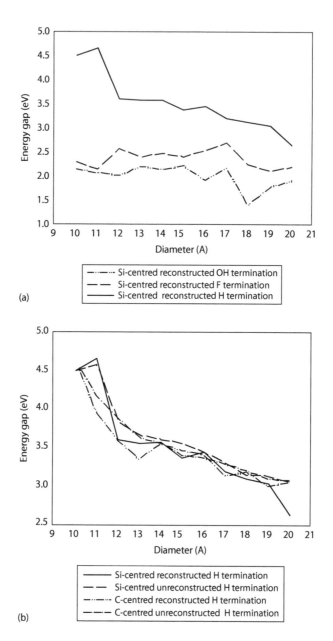

FIGURE 8.12

(a) Energy gap versus QD size with different surface termination groups and (b) energy gap of -H-terminated QD versus QD size by surface reconstruction and surface composition. Surface termination groups (OH, F, H) have a greater influence than size and surface reconstruction.

the onset of quantum confinement effects, as only those SiC-QDs large enough to be in region B in Figure 8.13 are excited.

Experimental findings show that the interplay between surface reconstructions and terminating species is critical below the threshold diameter of <30 Å. In the present context, the threshold size can be defined in a similar fashion, when the QD core–related JDoS start to dominate over those that are surface related.

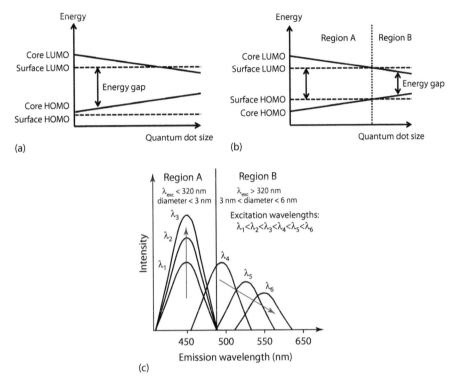

FIGURE 8.13
Illustration of core and surface states related to HOMO and LUMO as a function of size in surface reconstructed 4H-SiC-QD for (a) H-termination and (b) OH- or F-termination with the respective regions A and B. (c) A simplified dual-feature photoluminescence spectrum of SiC-QDs in regions A (when surface states dominate) and B (when core-related states dominate).

To illustrate the relative roles of the surface and core states as simulated in this study, the JDoS contours related to absorption onsets and at 10^3 eV^{-1} representative of the core (Figure 8.14) are plotted. Core atoms–related JDoS show a linear correlation with diameter^{-2} for both OH- and F-terminating groups, consistent with the particle in a sphere model [16,18]. However, the surface state–related JDoS do not increase. Based on the trend, it is predicted that the surface-related states will define the emission up to between 25 and 27 Å diameter for F- and OH-terminations (the intersections of the lines in Figure 8.14) [19]. For other surface groups, the intersection is expected to differ from these values, but the general feature of a dual-feature optical spectrum for systems involving main-group lone-pairs is expected.

8.3 Conclusions

In conclusion, 10–20 Å diameter SiC-QDs with OH-, F- and H-terminations have been investigated. In real experiments, the dual-feature PL was observed for all major polytypes comprising 3C-, 6H- and 4H-SiC. The SiC bulk bandgap increases with polytype in the order: 2.36 eV (525 nm), 3.00 eV (413 nm) and 3.23 eV (383 nm) for 3C, 6H and 4H,

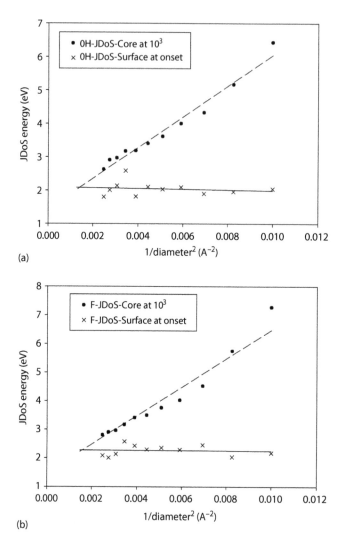

FIGURE 8.14
Change in energy of core-related and surface-related JDoS versus the inverse square of diameter for (a) OH-termination and (b) F-termination. Based on the trend, it is predicted that the surface-related states will define the emission up to between 25 and 27 Å diameter for F- and OH-terminations.

respectively. However, SiC-QDs derived from these polytypes exhibit a strikingly similar constant PL emission wavelength near 450 nm for sizes smaller than 3 nm, which is above bandgap for 3C but below bandgap for 6H and 4H-SiC. The role of polytypes was not observed in the optical properties of SiC-QDs particularly at smaller sizes. Beyond 3 nm, increasing the excitation wavelength results in red-shifting in the PL emission from 450 toward 560 nm, exhibiting PL that is characteristic of 3C-SiC regardless of the starting material. Due to these observations, polytypic phase transformation is frequently given as the explanation.

In this work, by examining the electronic structures of a wide range of compositions and surface treatments, it is concluded that the dual-feature PL spectra observed experimentally is best explained by considering the interplay of core and surface states, where surface states dominate at small scales (smaller than around 25 Å) and core states dominate at larger SiC-QDs.

It is found that the chemical nature of the surface termination groups is a critical factor in controlling the quantum confinement of 4H-SiC-QDs, and hence the optical properties. It is also noted that although surface reconstructions have an impact, they are generally much smaller than that arising from the choice of termination.

Finally, it is noted that SiC-QDs are fabricated using the wet etching methods employing HF, ethanol and water, which will naturally lead to a range of chemical terminations of the surfaces. It is therefore critical to understand the chemical nature of the surface termination, since control of the surface states will be essential in either exploiting or eliminating the size-dependent optical characteristics.

References

1. J. Fan, H. Li, J. Jiang, L. K. Y. So, Y. W. Lam and P. K. Chu, *Small*, 2008, 4, 1058–1062.
2. J. Y. Fan, X. L. Wu, H. X. Li, H. W. Liu, G. G. Siu and P. K. Chu, *Applied Physics Letters*, 2006, 88, 041909.
3. X. L. Wu, J. Y. Fan, T. Qiu, X. Yang, G. G. Siu and P. K. Chu, *Physical Review Letters*, 2005, 94, 026102.
4. X. Guo, D. Dai, B. Fan and J. Fan, *Applied Physics Letters*, 2014, 105, 193110.
5. J. Fan, H. Li, J. Wang and M. Xiao, *Applied Physics Letters*, 2012, 101, 131906.
6. X. L. Wu, S. J. Xiong, J. Zhu, J. Wang, J. C. Shen and P. K. Chu, *Nano Letters*, 2009, 9, 4053–4060.
7. D. Beke, Z. Szekrényes, I. Balogh, Z. Czigány, K. Kamarás and A. Gali, *Nanoscale*, 2015, 7, 10982–10988.
8. J. Zhu, S. Hu, W. W. Xia, T. H. Li, L. Fan and H. T. Chen, *Materials Letters*, 2014, 132, 210–213.
9. P. R. Briddon and R. Jones, *Physica Status Solidi (b)*, 2000, 217, 131–171.
10. M. J. Rayson and P. R. Briddon, *Physical Review B*, 2009, 80, 205104.
11. C. Hartwigsen, S. Goedecker and J. Hutter, *Physical Review B*, 1998, 58, 3641–3662.
12. J. P. Goss, M. J. Shaw and P. R. Briddon, in *Marker-Method Calculations for Electrical Levels Using Gaussian-Orbital Basis Sets*, ed. D. A. Drabold and S. K. Estreicher, 2007, vol. 104, pp. 69–93, Springer, Berlin.
13. D. R. Lide, *Handbook of Chemistry and Physics*, 2006, vol. 87, CRC Press, Boca Raton, FL.
14. M. Vörös, P. Deák, T. Frauenheim and A. Gali, *Applied Physics Letters*, 2010, 96, 051909.
15. X. H. Peng, S. K. Nayak, A. Alizadeh, K. K. Varanasi, N. Bhate, L. B. Rowland and S. K. Kumar, *Journal of Applied Physics*, 2007, 102, 024304.
16. G. Konstantatos and E. H. Sargent, *Colloidal Quantum Dot Optoelectronics and Photovoltaics*, 2013, Cambridge University Press, New York.
17. D. Dai, X. Guo and J. Fan, *Applied Physics Letters*, 2015, 106, 053115.
18. L. E. Brus, *The Journal of Chemical Physics*, 1983, 79, 5566–5571.
19. M. Rashid, A. K. Tiwari, J. P. Goss, M. J. Rayson, P. R. Briddon and A. B. Horsfall, *Physical Chemistry Chemical Physics*, 2016, 18, 21676–21685.

9

Molecular Beam Epitaxy of AlGaN/GaN High Electron Mobility Transistor Heterostructures for High Power and High-Frequency Applications

Yvon Cordier, Rémi Comyn, and Eric Frayssinet

CONTENTS

9.1 Introduction

Since it has been shown that an AlGaN/GaN heterostructure can generate a two-dimensional electron gas (2DEG) [1], GaN-based high electron mobility transistors (HEMTs) have been developed and are now established as the most interesting III-nitride electron devices for high-frequency power amplification as well as power switching. The reason for this is a combination of many factors [2]: the possibility of achieving high sheet carrier concentration in the 2DEG ($\sim 1 \times 10^{13}$/cm^2) with a high saturated velocity ($>1.5 \times 10^7$ cm/s), a quite high electron mobility (up to about 2000 cm^2/V.s at RT) and a breakdown electron field exceeding 3 MV/cm. Moreover, the chemical inertness and the wide energy bandgap of GaN guarantee the thermal stability of the devices.

The most commonly used growth techniques for III-nitride heterostructures are metal-organic vapor phase epitaxy (MOVPE) and molecular beam epitaxy (MBE), each of these techniques having its own advantages and drawbacks. While MOVPE is widely used due to its larger throughput and larger wafer size handling capability, MBE operates at lower temperatures under high vacuum with much less source products consumption and it is equipped with useful in situ inspection tools like reflection high energy electron diffraction (RHEED). MBE production tools for III-nitrides are rare, but many research reactors are used worldwide and are at the origin of demonstrations of optoelectronic and electronic devices. In the following sections, we will first discuss the advantages of using ammonia as the nitrogen source for MBE growth. This will be followed by a description of the growth of AlGaN/GaN HEMT heterostructures on GaN-on-sapphire templates and free-standing GaN, and on foreign substrates like silicon and silicon carbide. The behavior of transistor devices will be described and high-frequency power density results will be presented. For transistor applications on silicon, growth requires even more attention due to possible conductivity through the substrate. In this context, we will see the benefit of reducing the growth temperature of III-nitrides. Finally, we will describe a technological route for the monolithic integration of GaN with silicon electron devices in a complementary metal-oxide-semiconductor (CMOS) first approach.

9.2 Characteristics of Ammonia Source Molecular Beam Epitaxy

The main features of ammonia-MBE (NH_3–MBE) are the following. Ammonia thermally decomposes at the surface of the films at temperatures above 450°C [3]. The typical growth temperature for GaN is 800°C, while thick AlN films necessitate higher temperatures. Usually, AlGaN films can be grown in the 800°C–875°C temperature range, the optimum temperature depending on the Al content and the thickness of the film to be grown. Growth rates in the range of 0.5–1.5 µm/h are quite easy to achieve. However, compared to MOVPE (GaN grown typically at 1000°C), lateral growth is very limited for MBE and whenever a significant roughening develops, it is difficult to recover a flat surface, which is detrimental for electron transport in the channel of lateral devices such as HEMTs. The study described here has been carried out in a Riber Compact 21 MBE reactor equipped with an NH_3 gas injector [4,5] and a plasma source (ADDON RFN50/63) connected to an N_2 gas line [6,7]. The reactor configuration was optimized for uniform films on 2 in. diameter substrates [4]. Nevertheless, even with this configuration, the GaN thickness uniformity deviation is below 3% on 3 in. diameter substrates. As described by Vézian [8], when growth starts, ammonia-MBE-grown GaN exhibits a transition from a spiral growth mode to a mixed growth mode where two-dimensional nucleation is sufficiently active to give rise to kinetic roughening. According to this study, the spiral growth mode occurs at the beginning of the growth, thanks to the step flow growth mode in the presence of screw dislocations. As a result, a coarsening of growth mounds correlated with the decrease in the dislocation density is observed. Thick films (>2 µm) exhibit root mean square (RMS) roughness of the order of 4–5 nm (Figure 9.1a) and an increase in the correlation length saturating around 1 µm.

We discuss here the influence of the nitrogen source flow rate on the growth of GaN. The case of ammonia is quite simple. The optimum growth conditions are N-rich, i.e. they require a large ammonia flow rate (200 sccm in our MBE reactor). At the optimum growth temperature of 800°C and for a fixed Ga flux, the growth rate and the surface morphology slightly change while reducing the ammonia flow rate. Then, a first regime is reached with pits corresponding to the facet development of threading dislocations (TDs) opening at the surface. Moreover, a further reduction of the ammonia flow rate leads to a decrease in the growth rate as well as further development of the surface roughness due to the insufficient amount of available active nitrogen species compared to incoming Ga species [3,9]. The nitrogen plasma growth (PA-MBE) is very different. First, the most frequently reported growth conditions for smooth GaN films are temperatures near 720°C–730°C with fine-tuning of the nitrogen flow rate, radio frequency (RF) cell power and Ga flow rate in order to keep a thin metallic Ga film (2–3 monolayers) floating on the growing surface and a resulting surface roughness of typically 1 nm (Figure 9.1b). Out of this equilibrium, a rough film is obtained or an excess of gallium generates droplets at the surface [6]. In order to mitigate these effects, growth at a higher temperature (780°C–790°C) under a high nitrogen flow rate has been proposed [10], but the sensitivity to threading defects can make the growth more difficult to monitor. A crucial parameter for electron devices such as field-effect transistors is the residual doping level in the channel and the buffer resistivity. As seen in Figure 9.2, the ammonia flow rate has a crucial influence on the residual doping level [9]. Secondary ion mass spectroscopy (SIMS) indicates that silicon and oxygen are the main donors, even in the case of nitrogen plasma growth. Additionally, it seems that the flow rate itself has more influence than the nitrogen species, ammonia or nitrogen molecules. One last point concerns the growth of AlGaN alloys. The desorption of Ga and Al species is negligible when the HEMT AlGaN barrier is grown at 800°C with a large ammonia flow rate, so that composition and growth rate calibrations are facilitated.

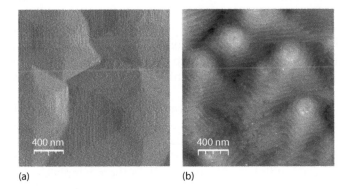

(a) (b)

FIGURE 9.1
Tapping mode AFM view of the surface of a GaN layer grown with ammonia (a) and with nitrogen plasma (b). Left picture is a derivative mode image in order to highlight the monolayer height steps present at the surface.

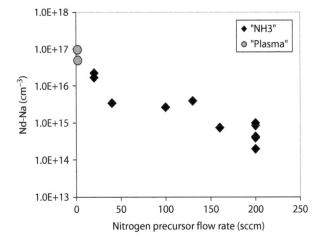

FIGURE 9.2
Donor concentration estimated by C-V as a function of the flow rate of the nitrogen precursor (NH_3 or nitrogen for plasma).

9.3 Homoepitaxy of GaN HEMTs

The availability of high resistivity GaN-on-sapphire templates and free-standing GaN substrates is very helpful for the development of HEMT structures on high crystal quality GaN. Unless significantly thicker GaN is regrown, the regrowth of GaN just replicates the substrate threading dislocation density (TDD) of roughly $1–5 \times 10^8/cm^2$ in the GaN-on-sapphire templates and $1–5 \times 10^7/cm^2$ in the free-standing substrates. Recently, bulk GaN substrates grown by the ammonothermal method with an ultralow dislocation density of $10^4 cm^{-2}$ and wafers up to 2 in. in diameter have been shown to be compatible with HEMT structures regrowth by MOVPE [11]. The growers have to face the problems related to the regrowth interface pollution. Even when the substrate/template is highly resistive/semi-insulating, the regrowth interface is contaminated with shallow donors such as silicon. The doping of GaN with elements such as carbon [12,13], beryllium [14], magnesium [15] and iron [16] is efficient to increase the resistivity of GaN. Therefore, the introduction of such elements in a regrown GaN layer is a solution to compensate the source of leakage in transistors. The incorporation rate can depend on the growth technique as well as the growth conditions. For instance, the incorporation of carbon from CBr_4 is highly dependent on temperature and not very efficient in the case of high-temperature MBE [17]. The ionization of methane is preferred for ammonia-MBE [13] since a high growth temperature is desirable.

However, such doping sources are not always available in growth reactors dedicated to HEMTs. Some authors proposed the growth of a thin AlN layer to upraise the conduction band at the regrown interface [18]. Another alternative is to develop GaN templates or substrates ready for epitaxial regrowth with a reduced electrical leakage. In our case, we have developed MOVPE GaN with iron doping [19,20]. In these templates, the amount of iron available at the regrowth interface is low enough to avoid unrecoverable surface roughening but sufficient to compensate the effect of silicon or oxygen contaminants. The transistors fabricated on such epi-ready insulating GaN-on-sapphire templates exhibit drain leakage currents as low as $10 \mu A/mm$ at $V_{ds} = 10 V$ when regrown by ammonia-MBE. As a high temperature favors the diffusion of iron, the leakage can be further reduced by about three orders of magnitude when regrown by MOVPE [20].

9.4 Heteroepitaxy of GaN HEMTs

The growth of GaN HEMTs on foreign substrates requires both insulating and stress-mitigating buffer layers. On sapphire, the difference in thermal expansion coefficients (TEC) induces a residual compressive strain in GaN, which is responsible for a noticeable convex bowing of the wafers. The large lattice parameter mismatch is responsible for a large number of dislocations, which helps in trapping the carriers related to the residual doping at the initial stages of growth. Further thickening of the GaN buffer drastically reduces the number of TDs. The growth of AlN is an alternative to increase the buffer layer resistivity because AlN is a wider bandgap material and the polarization electric field depletes the GaN/AlN regrowth interface from eventual free carriers. In the present study, we chose to grow with NH_3–MBE a HEMT structure on a 1 μm thick MOVPE AlN-on-sapphire template. Contrary to sapphire, GaN grown on SiC or on Si suffers a tensile strain induced by the TEC mismatch. To compensate this effect and the associated risk of layer cracking, GaN is grown on an AlN nucleation layer to benefit from an initial compressive strain related to the 2.5% lattice parameters mismatch strain between both materials. Furthermore, AlN appears to be a useful solution to avoid reactions between gallium and the silicon substrate [21]. But, if the high temperature (800°C) of the NH_3–MBE favors dislocation bending, interactions and elimination, it also promotes the relaxation of this strain [22] so that more complex structures with additional intercalated AlN layers have been grown to obtain 2 μm thick, crack-free HEMT structures on Si(111) [23] and SiC.

9.5 Electrical Properties

Figure 9.3 depicts the main kinds of structures we have grown. Table 9.1 summarizes the best results we obtained on these kinds of structures: the 2DEG carrier density and the low field electron mobility assessed by the Hall effect as well as the buffer residual donor density obtained by exploiting the capacitance-voltage (CV) measurements beyond pinch-off. The TDD assessed by atomic force microscopy (AFM) or x-ray diffraction (XRD) is

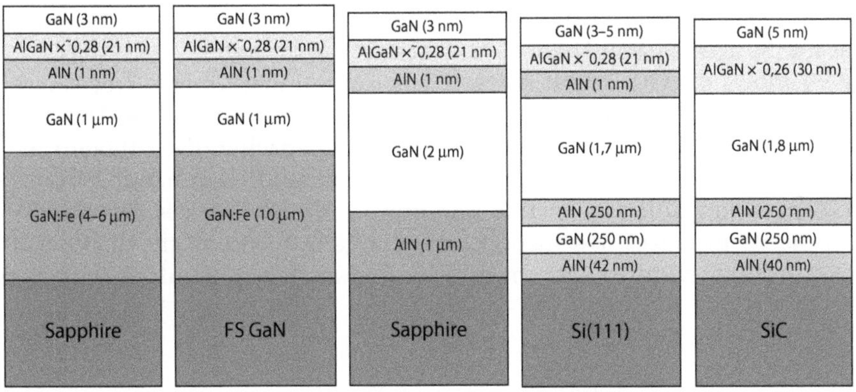

FIGURE 9.3
Schematic cross section of the typical HEMT structures grown by NH_3–MBE.

TABLE 9.1

2DEG Carrier Density and Mobility, Buffer Residual Donor Density and Threading Dislocation Density (TDD) in the Studied HEMTs

	GaN/Al$_2$O$_3$	GaN FS	AlN/Al$_2$O$_3$	H-SiC	Si(111)
Al content	28%	28%	28%	26%	28%
Ns @ 300 K (cm^{-2})	10^{13}	10^{13}	9×10^{12}	[a]8×10^{12}	9×10^{12}
μ @ 300 K (cm^2/Vs)	2,080	2,140	2,085	[a]1,769	2,000 ([a]1,780)
μ @ <10 K (cm^2/Vs)	30,000 ([a]12,500)	—	—	[a]8,740	12,700 ([a]7,880)
Nd-Na (cm^{-3})	3×10^{13}	3×10^{13}	$<1 \times 10^{13}$	3×10^{13}	$1-3 \times 10^{14}$
TDD (cm^{-2})	$0.4-1 \times 10^9$	$1-2 \times 10^7$	2.5×10^9	3×10^9	$3-4 \times 10^9$

[a] No AlN spacer between the AlGaN barrier and the GaN channel.

also reported. At room temperature, electron mobility is mainly limited by optical phonon scattering, interface roughness and alloy scattering. Due to the high carrier density, the screening of TD fields is quite efficient and reduces the influence of the TDD with respect to other scattering mechanisms. Furthermore, it is obvious that the insertion of a thin AlN spacer at the AlGaN/GaN interface enhances room temperature mobility by greater than 200 cm²/V.s. Mobility reaches 2000 cm²/V.s when the TDD is sufficiently low (below 5×10^9 cm⁻²). A consequence of the good crystal quality is that the carrier density in the 2DEG is mainly determined by the barrier thickness and the Al molar content as well as the GaN cap thickness. At low temperature, the TDD reduction has a noticeable impact on electron mobility enhancement.

9.6 Transistors Evaluation

In order to evaluate the potentialities of these structures, test devices including transmission line model (TLM) and isolation patterns, diodes and transistors have been fabricated by photolithography. The device process starts with mesa definition by a reactive ion etching (RIE) step in a $Cl_2/Ar/CH_4$ mixture. After a short RIE etching, TiAlNiAu stacks are deposited by e-beam evaporation. The ohmic contacts are achieved after rapid thermal annealing at 750°C for 30 s. NiAu films are then evaporated for the gate Schottky contact as well as for the access pads. These devices are not passivated.

Figure 9.4 shows the DC output characteristics of a transistor on a free-standing GaN substrate compared to one fabricated on a GaN-on-sapphire template. The $2.5 \times 150 \, \mu m^2$ gates are deposited in a nominal 11 μm source drain spacing. Even though the contact resistance and the sheet resistance are not very different (0.9 Ω.mm and 280 Ω/sq on sapphire, 0.7 Ω.mm and 270 Ω/sq on free-standing GaN), one notices that the I-V curves diverge while V_{gs} increases. We suspect here a large influence of the device self-heating on the electrical behavior. To confirm this, we first analyze the output characteristics of

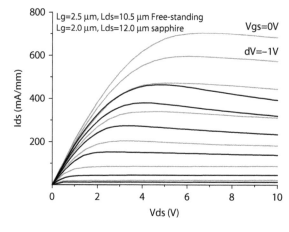

FIGURE 9.4
DC output characteristics of AlGaN/GaN HEMTs regrown on GaN/sapphire template and on a free-standing GaN substrate. Device dimensions are W = 150 μm, L_g = 2.0–2.5 μm and L_{sd} ~10.5–12.0 μm.

transistors fabricated with similar dimensions on the different kinds of substrates. We then define a relative DC drain current collapse R as

$$R = \left(Id_{knee} - Id_{20V}\right)/Id_{knee}$$

with Id_{knee} the maximum drain current and Id_{20V} the drain current obtained at $V_{ds} = 20\,V$ for a given gate bias V_{gs}. This is illustrated in the insert of Figure 9.5 on a transistor fabricated on Si(111).

When plotting the relative drain current collapse as a function of the drain current, one clearly sees the increase of the collapse with the drain current, but one also notices the similarity of the collapse for transistors on sapphire with AlN or GaN buffer. Furthermore, the transistors on SiC present the smaller collapse while an intermediate one is obtained on silicon and free-standing GaN. Such a collapse seems not related to the defect density reported in Table 9.1, but to the thermal conductivity of the substrate. However, one type of transistor developed on silicon presents a relative collapse as low as on SiC (dashed line in Figure 9.5). The peculiarity of this transistor is the unusual buffer layer thickness (0.5 μm GaN on a 0.5 μm AlN/GaN stress-mitigating stack).

To go further, pulsed measurements have been performed on transistors fabricated on sapphire and Si(111) [24]. Pulsed current-voltage measurements (600 ns, 100 Hz) performed on devices on sapphire and on silicon substrates confirm the primary importance of thermal effects (Figure 9.6). As expected, the pulsed drain currents obtained after biasing at $(V_{gs} = 0\,V, V_{ds} = 0\,V)$ quiescent point are systematically higher than when recorded under DC conditions. The difference increases dramatically on sapphire (Figure 9.6c). However, as shown in Figure 9.6b, it appears that the device on silicon with the thicker GaN buffer layer (here 1.7 μm GaN on a 0.5 μm AlN/GaN stress-mitigating stack) and therefore with a reduced TD density shows an increase of the pulsed drain current collapse with quiescent points at $(V_{gs} = -8\,V, V_{ds} = 0\,V)$ (gate lag) and $(V_{gs} = -8\,V, V_{ds} = 20\,V)$ (drain lag). Note that we

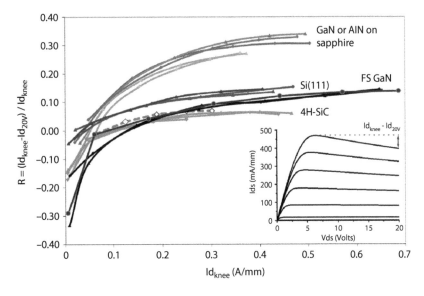

FIGURE 9.5
Relative drain current collapse measured in DC conditions. Device dimensions are $W = 150\,\mu m$, $L_g = 3–4\,\mu m$ and $L_{sd} = 9–12\,\mu m$.

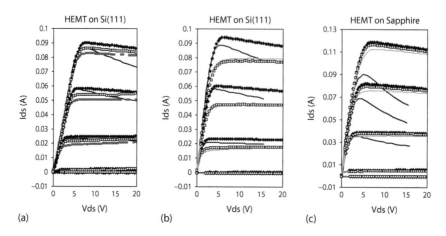

FIGURE 9.6
DC and pulsed measurements performed on HEMTs on silicon with thin buffer (a), thick buffer (b) and on sapphire substrate (c). On silicon, bias conditions are black line (DC, DC), diamonds (0 V, 0 V), squares (−8 V, 0 V), gray line (−8 V, 20 V). On sapphire, bias conditions are black line (DC, DC), diamonds (0 V, 0 V), squares (−4 V, 0 V), gray line (−4 V, 10 V).

previously observed a similar trend for the DC current collapse (Figure 9.5). Thus, it seems that in these bias conditions, the presence of dislocations helps in achieving a more stable device operation. However, GaN transistors generally operate under more severe bias conditions. The development of an efficient surface passivation drastically mitigates the current collapse. For instance, the combination of an N_2O plasma surface treatment with the deposition of SiN/SiO_2 passivation layers leads to small dispersions, as recently demonstrated on 0.25 μm gate transistors developed on similar structures with a resulting state-of-the-art $L_g \times Ft = 15$ GHz.μm product and a power density of 1.5 W/mm at 40 GHz [25].

A critical parameter for transistors and especially power transistors is the buffer resistivity and the breakdown voltage. Often, the buffer resistivity directly impacts the drain leakage current evidenced when the transistor is in the off-state. Sometimes, the gate leakage itself is the source of the leakage current in the off-state configuration. The buffer leakage current measured between ohmic contacts fabricated on HEMT active layers and separated by the mesa etching spacings superior to 10 μm are reported in Table 9.2. With such spacings, electric fields propagate down to the substrate so that all regions of the buffer layer can contribute. The drain leakage in transistors with gate lengths of 2–3 μm and source-to-drain spacings of 12–13 μm is also reported for a drain bias of 100 V and a gate bias V_{gs} of approximately 2 V beyond the threshold voltage. The transistor development W is 150 μm. It is clear from Table 9.2 that both the buffer leakage current and the drain leakage current are of the same order of magnitude, which confirms the primordial influence of the buffer resistivity. Nevertheless, a closer look shows that in some cases, the leakage

TABLE 9.2

Buffer and Transistor Leakage Currents Measured at $V_{ds} = 100$ V

Substrate	GaN/Al$_2$O$_3$	AlN/Al$_2$O$_3$	SiC	Si(111)	Si(111)
Buffer layer	GaN:Fe	GaN/AlN	GaN/AlN	GaN/AlN	AlGaN/AlN
Buffer leakage at 100 V	<10 μA/mm	<40 μA/mm	<200 μA/mm	20–100 μA/mm	20 μA/mm
Transistor W=150 μm leakage at 100 V	<70 μA	1 mA/mm @ 85 V	<30 μA	4–120 μA	<30 μA

currents of the transistors at pinch-off systematically overpass the buffer leakage currents. This is probably due to the increase in electric field crowding in the vicinity of the gate. The device on Fe-doped GaN-on-sapphire is very satisfying in terms of trade-off between crystal quality and drain leakage, contrary to the device on AlN-on-sapphire whose break-down voltage is below 100 V, as illustrated in Figure 9.7a.

The transistor leakage on hexagonal SiC better scales with buffer leakage, indicating a more stable behavior with respect to the electric fields and the self-heating. However, the buffer on SiC is not the most resistive, as shown in Figure 9.7a. This can be due to the possible diffusion of silicon from the substrate, as mentioned by Hoke [26] in the case of PA-MBE-grown AlN. The present buffer layer contains a thin AlN layer and a 0.5 μm thick GaN/AlN stack (Figure 9.3) and an increase by two orders of magnitude of the buffer leak-age current is obtained in the absence of the AlN interlayer. Thickening of the AlN layers

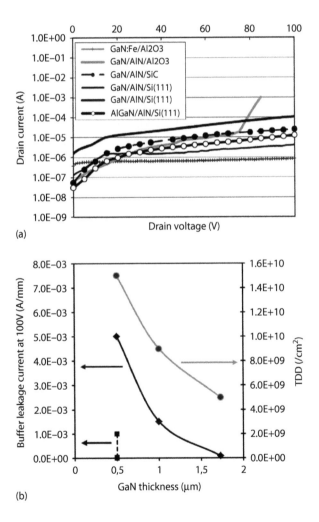

(a)

(b)

FIGURE 9.7
Transistors drain leakage current under pinch-off conditions (a). Buffer leakage current at 100 V and threading dislocation density in HEMT structures grown on Si(111) with different GaN buffer thicknesses (b). Circles represent TDD, diamonds represent the leakage for buffers on silicon with interlayers and squares represent the leakage for buffers on silicon without interlayers.

as well as reducing the growth temperature of these layers are possible ways to enhance the electrical resistance of buffer layers on SiC.

Our most studied devices are grown on Si(111). Figure 9.7a shows that depending on the growth conditions, the buffer resistivity can vary significantly. Moreover, the GaN thickness in structures like the one described in Figure 9.3 can influence the buffer resistivity in an expected way, increasing the leakage current while reducing the GaN thickness. As this occurs while the dislocation density increases, we suspect that the particular arrangement of dislocations (dislocation loops, bended dislocations, number of screw-type dislocations) in the GaN grown compressively strained on the AlN interlayer is responsible for this. Indeed, for 0.5 μm GaN thickness, the buffers grown without interlayers on Si(111) present much lower leakage currents (Figure 9.7b), but with clearly worse crystal quality (TDD > 2×10^{10}/cm²) and resulting electron mobility of the order of 1300 cm²/V.s. Showing the leakage current recorded at a given bias only is sometimes not enough to describe the behavior of a particular buffer layer. In [27], Leclaire et al. describe the effects of reducing the HEMT buffer layer thickness on the main current-voltage characteristics of transistors and Schottky diodes in order to evaluate the effect on the leakage both through the buffer and through the HEMT barrier layer. It is worth noting that what was deduced previously from the amplitude of the leakage current at 100 V was also true for the breakdown voltage defined at a leakage current of 1 mA/mm. Returning to thick buffer layers with interlayer stacks, a solution to stabilize the electrical resistivity can be the replacement of GaN in the buffer with a larger bandgap material such as AlGaN. However, the growth of such a layer on silicon is more difficult due to the smaller lattice parameter of AlGaN and the resulting stress. Nevertheless, we succeeded in growing up to 1.5 μm crack-free AlGaN with 5%–10% of Al and with a low crystal quality degradation [25,28]. This enabled more stable drain leakage currents below 30 μA at $V_{ds} = 100$ V in the studied devices.

These devices are not passivated so they are not able to sustain large drain and gate biases. Despite a drain leakage current that can be as low as a few microampere, the breakdown voltage obtained in air is around 200 V (234 V at best) with destruction of the metal contacts. The possible leakage current paths are presented in Figure 9.8. Electrons can be

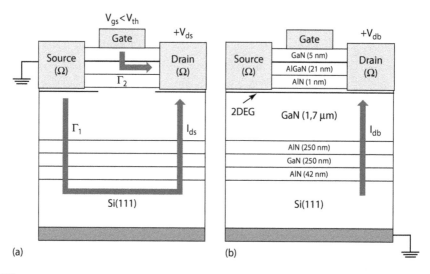

(a)   (b)

FIGURE 9.8
Leakage paths in the transistor on silicon under pinch-off conditions (a). Vertical leakage path in the device under vertical current configuration (b). (Adapted from Pérez-Tomás, A., et al., *J. Appl. Phys.*, 113, 174501, 2013.)

injected from the source to the substrate through the buffer layer and the nucleation layer, so that a current can flow via the buried part of the buffer or via the substrate (when not insulating) and then reach the positively biased drain contact. When the leakage arising from the gate (path Γ_2) is small enough, drain current leakage at pinch-off can follow path Γ_1, so that measuring the vertical leakage is helpful to identify the detailed mechanisms of the drain leakage.

The behavior of HEMT devices on Si, GaN-on-sapphire template and free-standing GaN substrates has been studied in this vertical configuration by Pérez-Tomás [29]. The drain is positively biased while the substrate is grounded. In this configuration, the leakage through the device on Si is a combination of a Poole–Frenkel (trap assisted) and a resistive conduction with an estimated resistance of 72 kΩ up to a soft breakdown at 420 V. We note here that the total buffer thickness is less than 2.3 μm, leading to an average breakdown field of more than 1.8 MV/cm. Due to the insulating sapphire, more than 350 V is necessary to notice a vertical leakage current with a resistive conduction (85 kΩ) up to 1 kV without any irreversible breakdown phenomena. This similar resistance is explained by the similar growth conditions of the thick GaN grown by MBE either on the silicon substrate or on the GaN-on-sapphire template. The difference in terms of TDD seems insufficient to have a noticeable effect on the vertical conduction. On the other hand, the device on the free-standing GaN substrate follows a resistive behavior only, with a resistance of 7 MΩ up to the destructive breakdown at 840 V. The absence of AlN layers in the buffer and the low TDD are efficiently compensated by the presence of iron in the 10 μm thick GaN layer. However, when the temperature is increased, the vertical leakage current on Si and free-standing GaN rapidly increases with an activation energy of 0.35 eV. In reverse bias conditions (when the drain is negatively biased), the leakage on the silicon substrate is larger with an activation energy of 0.1 eV and a resistance of 20 kΩ, due to the band lineup between the materials. A detailed analysis performed on silicon shows that for a broad range of drain bias, the gate-to-source leakage current is more than one order of magnitude lower than the leakage current between the drain and the source, which itself is the same as the vertical leakage current flowing from the substrate to the drain. This suggests that the preferential leakage path is Γ_1 and then any enhancement of the resistivity of the nucleation layer region will benefit the transistor power DC operation. Before we discuss this point, we come back to recent results we obtained after growing a HEMT with a buffer containing only AlGaN and AlN layers, the only GaN layers being part of the top active region composed of a 10 nm channel and a 0.5 nm cap, as shown in Figure 9.9. The devices are not passivated and $2 \times 150\,\mu m^2$ gate transistors show DC characteristics with a leakage current limited to 1 μA/mm. On the other hand, a monotonous increase of the vertical leakage current is noticed in these devices until the destructive breakdown at bias that can reach 740 V (Figure 9.10), which is particularly high for a structure with a total thickness below 2 μm and results in an average breakdown field of 3.8 MV/cm.

The present results confirm that MBE can provide materials able to sustain high electric fields even when grown on silicon, and show that small Al molar fractions in the buffer layer drastically enhance the breakdown voltage. Compared to GaN, this is probably due to the larger electrical resistivity of AlGaN linked to the increase in bandgap energy, but it may also be linked to the surface quality that presents a lower roughness as evidenced by AFM (generally with RMS below 2 nm vs 4–5 nm for GaN) and which may translate into less propensity to develop defects like hillocks and V-shaped defects, often cited as probable current leakage paths.

As evocated previously, any enhancement in the properties of the AlN nucleation layer on silicon should result in a benefit for the transistors. An interesting example concerns the

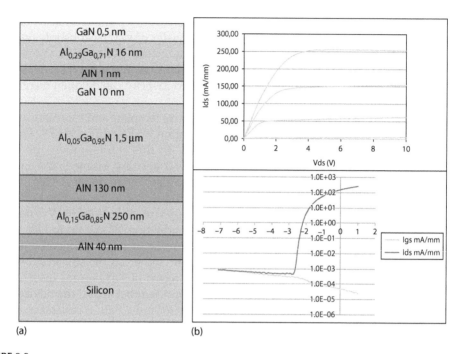

(a)   (b)

FIGURE 9.9

Schematic cross section of a HEMT structure grown on silicon without any GaN layer in the buffer layer (a); DC output (V_{gs}: 0 V; –4 V by –1 V steps) and transfer characteristics of a transistor (b).

effect of the AlN growth temperature. It is well known that compared to MOCVD, MBE is able to produce compact and smooth AlN nucleation layers at lower growth temperatures. As illustrated by Freedsman et al. [30], it seems more delicate to fabricate reliable, smooth, insulating AlN layers with MOCVD, especially at low temperature. To illustrate the benefits of MBE, the surface morphology of AlN nucleation layers grown on silicon is shown in Figure 9.11. All three samples have been grown well below 1000°C (920°C for

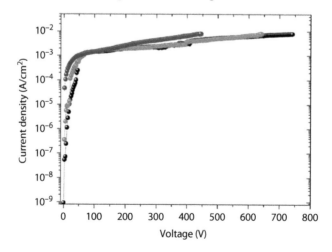

FIGURE 9.10

Leakage current recorded in the vertical configuration for a HEMT structure grown on silicon without any GaN layer in the buffer layer. (Courtesy of Dogmus E., et al., IEMN-CNRS.)

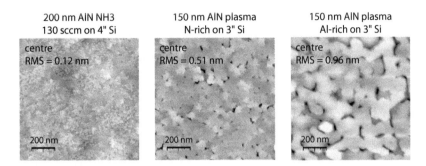

FIGURE 9.11

Tapping mode AFM view of the surface of AlN layers grown with ammonia-MBE and plasma-assisted MBE in the N-rich and Al-rich regimes.

ammonia-MBE and 830°C for N-rich and Al-rich plasma-assisted MBE) and present rather smooth surfaces with RMS roughness below 1 nm without any pits. The thickness is in the 100–200 nm range but similar morphology is obtained for films as thin as 10 nm. However, if temperatures higher than 900°C permit the slight enhancement of the crystal quality in terms of XRD line widths, other phenomena like impurities diffusion/incorporation can mitigate the enthusiasm to grow AlN at higher temperatures; on the contrary, it leads to considering the feasibility of growth at lower temperatures, as Comyn did [31]. The latter demonstrated that within HEMT buffer layer structures grown by ammonia-MBE, it was possible to grow AlN layers at 830°C with limited degradation of the crystal quality. Within a thin buffer HEMT structure (0.5 μm GaN/0.2 μm AlN), Comyn noticed a decrease of the vertical buffer leakage current by two orders of magnitude (Figure 9.12) and an increase of the lateral breakdown voltage while reducing the AlN growth temperature from 920°C to 830°C.

A SIMS analysis of the sample with GaN grown at 800°C and AlN grown at 920°C is shown in Figure 9.13a, while the same structure with AlN grown at 830°C is shown in Figure 9.13b. The only noticeable difference is the silicon profile, which drops more sharply within the AlN nucleation layer grown at low temperature to stabilize at a level about half that measured for AlN grown at 920°C. Despite the huge bandgap of AlN (6.1 eV), silicon

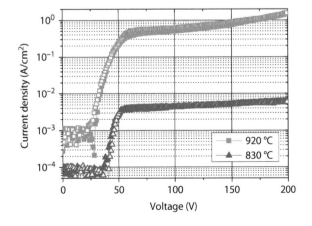

FIGURE 9.12

Leakage current recorded in the vertical configuration for the 500 nm GaN buffer layers with AlN grown at 830°C and 920°C.

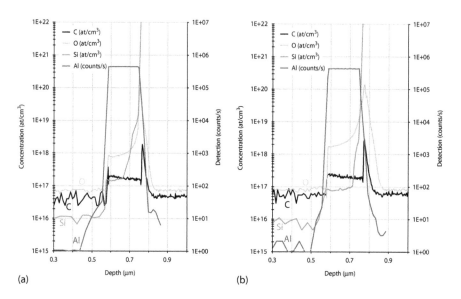

FIGURE 9.13
SIMS profiles recorded for carbon, oxygen, silicon and aluminum elements within buffer layer structures with AlN grown at (a) 920°C and (b) 830°C.

is still an active donor in this material [32], which may explain the difference in electrical resistivity. Another possibility is that AlN grown at a reduced temperature may contain a larger number of punctual defects able to trap any carrier.

9.7 RF Devices

Both the buffer and the substrate resistivity are crucial for RF performances. Even when it is possible to extract the losses in access regions of transistors (de-embedding) [33], performing functional devices and exploiting them within monolithic circuits require minimal losses especially at frequencies of several tens of gigahertz (GHz). For these reasons, substrates with high resistivity are used (typically superior to 3 kΩ.cm). However, the presence of Al atoms in the early stage of the growth and the high temperature usually employed to obtain a good crystal quality film can reduce the overall substrate resistivity. In order to limit the capacitive coupling of the device with the substrate, a thick buffer layer and maybe a lower temperature growth technique like MBE are preferable. For instance, Lecourt et al. measured RF losses of 0.4 dB/mm at 50 GHz on a 2.5 μm thick structure grown on resistive Si(111) (ρ > 1kΩ.cm) by ammonia-MBE [34]. More, it is interesting to note that the propagation losses measured on thin GaN buffer layers grown on AlN are not so sensitive to the growth temperature provided it is below 1000°C. As shown in Figure 9.14, propagation losses below 0.5 dB/mm are obtained up to 70 GHz in 0.5 μm GaN with 0.2 μm AlN grown by MBE at 830°C and 920°C on Si(111) (ρ > 3kΩ.cm). On the other hand, the effect of the growth temperature is more critical when GaN/AlN structures are regrown by MOCVD.

Figure 9.15 shows the state of the art in terms of power densities obtained at 40 GHz for devices with varying gate lengths. These results have been obtained with bias points

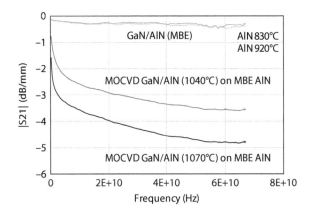

FIGURE 9.14
RF propagation losses in GaN/AlN buffer layers grown on silicon. (Courtesy of Defrance N., et al., IEMN-CNRS.)

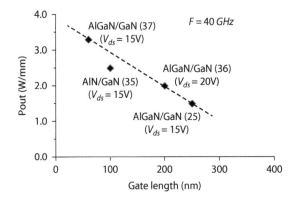

FIGURE 9.15
Gate length dependence of the state-of-the-art output power densities obtained at 40 GHz with Al(Ga)N/GaN HEMTs on silicon substrates.

$V_{ds} = 15$–20 V and the dependence of the performance is almost linear. The power density results of 1.5 W/mm with 250 nm gates [25] and 3.3 W/mm with 60 nm gates [37] have been obtained with AlGaN/GaN HEMTs grown by ammonia-MBE on silicon. The two intermediate results [35,36] have been obtained with MOVPE-grown structures. As expected, the achievable output power density enhances while lowering the operation frequency (Table 9.3). Results on GaN substrates are still quite rare but at least as good as on silicon. Devices on SiC substrates clearly benefit from a better thermal conductivity, which is crucial for high-frequency power devices. It is often difficult to find direct comparisons between devices made with the same technology on structures grown with different techniques. Palacios did this for HEMT structures grown by PA-MBE and MOVPE on SiC [45]. Ammonia-MBE and PA-MBE produce similar results that are not very far from those obtained using MOVPE. However, the long-term reliability of HEMTs is still an issue, in particular the degradation induced by the electrical stress. An interesting study has been reported by Puzyrev [46] on HEMTs grown either by PA-MBE under Ga-rich and N-rich conditions or by ammonia-MBE or MOVPE (both N-rich conditions). According to this study, hydrogenated Ga-vacancies loosing hydrogen atoms during the stress with hot electrons are responsible for a positive threshold voltage shift and a transconductance

TABLE 9.3

High-Frequency Power Density Results Reported on Silicon, GaN and Silicon Carbide Substrates with Different Growth Techniques

Substrate	Growth Technique	Gate Length (μm)	Frequency (GHz)	Drain Bias (V)	Output Power Density (W/mm)	References
Si	NH₃-MBE	0.25	40	15	1.5	[25]
Si	NH₃-MBE	0.06	40	15	3.3	[37]
Si	NH₃-MBE	0.25	18	35	5.1	[38]
Si	NH₃-MBE	0.3	10	40	7	[39]
GaN	MOVPE	0.15	10	25	4.8	[40]
				50	9.4	
GaN	PA-MBE	0.5	10	25	4.8	[41]
SiC	NH₃-MBE	0.6	10	30	6.3	[42]
				48	11.2	
SiC	PA-MBE	0.25	10	31	6.3	[43]
SiC	NH₃-MBE	0.2	30	40	6.5	[44]
SiC	PA-MBE	0.16	40	30	8.6	[45]
SiC	MOVPE	0.16	40	30	10.5	[45]

degradation in samples grown by PA-MBE (Ga-rich and N-rich). On the other hand, in the case of ammonia-MBE, the author proposes that the dehydrogenation of N-antisite defects is responsible for a negative threshold voltage shift and a reduced transconductance change due to the lower charge state of the defects. The degradation of MOCVD-grown devices is similar to that of ammonia-MBE-grown devices, showing that ammonia favors antisite-related defects themselves passivated by hydrogen.

9.8 Monolithic Integration with Si CMOS Technologies

Compared with MOVPE, another advantage of MBE is the possibility to grow structures at temperatures compatible with the monolithic integration of already processed devices like silicon MOSFETs. This is an alternative to approaches relying on the transfer of GaN epilayers [47] or even devices from the growth substrate to the silicon wafer [48]. In this CMOS first approach, the Si devices can be fabricated in a conventional silicon device process line with no risk associated with contamination of III-V elements. Metal connections for CMOS circuits, e.g. Al, are not present in this step to avoid fusion and are fabricated later. The silicon devices are protected with dielectric masks, and windows are opened to define GaN growth areas.

Questions arise on the choice of the silicon substrate crystal orientation, which has to be compatible with both silicon CMOS device fabrication and good quality III-nitrides. A first possibility is a special silicon-on-insulator (SOI) substrate with an Si(100) top layer for CMOS and a thick Si(111) bulk substrate for IIII-nitride growth [49]. Another possibility is the use of an Si(100) substrate with a perfectly controlled off-cut in order to grow IIII-nitrides with a single in-plane crystal orientation via the formation of only one kind of surface steps (double steps). Even if it is permitted to fabricate high RF performance GaN HEMTs on Si(100) [50], it seems difficult to avoid a high-temperature surface preparation to stabilize the double steps, which can compromise the properties of the silicon devices.

Recently, progress has been made in reducing the thermal budget of preparation for the growth of GaAs on Si(100) [51], but the efficiency of such approaches remains to be demonstrated for GaN on silicon. The growth thermal budget can have a significant impact on the diffusion of dopants in silicon so trade-offs have to be made for monolithic integration. The trade-off we have chosen is the fabrication of both CMOS and GaN HEMTs on Si(111) or on Si(110). The latter allows the growth of III-nitrides with high crystal quality [52] and device performance [50], at least as good as on Si(111), the price to pay being an increased density of interface states at the SiO_2/Si interface of the MOS devices.

The impact of the growth of AlN (at 920°C) and GaN (at 800°C) on the dopants diffusion in silicon is shown in Figure 9.16a. It is clear from the figure that the growth of AlN for the nucleation layer and the stress-mitigating stack is the main contributor to the thermal budget of diffusion. According to simulations, dopants like phosphorus, boron or arsenic should not suffer a huge diffusion if the thermal budget associated with the growth is well below 900°C for several hours, meaning that AlN present in nucleation and stress-mitigating layers has to be grown at a reduced temperature [49,53]. Plasma-assisted MBE usually satisfies such criteria with AlN grown below 830°C and GaN grown at 700°C–730°C. As shown in previous sections, ammonia-MBE can grow AlN with acceptable crystal quality at temperatures as low as 830°C while GaN and AlGaN layers are grown at 800°C.

Another critical aspect of the integration is the mechanical stability of the mask used to perform the local area growth. Indeed, contrary to GaN, the growth of AlN on dielectric masks like SiO_2 or Si_3N_4 is not selective, and consequently the following GaN-based layers stick to these masks. The large tensile thermal mismatch strain between GaN or AlN and silicon rapidly generates cracks and delamination once a critical thickness is reached (Figure 9.16b). To overcome this problem, we have developed a mask covered with a very rough quasi-porous GaN layer on SiO_2 [53]. After the growth of HEMT structures, this sacrificial layer provides easy paths to delamination, preserving the integrity of the underlying SiO_2/Si layers. The delamination is mostly spontaneous on the sample when cooling down and it is totally achieved after cleaning in solvents and buffered hydrofluoric (HF) solution with ultrasonic agitation (Figure 9.16c).

The crystal quality assessed from XRD rocking curves is unchanged compared to planar growth. Furthermore, the chemical stability of the masks used for selective area growth has been confirmed by SIMS measurements showing levels of silicon, oxygen and carbon contaminations that compare well with planar growth. Transistors fabricated on such GaN HEMT structures present low leakage currents and normal output DC characteristics, as

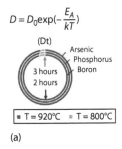
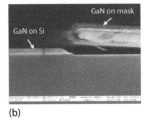
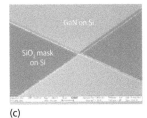

(a)   (b)   (c)

FIGURE 9.16
(a) Growth process impact for thick GaN HEMT on dopant diffusion according to the thermal budgets (D × time); (b) cross-section view of a delaminated part of the dielectric mask on silicon; (c) crack-free local area growth process after polycrystalline GaN removal.

shown in Figure 9.17. This confirms the quality of the local area low-temperature growth process developed for integration.

As mentioned earlier, the necessary trade-off for this monolithic integration involves the degradation of dopants profiles (Figure 9.18) and consequently the threshold voltage and the drain current in silicon CMOS devices. To evaluate the impact of the GaN HEMT growth on these parameters, silicon NMOS have been processed on p-doped Si(110) and Si(100) substrates with a resistivity of $1–10\,\Omega.cm$. A polysilicon gate was defined on a 30 nm thick SiO_2 thermal oxide. The transistor access regions were implanted with phosphorus while the threshold voltage was adjusted with a boron implantation, followed by an activation step in an N_2 ambient for 1 h at 950°C. For the test, the device was covered by a plasma-enhanced chemical vapor deposition (PECVD) SiO_2 film of thickness greater than 200 nm. The wafers were then cut in two pieces; the first halves were annealed in the MBE chamber under NH_3 flow to simulate the different growth process steps while the others halves were used as references. Finally, electrical contacts were fabricated on all samples, consisting of Al pads deposited after opening vias into the SiO_2 layers. Figure 9.18c compares the output characteristics of a 3 µm gate transistor fabricated on a reference Si(100) wafer with a transistor from the same wafer after NH_3 annealing corresponding to the growth of a thick GaN HEMT structure with low-temperature-grown AlN. It is clear that the output current capability is preserved as well as the normally off behavior, but with a change of threshold voltage and source drain resistance.

One proposed mechanism to explain such changes is the oxynitridation-enhanced diffusion (ONED) of dopants due to NH_3. The changes in transistor characteristics could be due to a reduction of the effective gate length (L_{eff}) as described in Figure 9.18b, accompanied by a positive shift of the threshold voltage due to a large increase in the subthreshold slope (SS). However, the latter is significantly reduced by annealing in N_2 at 400°C, which is also necessary in a standard MOS process in order to obtain low SS. Furthermore, the effective gate length variations (Figure 9.19a) determined using the method described in [54] are very large compared to diffusion length simulations. The noticeably larger threshold voltage shift (Figure 9.19b) and SS (Figure 9.19c) obtained on Si(110) compared to Si(100) (Figure 9.19c) leads to the assumption that ONED mainly affects the SiO_2/Si interface properties. In addition, since no significant diffusion length variation occurs when growing thicker GaN buffers (at 800°C), we assume that ONED is activated in the 830°C–850°C temperature range (Figure 19a). According to [55,56], ONED is favored by the presence of

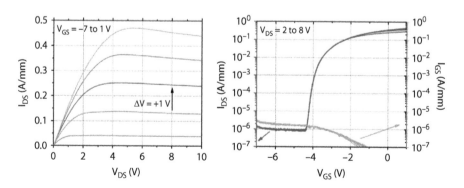

FIGURE 9.17
DC output characteristics and transfer characteristics at $V_{ds} = 2–8\,V$ of a transistor fabricated on an AlGaN/GaN HEMT structure after local area growth on silicon. The gate is 3 µm long and centered in a 13 µm source drain spacing.

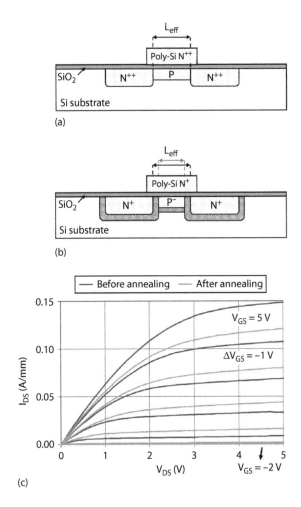

FIGURE 9.18
Cross section of the NMOS before (a) and after (b) annealing; (c) evolution of silicon NMOS output characteristics ($L_g = 3\,\mu m$) after exposure to the process corresponding to the thick structure with low-temperature AlN grown at 850°C.

hydrogen in NH_3; therefore, an alternative mask, e.g. SiN, has to be deposited on SiO_2 in order to effectively inhibit the diffusion of NH_x species, thereby preventing such a drift of MOS device characteristics.

9.9 Conclusion

MBE is able to produce device-quality materials for field-effect transistors such as HEMTs. Compared with ammonia-MBE and MOVPE, the low growth rate of PA-MBE (0.3–0.5 μm/h) has been considered for a long time as a drawback for production. A first attempt to solve this has been to use several nitrogen plasma cells simultaneously allowing to reach more than 0.8 μm/h growth rate [57]. More recently, new developments on plasma cells have enabled to reach more than 2 μm/h growth rate for GaN with unchanged surface

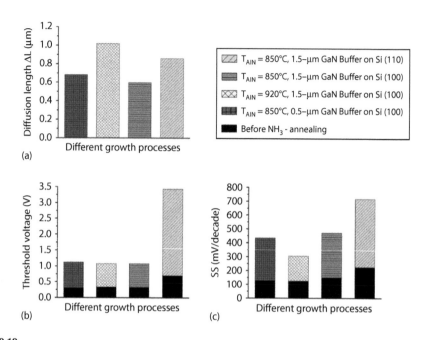

FIGURE 9.19

(a) Extracted effective gate length in NMOS transistors on Si(100) and Si(110) with and without annealing in NH$_3$ to simulate the HEMT growth processes. (b) Threshold voltage of the same. (c) Sub-threshold slope (SS).

morphology and optical properties [58–60]. However, uniformity has not yet been demonstrated. With a growth temperature intermediate between plasma-assisted MBE and MOVPE, ammonia-MBE presents the advantage of combining a simple N-rich growth regime with an acceptable growth rate and uniformity [4,9]. Moreover, it seems that ammonia-MBE offers the possibility to obtain a better purity material with a lower residual donor concentration, making the incorporation of additional compensation species not always necessary. Compared with MOVPE, one advantage of MBE is the possibility to develop structures at temperatures compatible with the integration of already processed devices, e.g. silicon MOSFETs. Furthermore, the possibility to grow smooth and good crystal quality AlN nucleation layers on silicon at noticeably lower temperatures is an advantage for producing high resistivity AlN/Si interfaces as well as buffer layers with high breakdown voltages and low RF propagation losses. All these points explain why MBE, and in our case why ammonia-source MBE, is so useful to develop electron device structures such as HEMTs.

Acknowledgments

The authors would like to thank all their colleagues in France, Spain, Belgium and Canada who contributed to the results presented here and who will recognize themselves in the references. They are also grateful to RIBER SA and to the agencies that supported a part of these studies. Among these agencies, the French Délégation Générale de l'Armement and the National Research Agency are deeply acknowledged.

References

1. M. Asif Khan, J.N. Kuznia, J.M. Van Hove, N. Pan, and J. Carter, *Appl. Phys. Lett.*, 60, 3027 (1992).
2. B. Gill (Ed.), *III-Nitride Semiconductors and their Modern Devices*, Oxford Science Publications, Oxford University Press, Oxford (2013).
3. M. Mesrine, N. Grandjean, and J. Massies, *Appl. Phys. Lett.*, 72, 350 (1998).
4. Y. Cordier, F. Pruvost, F. Semond, J. Massies, M. Leroux, P. Lorenzini, and C. Chaix, *Phys. Stat. Sol C*, 3, 2325 (2006).
5. Y. Cordier, F. Semond, J. Massies, M. Leroux, P. Lorenzini, and C. Chaix, *J. Crystal Growth* 301–302, 434 (2007).
6. F. Natali, Y. Cordier, C. Chaix, and P. Bouchaib, *J. Crystal Growth*, 311, 2029 (2009).
7. F. Natali, Y. Cordier, J. Massies, S. Vezian, B. Damilano, and M. Leroux, *Phy. Rev. B*, 79, 035328 (2009).
8. S. Vézian, F. Natali, F. Semond, and J. Massies, *Phys. Rev. B*, 69, 125329 (2004).
9. Y. Cordier, F. Natali, M. Chmielowska, M. Leroux, C. Chaix, and P. Bouchaib, *Phys. Status Solidi C*, 9 (3–4), 523 (2012).
10. G. Koblmüller, F. Reurings, F. Tuomisto, and J.S. Speck, *Appl. Phys. Lett.*, 97, 191915 (2010).
11. P. Kruszewski, P. Prystawko, I. Kasalynas, A. Nowakowska-Siwinska, M. Krysko, J. Plesiewicz, J. Smalc-Koziorowska, R. Dwilinski, M. Zajac, R. Kucharski, et al., *Semicond. Sci. Technol.*, 29, 075004 (2014).
12. C. Poblenz, P. Waltereit, S. Rajan, S. Heikman, U.K. Mishra, and J.S. Speck, *J. Vac. Sci. Technol. B*, 22 (3), 1145 (2004).
13. J.B. Webb, H. Tang, S. Rolfe, and J.A. Bardwell, *Appl. Phys. Lett.*, 75, 953 (1999).
14. D.F. Storm, D.S. Katzer, J.A. Mittereder, S.C. Binari, B.V. Shanabrook, X. Xu, D.S. McVey, R.P. Vaudo, and G.R. Brandes, *J. Crystal Growth*, 281, 32 (2005).
15. T.M. Kuan, S.J. Chang, Y.K. Su, J.C. Lin, S.C. Wei, C.K. Wang, C.I. Huang, W.H. Lan, J.A. Bardwell, H. Tang, et al., *J. Crystal Growth*, 272, 300 (2004).
16. A. Corrion, F. Wu, T. Mates, C.S. Gallinat, C. Poblenz, and J.S. Speck, *J. Crystal Growth*, 289, 587 (2006).
17. S.W. Kaun, M.H. Wong, U.K. Mishra, and J. Speck, *Semicond. Sci. Technol.*, 28, 074001 (2013).
18. Y. Cao, T. Zimmermann, H. Xing, and D. Jena, *Appl. Phys. Lett.*, 96, 042102 (2010).
19. Y. Cordier, M. Azize, N. Baron, S. Chenot, O. Tottereau, and J. Massies, *J. Crystal Growth*, 309, 1 (2007).
20. Y. Cordier, M. Azize, N. Baron, Z. Bougrioua, S. Chenot, O. Tottereau, J. Massies, and P. Gibart, *J. Crystal Growth*, 310, 948 (2008).
21. A. Watanabe, T. Takeuchi, K. Hirosawa, H. Amano, K. Hiramatsu, and I. Akasaki, *J. Crystal Growth*, 128, 391 (1993).
22. Y. Cordier, N. Baron, S. Chenot, P. Vennéguès, O. Tottereau, M. Leroux, F. Semond, and J. Massies, *J. Crystal Growth*, 311, 2002 (2009).
23. N. Baron, Y. Cordier, S. Chenot, P. Vennéguès, O. Tottereau, M. Leroux, F. Semond, and J. Massies, *J. Appl. Phys.*, 105, 033701 (2009).
24. Y. Cordier, N. Baron, F. Semond, M. Ramdani, M. Chmielowska, E. Frayssinet, S. Chenot, H. Tang, C. Storey and J.A. Bardwell, *Proceedings of the 35th Workshop on Compound Semiconductor Devices and Integrated Circuits*, Catania (Italy), May 29–June 1, pp. 89–90 (2011).
25. S. Rennesson, F. Lecourt, N. Defrance, M. Chmielowska, S. Chenot, M. Lesecq, V. Hoel, E. Okada, Y. Cordier, and J.-C. De Jaeger, *IEEE Trans. Electron. Dev.*, 60, 3105 (2013).
26. W.E. Hoke, A. Torabi, J.J. Mosca, R.B. Hallock, and T.D. Kennedy, *J. Appl. Phys.*, 98, 084510 (2005).
27. P. Leclaire, S. Chenot, L. Buchaillot, Y. Cordier, D. Theron, and M. Faucher, *Semicond. Sci. Technol.*, 29, 115018 (2014).
28. Y. Cordier, F. Semond, M. Hugues, F. Natali, P. Lorenzini, H. Haas, S. Chenot, M. Laügt, O. Tottereau, and P. Vennegues, *J. Crystal Growth*, 278/1-4, 393 (2005).

29. A. Pérez-Tomás, A. Fontserè, J. Llobet, M. Placidi, S. Rennesson, N. Baron, S. Chenot, J.C. Moreno, and Y. Cordier, *J. Appl. Phys.*, 113, 174501 (2013).

30. J.J. Freedsman, A. Watanabe, Y. Yamaoka, T. Kubo, and T. Egawa, *Status Solidi A*, 213, 424 (2016).

31. R. Comyn, Y. Cordier, V. Aimez, and H. Maher, *Phys. Status Solidi A*, 212, 1145 (2015).

32. S. Contreras, L. Konczewicz, J. Ben Messaoud, H. Peyre, M. Al Khalfioui, S. Matta, M. Leroux, B. Damilano, and J. Brault, *Superlattices Microstruct.*, 98, 253 (2016).

33. E.M. Chumbes, A.T. Schremek, J.A. Smart, Y. Wang, N.C. MacDonald, D. Hogue, J.J. Komiak, S.J. Lichwalla, R.E. Leoni and J.R. Shealy, *IEEE Trans. Electron. Dev.*, 48, 420426 (2001).

34. F. Lecourt, Y. Douvry, N. Defrance, V. Hoel, Y. Cordier, and J.C. De Jaeger, *Proceedings of the 5th European Microwave Integrated Circuits Conference (EuMIC)*, Paris, September 27–28, pp. 33–36 (2010).

35. F. Medjdoub, M. Zegaoui, B. Grimbert, D. Ducatteau, N. Rolland, and P. Rolland, *IEEE Electron. Device Lett.*, 33, 1168 (2012).

36. D. Marti, S. Tirelli, A. Alt, J. Roberts, and C. Bolognesi, *IEEE Electron. Device Lett.*, 33, 1372 (2012).

37. A. Soltani, J.-C. Gerbedoen, Y. Cordier, D. Ducatteau, M. Rousseau, M. Chmielowska, M. Ramdani, and J.-C. De Jaeger, *IEEE Electron. Device Lett.*, 34, 490 (2013).

38. D. Ducatteau, A. Minko, V. Hoel, E. Morvan, E. Delos, B. Grimbert, H. Lahreche, P. Bove, C. Gaquiere, J.C. De Jaeger, et al., *IEEE Electron. Device Lett.*, 27, 7 (2006).

39. D.C. Dumka, C. Lee, H.Q. Tserng, P. Saunier, and R. Kumar, *Electron. Lett.*, 40 (16), 1023 (2004).

40. K.K. Chu, P.C. Chao, M.T. Pizzella, R. Actis, D.E. Meharry, K.B. Nichols, R.P. Vaudo, X. Xu, J.S. Flynn, J. Dion, et al., *IEEE Electron. Device Lett.*, 25 (9), 596 (2004).

41. D.F. Storm, J.A. Roussos, D.S. Katzer, J.A. Mittereder, R. Bass, S.C. Binari, D. Hanser, E.A. Preble, and K. Evans, *Electron. Lett.*, 42 (11), 663 (2006).

42. C. Poblenz, A.L. Corrion, F. Recht, S.S. Chang, R. Chu, L. Shen, J.S. Speck, and U.K. Mishra, *IEEE Electron. Device Lett.*, 28 (11), 945 (2007).

43. N.X. Nguyen, M. Micovic, W.-S. Wong, P. Hashimoto, L.-M. McCray, P. Janke, and C. Nguyen, *Electron. Lett.*, 36 (5), 468 (2000).

44. Y. Pei, C. Poblenz, A.L. Corrion, R. Chu, L. Shen, J.S. Speck, and U.K. Mishra, *Electron. Lett.*, 44 (9), 598 (2008).

45. T. Palacios, A. Chakraborty, S. Rajan, C. Poblenz, S. Keller, S.P. DenBaars, J.S. Speck and U.K. Mishra, *IEEE Electron. Device Lett.*, 26 (11), 781 (2005).

46. Y.S. Puzyrev, T. Roy, M. Beck, B.R. Tuttle, R.D. Schrimpf, D.M. Fleetwood, and S.T. Pantelides, *J. Appl. Physics*, 109, 034501 (2011).

47. J.W. Chung J.-K. Lee, E.L. Piner, and T. Palacios, *IEEE Electron. Device Lett.*, 30, 1015 (2009).

48. Y.-C. Wu, M. Watanabe, and T. LaRocca, *Proceedings of the Radio Frequency Integrated Circuits Symposium (RFIC)*, Phoenix, AZ, May 17–19 (2015).

49. W.E. Hoke, R.V. Chelakara, J.P. Bettencourt, T.E. Kazior, J.R. Laroche, T.D. Kennedy, J.J. Mosca, A. Torabi, A.J. Kerr, H.-S. Lee, et al., *Sci. Technol. B*, 30, 02B101 (2012).

50. A. Soltani, Y. Cordier, J.-C. Gerbedoen, S. Joblot, E. Okada, M. Chmielowska, M.R. Ramdani and J.-C. De Jaeger, *Semicond. Sci. Technol.*, 28, 094003 (2013).

51. R. Alcotte, M. Martin, J. Moeyaert, R. Cipro, S. David, F. Bassani, F. Ducroquet, Y. Bogumilowicz, E. Sanchez, Z. Ye, et al., *APL Mater.*, 4, 046101 (2016).

52. Y. Cordier, J.-C. Moreno, N. Baron, E. Frayssinet, J.-M. Chauveau, M. Nemoz, S. Chenot, B. Damilano, and F. Semond, *J. Crystal Growth*, 312, 2683 (2010).

53. R. Comyn, Y. Cordier, S. Chenot, A. Jaouad, H. Maher, and V. Aimez, *Phys. Status Solidi A*, 213, 917 (2016).

54. K. Terada and H. Muta, *Jpn. J. Appl. Phys.*, 18, 953 (1979).

55. B. Balland and A. Glachant, Silica, silicon nitride and oxynitride thin films: An overview of fabrication techniques, properties and applications. In *Instabilities in Silicon Devices*, G. Barbottin and A. Vapaille (Eds), North-Holland, Amsterdam, pp. 3–144 (1999).

56. P. Fahey, R. Dutton, and M. Moslehi, *Appl. Phys. Lett.*, 43, 683 (1983).

57. R. Aidam, E. Diwo, N. Rollbühler, L. Kirste, and F. Benkhelifa, *J. Appl. Phys.*, 111, 114516 (2012).
58. B.M. McSkimming, F. Wu, T. Huault, C. Chaix, and J.S. Speck, *J. Crystal Growth*, 386, 168 (2014).
59. Y. Kawai, S. Chen, Y. Honda, M. Yamaguchi, H. Amano, H. Kondo, M. Hiramatsu, H. Kano, K. Yamakawa, and S. Den, *Phys. Status Solidi C*, 8, 2089 (2011).
60. Y. Cordier, B. Damilano, P. Aing, C. Chaix, F. Linez, F. Tuomisto, P. Vennéguès, E. Frayssinet, D. Lefebvre, M. Portail, et al., *J. Crystal Growth*, 433, 165 (2016).

10

Silicon Carbide Oscillators for Extreme Environments

Daniel R. Brennan, Hua-Khee Chan, Nicholas G. Wright, and Alton B. Horsfall

CONTENTS

10.1 Introduction

Silicon carbide (SiC)–based electronics have the advantage of being highly radiation resistant and capable of operating at temperatures not possible with conventional, silicon-based technologies [1]. This allows for their utilization as sensor circuits for applications located in hazardous environments, which are often inaccessible or dangerous to personnel. For this reason, it is desirable for the circuit to include some form of wireless communication capability that is realized with SiC-based components. In electronics, there is a standard requirement for repetitive waveforms to perform multiple functions, including the carrier waves required for communications systems. In order to meet this requirement, oscillator circuits are required that, depending on their complexity, can produce any waveform including high-frequency radio waves. These high-frequency radio signals form the basis of wireless communication systems. The challenge in realizing these high-frequency oscillators for deployment in extreme environments is the lack of process maturity in SiC and the shift in performance with temperature. This chapter provides an overview of a range of oscillator circuit topologies that are compatible with analog circuit designs and hence existing SiC technology, before focusing on the realization of a high-temperature Colpitts oscillator, which is used to demonstrate a high-temperature prototype communication circuit. The drift in the central frequency with the temperature of SiC-based oscillators is reported and described in terms of the parametric characteristics of the devices. The circuit is then expanded to enable the first SiC Colpitts oscillator–based voltage-controlled oscillator (VCO), which is presented as a method of enabling the closed loop control of the frequency of an oscillator over a wide temperature range. The chapter concludes with demonstrations of both the amplitude and frequency modulation (FM) of carrier waves from a circuit operating at 300°C, which are subsequently discussed with a view to creating a high-temperature wireless communication system.

10.2 Silicon Carbide Technology Overview

The development of SiC technology has become increasingly rapid in recent years. With significant improvements in wafer growth technology and materials processing, SiC devices have become part of the mainstream in power electronic applications [2,3] and a number of conferences are running sessions dedicated to the deployment of this technology in mainstream applications. Further advantages in terms of small, signal-level devices [4] and sensors [5] have resulted in significant improvements in capabilities; however, these are still only available at research level, where a number of groups are active. The underlying rationale for the continued investment in SiC technology is the excellent material properties, which derive from the high chemical stability of the Si–C bond. SiC is the only technologically relevant semiconductor that has a stable thermal oxide (SiO_2) [6] and can exist in a large number of polytypes – different crystal structures built from the same Si-C sub-unit organized into different stacking sequences. There are over 100 of these polytypes known; however, the vast majority of the research has focused on three: 3C, 6H and 4H. Of these, the 4H polytype is the most common for electronic devices, and commercially available power electronic devices are manufactured almost exclusively from this polytype due to its overall superior material properties. The bandgap of 4H SiC is 3.23 eV at room temperature (compared to 1.12 eV for silicon), which dramatically reduces the intrinsic carrier concentration in comparison to semiconductors such as silicon or gallium arsenide, allowing devices to theoretically operate at temperatures up to 1000°C [7]. SiC also has a high saturation electron velocity, $2 \times 10^7 \, \mathrm{cm \, s^{-1}}$, a thermal conductivity in excess of copper at room temperature and a critical electric field that is almost an order of magnitude higher than that of silicon. These parameters have been exploited in the high-performance power metal-oxide-semiconductor field-effect transistors (MOSFETs) and diodes that are commercially available, but also have the potential to realize high-performance, high-frequency oscillators.

10.3 The Electronic Oscillator

An electronic oscillator is a circuit that produces a repetitive, oscillating signal, converting a direct current (DC) power supply into an alternating current (AC) signal. The shape of the required signal waveform depends on the application for which the oscillator is designed and typical examples include the generation of clock pulses for digital electronics [8], pulse width–modulated switching waveforms for switch mode power supplies [9] and radio frequency (RF) signals for wireless electronic communications [10]. A significant number of oscillator circuits have been realized; however, they can be categorized into two main types: the nonlinear oscillator and the linear or harmonic oscillator.

10.3.1 The Nonlinear Oscillator

The nonlinear oscillator (sometimes referred to as a relaxation oscillator) produces a waveform that is nonsinusoidal, such as a square wave, sawtooth or triangular waveform. The circuit topology comprises an energy storage component in conjunction with a nonlinear switching circuit. The energy storage component is often a capacitor; however, circuits based on the use of inductors have been demonstrated. Alternative implementations of this circuit comprise those that include a negative resistance device that periodically charges and discharges the energy in the storage element, resulting in abrupt changes in the output voltage of the circuit. Square wave relaxation oscillators are used extensively in digital electronics to provide the clock signal for sequential logic circuits such as timers and counters, whereas sawtooth oscillators are used specifically in time base circuits. An example of a single active device relaxation oscillator can be constructed using a unijunction transistor (UJT) device. A UJT comprises one p-n junction, making it similar in structure to a p-n diode; however, it differs in the fact that it has three terminals and so it is also referred to as a double base diode. It is worth noting that while the UJT has a similar construction to a junction field-effect transistor (JFET), the operation in a circuit is significantly different. In a UJT, the current through the device is modulated by the bipolar current injected through the gate under forward bias, rather than by the influence of the depletion region formed in the channel [11] with the gate under reverse bias.

As can be seen from Figure 10.1, the structure of a UJT does show some similarity to that of an N-channel JFET. The device consists of a lightly doped n-type substrate with three p-type terminals, which are labeled as base one (B1), base two (B2) and an emitter (E). B1 and B2 are ohmic contacts formed on the n-epilayer, whereas the E is formed from a heavily doped p-type region. The main physical difference between the UJT and the JFET is that the p-type region completely surrounds the n-type region in the JFET and the surface area of the JFET gate is significantly larger than the area of the emitter junction in the UJT. In a typical UJT, the doping of the emitter region is high, while the dopant concentration in the n-type region is low. This results in the resistance between the base terminals being significantly higher than is typical for a JFET, being in the region of 4–10 kΩ when the emitter is open circuit. In practical devices, the n-type channel has a high resistance and the resistance between the emitter and contact B1 is larger than that between the emitter and B2, resulting from the physical asymmetry in the structure, i.e. the emitter is physically closer to B2 than B1. A UJT does not have the ability to amplify; however, it is capable of controlling a large AC power with a small signal applied to the gate. In operation, the UJT is biased with a positive voltage between the two bases,

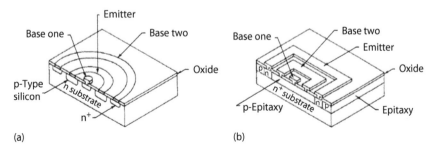

FIGURE 10.1
Example of the physical construction of a UJT. (a) Diffused planar structure and (b) epitaxial planar structure. (From Andronov, A.A., et al., *Theory of Oscillators*, Dover Publications, 2009 [12].)

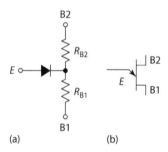

FIGURE 10.2
(a) Schematic representation of a UJT, (b) circuit symbol of a UJT.

whereby B2 is biased more positively than B1, resulting in a potential drop created along the length of the device. This can be seen in the schematic representation of a UJT shown in Figure 10.2a. When the emitter voltage is greater than the voltage applied to B2 plus the built-in potential of the p-n junction (typically 0.7 V for silicon and 2.8 V for SiC [13]), current flows from the emitter into the base region. Because the base region is very lightly doped, the additional charge carriers in the base region, which are injected by the emitter current, result in conductivity modulation. This excess charge reduces the resistance of the base region between the emitter junction and the B2 contact. This reduction in resistance results in the potential difference between the emitter and B1 increasing, which means that the emitter junction is more forward biased, and so even more current is injected. Overall, the effect appears to be similar to a negative resistance at the emitter terminal. This apparent negative resistance is the property that makes the UJT useful as the active component in simple oscillator circuits.

With the emitter open circuit, it is possible to determine the total resistance of the channel, also known as interbase resistance (R_{BB0}), as

$$R_{BB0} = R_{B1} + R_{B2} \tag{10.1}$$

As described previously, practical devices are nonsymmetrical and therefore it is possible to express the ratio between the resistance of the emitter to B_1 to that of the interbase resistance R_{BB0} as the intrinsic standoff ratio, which can be calculated using

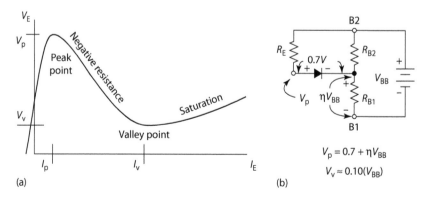

FIGURE 10.3
(a) Emitter characteristic curve, (b) model for V_P.

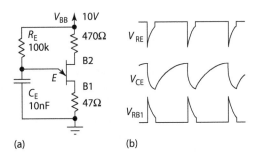

FIGURE 10.4
(a) Nonlinear electronic oscillator circuit using a UJT, (b) waveforms seen at respective points of the circuit.

$$\eta = \frac{R_{B1}}{R_{B1} + R_{B2}} = \frac{R_{B1}}{R_{BB0}} \qquad (10.2)$$

Figure 10.3a depicts the emitter current vs voltage characteristic curve for a typical UJT.

As the emitter voltage, V_E, increases further, the emitter current, I_E, increases until the peak point, V_P, beyond which the high level of current injection results in a negative resistance region, as can be seen from Figure 10.3.

The emitter voltage reaches a minimum at the valley point, V_V, where the emitter resistance is minimized. By biasing the UJT in the negative resistance region, it is possible to construct a simple electronic oscillator circuit, as shown schematically in Figure 10.4.

The operation of the oscillator circuit shown in Figure 10.4a can be described as follows: the external capacitor, C_E, charges based on the current flowing through the resistor, R_E, until the voltage at the emitter contact reaches the peak point, V_P. The reduction in the UJT resistance between the emitter and base 1 contacts (due to the negative resistance characteristics) discharges the capacitor. This can be seen in the waveform characteristics in Figure 10.4b. Once the potential across the capacitor has discharged below the valley point, the resistance between the emitter and base 1 increases and the capacitor is free to charge again. This results in a repeating oscillation, where the frequency can be controlled by the selection of the external components. Neglecting any parasitics in the circuit, the oscillation frequency of a UJT oscillator can be expressed as

$$f = \frac{1}{RC\ln\left(\frac{1}{1-\eta}\right)} \qquad (10.3)$$

To date, no UJT structures have been realized in SiC technology, which precludes this topology from high-temperature operations. The requirement for negative resistance in the device operation places significant limits on the carrier lifetime in the semiconductor. At present, the typical carrier lifetime in SiC is of the order of 1 μs [14], which is significantly lower than in the state-of-the-art silicon or gallium arsenide wafers, where values closer to 1 ms are commonplace.

10.3.2 The Linear Oscillator

In contrast to the nonlinear oscillator described in the previous section, the linear or harmonic oscillator produces a sinusoidal waveform. Two fundamental types of linear oscillator are possible: the negative resistance oscillator and the feedback oscillator.

10.3.2.1 The Negative Resistance Oscillator

The negative resistance oscillator uses an active component that exhibits negative resistance, such as a magnetron tube or a diode. Previous research has shown the possibility of an impact ionization avalanche transit-time (IMPATT) diode in SiC, which has suitable characteristics for the realization of high-frequency oscillators [15,16].

Similar to the nonlinear oscillator described in the previous section, the operation of these oscillators relies on the concept of negative resistance. Considering the simple tank circuit shown in Figure 10.5a, the initial current pulse is applied at I_{in}. The LCR tank responds with a decaying oscillatory behavior because in every oscillation a fraction of the energy that reciprocates between the capacitor and the inductor is lost as heat in the resistor.

If a device or a circuit that exhibits negative resistance is placed in parallel with R_P, it can be modeled as $-R_P$. If the magnitude of the negative resistance is equal to the parasitic resistance in the tank circuit, R_P, then the circuit continues to oscillate indefinitely, as shown schematically in Figure 10.5b. Such oscillators are used in the production of high frequencies, typically in the microwave region (>300 MHz) and above. The advantage of SiC in such circuits is its ability to operate at higher temperatures than an equivalent silicon device, thus allowing designers to dissipate much higher levels of power in a circuit [18].

Figure 10.6a depicts an example oscillator circuit utilizing an IMPATT diode as the negative resistance element. The circuit topology in the figure contains a resonant circuit, in this example an inductor-capacitor (LC) circuit, but alternatives including those in which a crystal or cavity resonator is connected across an IMPATT diode have identical characteristics.

As shown in Figure 10.5a, a resonant circuit has the possibility to operate as an oscillator, in that it has the ability to store energy in the form of electrical oscillations if excited. However, it will have some form of internal resistance and, as a result, the magnitudes of the oscillations are damped and eventually become zero. The negative resistance of the active device cancels the internal resistance in the resonator, thus creating a resonator where the magnitude of the oscillations shows no evidence of damping. This results in the

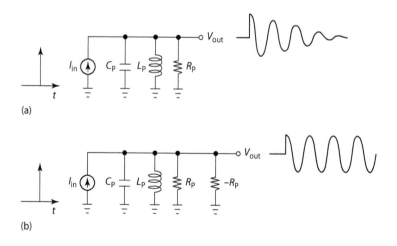

FIGURE 10.5

(a) Decaying impulse response of a tank circuit, (b) the addition of negative resistance to cancel loss in R_P. (From Senhouse, L.S., *IEEE Trans. Elect. Dev.*, 16, 1969, [17].)

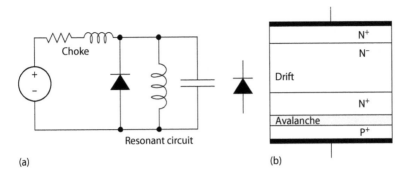

FIGURE 10.6
(a) Circuit of an IMPATT diode–based negative resistance oscillator, (b) structure of an IMPATT diode.

creation of oscillations with a constant magnitude at the resonant frequency, which can be determined using Equation 10.5.

10.3.2.2 The Feedback Oscillator

The feedback oscillator is the most common form of linear oscillator. It requires the use of an active component, typically an electronic amplifier based on a single transistor. The output from the amplifier is connected in a feedback loop to the input through a frequency selective electronic filter, resulting in positive feedback. For the oscillator to operate, the gain of the two stages of the amplifier needs to fulfill the requirement that the product of the gain for the two individual stages must exceed unity, i.e. $BA > 1$. This limitation can be considered in terms of the energy required to ensure a constant magnitude of the output waveform. A schematic representation of the feedback oscillator is shown in Figure 10.7.

When power is first supplied to the amplifier, the electronic noise in the circuit provides the initial signal to initiate the oscillations. The noise travels around the feedback loop, where it is amplified and filtered until it becomes a sine wave at a single specific frequency [20].

Feedback oscillator circuits can be classified according to the type of frequency selective filter they employ in the feedback loop. There are three specific types: crystal oscillators, resistor-capacitor (RC) oscillators and LC oscillators each with their own merits and limitations.

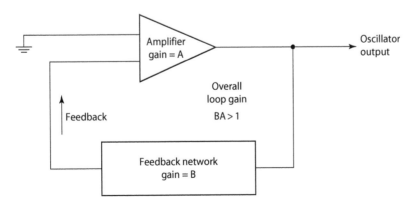

FIGURE 10.7
Schematic representation of a feedback oscillator.

10.3.2.2.1 Crystal Oscillators

Crystal oscillators utilize a piezoelectric crystal as the filter in the feedback loop. The crystal mechanically vibrates as a resonator and the frequency of the vibration determines the oscillation frequency of the electronic circuit. A common example circuit for a crystal oscillator is the Pierce oscillator, as shown schematically in Figure 10.8.

Crystals for electronic circuits are typically manufactured from quartz. Quartz mechanical resonators typically have a very high Q-factor, resulting in high selectivity to the desired frequency in the feedback loop. In addition, the oscillator frequency shows a weak temperature dependence in comparison to tuned circuits. These two characteristics mean that crystal oscillators have significantly better frequency stability than either LC or RC oscillators. However, quartz is not suitable for the development of high-temperature oscillators because it undergoes a phase transition at 573°C, referred to as the quartz inversion [19]. Above this temperature, the crystals do not operate as electronic filters and the oscillator produces a wide range of frequencies simultaneously. At the current time, no other material has been identified that can demonstrate the mechanical properties of a quartz resonator and operate at temperatures beyond this temperature. While crystal oscillators provide greater frequency stability with variations in temperature, the availability and complexities associated with incorporating a high-temperature crystal component were viewed as a disadvantage in comparison with the simplicity of prototype high-temperature LC oscillator–based circuits. Hence, crystal oscillators have not been experimentally investigated further.

10.3.2.2.2 RC Oscillators

RC oscillators utilize a filter network composed of discrete resistors and capacitors to produce lower frequencies (in comparison to the microwave frequencies generated using the negative resistance oscillators described previously), typically in the kilohertz frequency range. A common example circuit topology is the phase shift oscillator, as shown in Figure 10.9.

It can be seen from Figure 10.9 that the JFET is operating as an inverting amplifier, producing an output signal that is phase shifted by 180° in comparison to the signal at the input. In this topology, the RC network shown in the lower part of the image acts as a

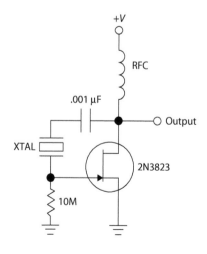

FIGURE 10.8
A Pierce crystal oscillator utilizing a JFET as an active device.

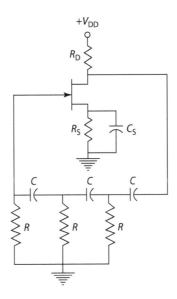

FIGURE 10.9
Phase shift oscillator utilizing a JFET. (From Sanitram, K., *Basic Electronics: Devices, Circuits and Fundamentals*, Dover Publications, 1972 [24].)

phase shift network. This phase shift in this part of the circuit matches that of the transistor, resulting in a fully in-phase signal. The phase shift is generated by the three RC sections in cascade, each of which produces a 60° phase shift. In general, each of the three sections shares common values for the resistance and capacitance of the components. In this case, the frequency of oscillation excluding parasitics can determined from

$$f = \frac{1}{2\pi RC\sqrt{6}} \tag{10.4}$$

Monolithically integrated RC oscillators (where the resistors and capacitors are fabricated on the same semiconductor die as the JFET) are possible. Because of the limited values of capacitance and resistance that can be realized, these circuits typically operate in the megahertz regime; however, the gate-source capacitance and channel resistance of the JFET become comparable to the component values, limiting the maximum oscillation frequency.

10.3.2.2.3 LC Oscillators

In an LC oscillator, the filter is a tuned circuit commonly referred to as a tank circuit, which consists of an inductor and a capacitor connected in parallel. During operation, charge flows back and forth between the plates of the capacitor through the inductor. This transfer of charge occurs at the resonant frequency and so the tank circuit acts as a filter, where the frequency is determined by the capacitance and inductance values. The value is often expressed in terms of the resonant angular frequency, which can be calculated from

$$\omega_0 = \frac{1}{\sqrt{LC}} \tag{10.5}$$

where L is the inductance in Henries and C is the capacitance in Farads.

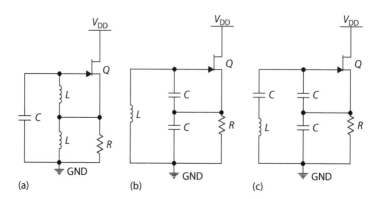

FIGURE 10.10
(a) Hartley, (b) Colpitts and (c) Clapp oscillators.

The internal losses within the tank circuit are compensated by the amplifier, which draws energy from the DC supply used in the circuit, resulting in a constant oscillation magnitude. LC oscillators are most commonly used at radio frequencies, where a tunable frequency source is required. Three commonly used circuit configurations are the Hartley, Colpitts and Clapp oscillators, as shown in Figure 10.10a–c, respectively.

As can be seen from the schematic circuits shown in Figure 10.10, the Colpitts and Clapp oscillators lend themselves more easily to miniaturization than the Hartley oscillator circuit, as they only utilize a single inductor in the design. High-temperature inductors tend to be physically large, as the permittivity and saturation magnetization of magnetic materials reduces with increasing temperature [21] and so the Hartley oscillator will be significantly larger than either the Colpitts or Clapp circuits. In this work, a Colpitts oscillator design was selected due to the ability of the oscillations to self-start utilizing components with lower Q values than the slightly more complicated Clapp design and this offers a more relaxed set of design criteria.

10.4 The Colpitts Oscillator

The Colpitts oscillator is a common form of LC oscillator that utilizes an LC tank circuit and an active device to counteract the damping effect caused by parasitic resistances. The Colpitts oscillator can be realized using a single transistor acting as an amplifier with the addition of a tank circuit. The circuits considered here are based on the use of a depletion mode SiC JFET; however, the analysis can be expanded to include a wide range of alternative transistor families, including MOSFETs and bipolar junction transistors (BJTs) as well as thermionic valves. By feeding the signal back from the output of the amplifier to the input through this LC tank to select a single frequency, it is possible to commission a Colpitts oscillator circuit, as shown by the circuit given in Figure 10.11.

The analysis of operation for a Colpitts oscillator is often performed using a linear systems feedback approach, as described in a number of references including [22]. The main outcome from this approach is a set of expressions for both frequency and the minimum JFET gain in order for the oscillations to maintain a constant magnitude with time. These conditions are described in Equations 10.6 and 10.7, respectively.

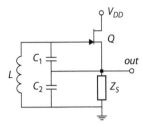

FIGURE 10.11
Circuit schematic of a Colpitts oscillator.

$$f_0 = \frac{1}{2\pi\sqrt{L\dfrac{C_1 C_2}{C_1 + C_2}}} \tag{10.6}$$

$$g_m r_o = \frac{C_2}{C_1} \tag{10.7}$$

This approach typically results in higher calculated frequencies than those predicted using computer-based simulation tools (for example, simulation program with integrated circuit emphasis [SPICE]) or obtained experimentally due to the exclusion of the gate-source, gate-drain and source-drain capacitances within the JFET. It is also possible to model the characteristics of the Colpitts oscillator using an alternative approach that is based on the concept of negative resistance [9] in one-port oscillators such as the SiC IMPATT diode, replacing the transistor with an ideal entire circuit that has the parasitics explicitly defined as external components, as shown in Figure 10.12.

From Figure 10.12, it can be see that in order to overcome the energy loss from a finite R_P, the active circuit must form a small-signal negative resistance ($R_P < 0$), which is required to replenish the energy lost in every oscillation. Hence, the negative resistance can be interpreted as a source of energy. In this example, R_P denotes the equivalent parallel resistance of the tank, and for oscillations to be self-starting it is necessary that $R_P + (-R_P) > 0$. As the oscillation amplitude increases, the amplifier will start to saturate, thus decreasing the gain from the feedback loop until it reaches unity. This steady-state condition satisfies the Barkhausen criterion [23] and the oscillations continue. Once the circuit is operating in the steady state, the two resistances, R_P and $-R_P$, must be of equal amplitude. This analytical approach allows for the inclusion of the parasitic capacitances that are inherent in the JFET structure, thus yielding a far more accurate calculation of the oscillation frequency of the circuit. Neglecting the effect of the inductor in the circuit, the AC

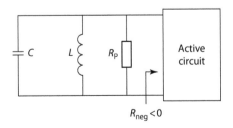

FIGURE 10.12
RLC tank with negative resistance created by the oscillator active network.

equivalent circuit in Figure 10.12 can be used to determine the input impedance, as shown in Figure 10.13. The input impedance of the circuit can be determined from Equation 10.8:

$$|Z_S| \gg \left| \frac{1}{i\omega(C_2 + C_{DS})} \right| \tag{10.8}$$

The parasitic capacitances of the JFET (denoted by Q) have been included in Figure 10.13.

These are the gate-source, C_{GS}, drain-source, C_{DS}, and gate-drain, C_{GD}, capacitances. By assuming the following conditions, it is possible to describe the JFET by means of a small-signal model, which can be utilized within SPICE to accurately describe the behavior of the circuit:

$$r_{DS} \gg \left| \frac{1}{i\omega(C_2 + C_{DS})} \right| \tag{10.9}$$

$$r_{GS} \gg \left| \frac{1}{i\omega(C_1 + C_{GS})} \right| \tag{10.10}$$

where r_{DS} and r_{GS} are the small-signal drain-source and gate-source resistances of the JFET. By replacing the SiC JFET with the simplified small-signal model, it is possible to obtain the equivalent circuit of the oscillator shown in Figure 10.14, which is the small-signal transconductance of the SiC JFET.

From results previously published in the literature and utilizing standard circuit theory, it is possible to determine the limits for the negative resistance required to achieve a steady-state operation for the circuit shown in Figure 10.14:

$$r_{neg} = -\frac{g_m}{\left[\omega^2 (C_1 + C_{GS})(C_2 + C_{DS}) \left(1 + \frac{C_{GD}}{C_1 + C_{GS}} + \frac{C_{GD}}{C_2 + C_{DS}} \right) \right]} \tag{10.11}$$

and the equivalent total capacitance in the circuit C_t can be expressed as

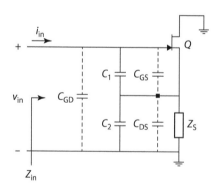

FIGURE 10.13

An AC equivalent circuit of a Colpitts oscillator ignoring the inductor L and including the parasitic capacitances inherent in the JFET transistor structure.

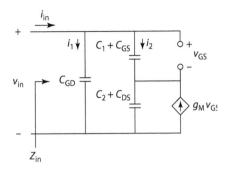

FIGURE 10.14
Small-signal equivalent of the Colpitts oscillator.

$$C_t = \frac{(C_1 + C_{GS})(C_2 + C_{DS})}{C_1 + C_2 + C_{GS} + C_{DS}}\left(1 + \frac{C_{GD}}{C_1 + C_{GS}} + \frac{C_{GD}}{C_2 + C_{DS}}\right) \tag{10.12}$$

It should be noted that combining Equation 10.12 with Equation 5.5 results in the reduction to a well-known expression for the negative resistance [17,24]:

$$r_{neg} = \frac{g_m}{\omega^2 C_1 C_2} \tag{10.13}$$

Utilizing these values and reconnecting the inductor, L, along with a series total loss resistance, r_t, to the input impedance, Z_{in}, it is possible to obtain a series equivalent circuit of the Colpitts oscillator, shown schematically in Figure 10.15.

The total series loss resistance, r_t, can be calculated from the series resistance of both the capacitors and inductors using

$$r_t = r_L + r_{C_1} + r_{C_2} \tag{10.14}$$

Hence, the equivalent parallel resistance of the Colpitts oscillator, R_P, including contributions from the tank circuit can be determined from

$$R_P = \frac{L}{r_t \frac{(C_1 + C_{GS})(C_2 + C_{DS})}{C_1 + C_2 + C_{GS} + C_{DS}}\left(1 + \frac{C_{GD}}{C_1 + C_{GS}} + \frac{C_{GD}}{C_2 + C_{DS}}\right)} \tag{10.15}$$

By substituting Equation 10.12 into Equation 5.5, the oscillation angular frequency for a circuit including the parasitic components can be determined using

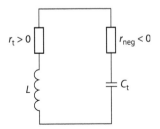

FIGURE 10.15
High-temperature Colpitts oscillator circuit diagram.

$$\omega = \sqrt{\left(L \frac{(C_1 + C_{GS})(C_2 + C_{DS})}{C_1 + C_2 + C_{GS} + C_{DS}} \left(1 + \frac{C_{GD}}{C_1 + C_{GS}} + \frac{C_{GD}}{C_2 + C_{DS}} \right) \right)^{-1}} \qquad (10.16)$$

Substituting Equation 10.16 into Equation 10.11 results in

$$r_{neg} = -\frac{g_m L}{C_1 + C_2 + C_{GS} + C_{DS}} \qquad (10.17)$$

Finally, referring back to Figure 10.13, it is possible to determine the negative resistance required for the Colpitts oscillator to self-start, resulting in

$$r_{neg} = -\frac{(C_1 + C_2 + C_{GS} + C_{DS})^2}{g_m (C_1 + C_{GS})(C_2 + C_{DS}) \left(1 + \frac{C_{GD}}{C_1 + C_{GS}} + \frac{C_{GD}}{C_2 + C_{DS}} \right)} \qquad (10.18)$$

Equation 10.18 shows that for the circuit to operate, the JFET needs to have a high transconductance, g_m, and low parasitic capacitances, C_{GS} and C_{GD}. This identifies the challenge of designing high-temperature oscillators, to the extent that in addition to the external capacitance values remaining unchanged (the parasitic capacitances within the transistor have a weak temperature dependence), the transconductance of the JFET will limit the upper operating temperature of the circuit. The transconductance of a JFET can be expressed as

$$g_m = \frac{\Delta(I_{DS})}{\Delta(V_{GS})} = \frac{2 I_{DSS}}{|V_P|} \left(1 - \frac{V_{GS}}{V_P} \right) \qquad (10.19)$$

10.4.1 High-Temperature Colpitts Oscillator

To demonstrate the importance of an accurate model in determining the resonant frequency of an oscillator, a comparison between the different circuit topologies was performed. Experimentally determined parameters for SiC devices operating at high temperatures were used in simulations using both SPICE-based simulations and the theoretical analysis described in the previous section. The characterized SiC components were then packaged into a hybrid module circuit board that was fabricated on an aluminum oxide ceramic substrate. Figure 10.16 shows a schematic representation of the high-temperature Colpitts oscillator, with the room temperature component values shown. The hybrid module was placed inside a Carbolite oven with electronic temperature control during the measurements and allowed to settle for 20 min at each temperature prior to any measurements being taken. The frequency spectrum of the oscillator was measured using an Agilent E4403B spectrum analyzer, which was coupled to the RF output of the circuit under test.

10.4.2 High-Temperature Voltage-Controlled Oscillator

The data in Figure 10.17 compare the calculated, simulated and experimental frequencies of a high-temperature Colpitts oscillator at temperatures up to 573 K. It can be seen from the experimental data that the frequency of the oscillator reduces with increasing temperature. This reduction is due to both the increased capacitance density of the AlN dielectric capacitors, which results from the increasing dielectric constant and the increasing

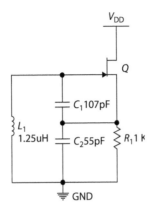

FIGURE 10.16
Comparison of calculated, simulated and experimental frequencies as a function of ambient temperature.

parasitic capacitance of the JFET p-n junction. As can be seen from the data, the predicted oscillator frequency using traditional theory is significantly higher than that obtained experimentally, predominantly because of the parasitic capacitances and inductances in the fabricated circuit. Estimating the parasitic values for the circuit results in modified values, as shown in Figure 10.17; however, the variation of the modified values with temperature does not match the experimental data well. This indicates that the performance of the oscillator circuit is dominated by the shift in the characteristics of the AlN capacitors. The variation in the capacitance density of AlN capacitors has been extracted from the capacitance-voltage characteristics [25] and this can be included as a temperature dependence in the SPICE model of the circuit, resulting in the data set labeled 'Frequency Simulated'

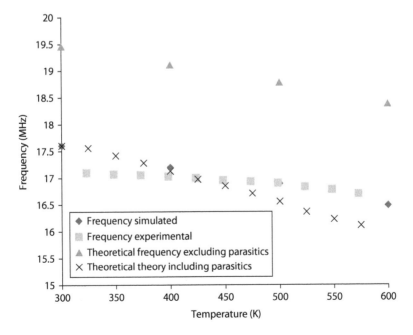

FIGURE 10.17
Comparison of calculated, simulated and experimental frequencies of a silicon carbide Colpitts oscillator as a function of temperature.

in Figure 10.17. This shows a better agreement with the experimental data for temperatures above 400 K; however, the low-temperature frequency is overestimated, similar to the modified theoretical predictions.

A working communication system requires receiver electronics that are tuned to the specific frequency of the carrier waveform generated by the oscillator circuit. While it is possible to create complicated receiver electronics capable of tracking a drifting signal, the challenge of then demodulating the information stored in the carrier signal becomes significantly more difficult. For this reason, it is far more desirable to directly control the frequency of the oscillator in the transmission system. For oscillator circuits based on LC tanks, this is commonly achieved with a device known as a varactor. A varactor is a p-n diode that has been optimized to give a significant capacitance under reverse bias conditions, rather than a conventional p-n diode structure where reverse bias capacitance is minimized to increase the switching speed. Since p-n diodes under reverse bias exhibit a depletion region, the width of which varies with voltage, the capacitance of these devices is inversely proportional to the square root of the applied voltage. Although a small number of reports on SiC varactors can be found in the literature [26,27], it is worth noting that all diodes, including SiC Schottky diodes, exhibit an identical change in capacitance with applied bias. For this reason, it is possible to modify the high-temperature Colpitts oscillator circuit shown in Figure 10.15 to form a high-temperature VCO by the inclusion of a Schottky diode, resulting in the circuit shown in Figure 10.19.

Neglecting the effects of the JFET junction capacitance and any stray capacitance from the circuit on the frequency of the oscillations, the addition of a reverse-biased SiC Schottky diode acting as a varactor can be considered as adding two capacitors, denoted by C_2 and D_1, in series as part of the tank circuit. This results in a capacitance that is voltage dependent in the lower branch of the tank circuit and the oscillator frequency (ignoring the influence of parasitics in the circuit) can now be approximated by

$$f = \frac{1}{2\pi\left(\sqrt{L\frac{C_1 C_{TOT}}{C_1 + C_{TOT}}}\right)} \tag{10.20}$$

where C_{TOT} is the capacitance of the Schottky diode and C_2 in series.

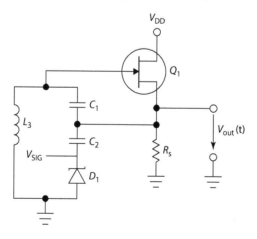

FIGURE 10.18
Circuit schematic of a high-temperature, voltage-controlled oscillator.

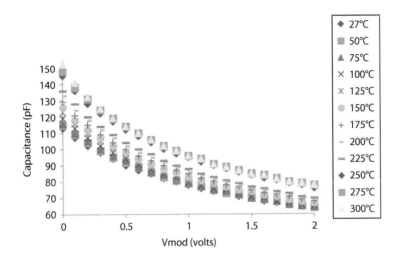

FIGURE 10.19
Capacitance of an SiC Schottky diode with different applied voltage bias as a function of temperature.

The capacitance of the Schottky diode can be decreased by the application of an external bias to the V_{SIG} port shown in Figure 10.18 and can be expressed as

$$C = A \left(\frac{q\varepsilon}{2} \right)^{1/2} \left(\frac{N_D}{\phi_B - V_A} \right)^{1/2} \tag{10.21}$$

where A is the diode area, E is the dielectric constant of the SiC, q is the electronic charge, N_D is the dopant concentration in the n-type region (assuming the abrupt junction approximation), ϕ_B is the barrier height and V_A is the applied bias.

The data in Figure 10.19 show the variation in the capacitance of an SiC Schottky diode as a function of reverse bias for the range of temperatures of interest. The observed decrease in capacitance with increasing temperature is linked to the increasing depletion width formed in the device, which results in the decreasing capacitance of the diode, as is expected from parallel plate capacitor theory [14]. Note that the capacitance of D_1 is in series with the capacitance of C_1 and so it is possible to control C_{TOT} by applying a voltage to D_1. This results in the decrease of C_{TOT} and hence an increase in the resonant frequency of the oscillator, as described by Equation 10.20 and can be seen from the data plotted in Figure 10.20.

As shown by the data in Figure 10.20, it is possible control the frequency of an SiC Colpitts-based VCO across a wide temperature range. The data show the potential to create an electronic feedback loop to stabilize the resonant frequency of the circuit through the application of a bias at the V_{SIG} port of the circuit. This technique can be used to nullify the temperature dependence of the oscillator for temperatures between 273 K and 573 K. This technique can also be used to achieve direct FM of the oscillations and hence open up the possibility of generating modulated signals that can be utilized for data transmission in extreme environments.

10.4.3 Modulation

Modulation is the term used to describe the process of varying one or more properties of a periodic waveform. The high-frequency carrier signal is modulated with a lower-frequency

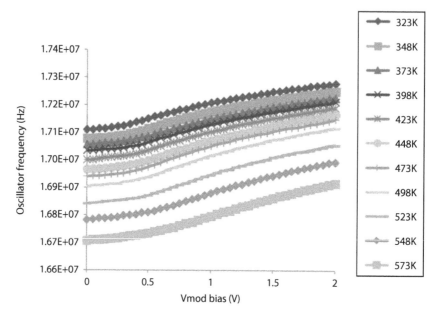

FIGURE 10.20
Frequency range of SiC voltage–controlled oscillator.

signal (often referred to as the secondary signal), which typically contains the information to be transmitted [12]. In principle, three basic methods of modulating the carrier waveform are practical and they can be implemented using either analog or digital methods. Figure 10.21a–c schematically shows examples of amplitude modulation (AM), FM and phase modulation (PM) schemes, respectively.

Considering the AM scheme, shown in Figure 10.21, the carrier waveform of a set frequency is combined with an analog information signal by modulating the amplitude of the carrier signal. This is one of the simplest forms of communication systems; however, data fidelity can be compromised by temporary changes in the efficiency of the communications channel between the transmitter and the receiver, such as changes in weather or people passing by. It is possible to utilize this method in an engineering context for the transmission of low data rate, noncritical digital signals for short distances; however, in order to achieve the performance required for a wireless sensor node, alternative modulation schemes are required. Figure 10.21b is an example of an FM scheme implemented using digital techniques, where the amplitude of the carrier waveform remains constant, while the frequency is varied between two distinct states. Each of these two states is used to represent a binary number. While this modulation system has a greater level of complexity for both the transmitter and the receiver, the significant increase in data fidelity has made this modulation scheme extremely popular for the transmission of data from wireless nodes. An example of a digital PM scheme is depicted in Figure 10.21c. In this scheme, the magnitude and frequency of the signal remains constant, while the phase of the waveform is shifted by 180° to indicate the different binary numbers. This scheme offers the potential of the greatest information density, but with the highest level of complexity – often using microprocessors to control the transmitter and receiver in real time.

Figure 10.22 shows a schematic representation of a simple analog AM communications system that is suitable for implementation in SiC technology. In this example, the analog signal used to modulate the carrier waveform could be produced using a sensor either

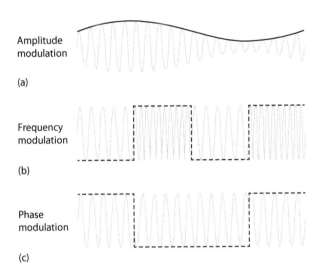

FIGURE 10.21
Example waveforms showing (a) amplitude modulation, (b) frequency modulation and (c) phase modulation.

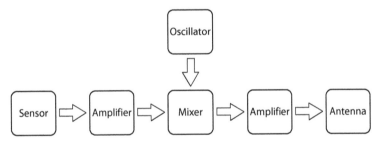

FIGURE 10.22
Block diagram of a simple analog AM communications system.

directly or through previous amplification electronics. The amplitude of the modulation signal is directly proportional to the signal on the sensor and so AM modulation can be achieved by means of a mixer circuit. The mixer ensures that the output amplitude of the oscillator is modulated by the magnitude of the sensor signal. This is the basis of a simple analog communications system where the amplitude of the carrier signal can be directly related to the sensor. The amplifier needs to demonstrate a high linearity at a single frequency to ensure that the data fidelity is maintained and to avoid complex deconvolution at the receiver.

Previously in this chapter, it was shown that by utilizing SPICE models for the SiC JFET and Schottky diode along with the characteristics of the high-temperature passive components, it is possible to design, simulate and commission a high-temperature Colpitts oscillator using SiC devices. Here, this is extended to show that the amplitude of the carrier signal can be directly modulated at 280°C [10]. Using the schematic circuit shown in Figure 10.15 as the Colpitts oscillator, capacitors fabricated from AlN dielectric C_1 and C_2 were selected to have values of 68 and 82 pF, respectively. The choice of AlN as the dielectric was to minimize the leakage current through the capacitors and to enable the integration of the capacitors with the dielectric films commonly used in the fabrication of high-temperature packages. A high-temperature printed circuit board was commissioned from a thick gold film on a ceramic substrate. The approximate frequency of the LC tank was determined from the physical size of the inductor, L. Based on Equations 10.20 and 10.22, the frequency of the oscillations was predicted to be approximately 22 MHz for an inductor value of 1.4 μH. The load resistor in the circuit was set to 1 kΩ, and 12 V DC was supplied to the circuit at the drain of the JFET denoted by Q. The inductance of a planar spiral can be determined from

$$L = \frac{N^2\left(\dfrac{D_i + N(W+S)}{2}\right)}{30\left(\dfrac{D_i + N(W+S)}{2}\right) - 11D_i} \tag{10.22}$$

where N is the number of turns on the inductor, D_i is the diameter of the innermost turn, N is the number of turns, W is the width of the tracks and S is the spacing between the turns.

Simulations of the oscillator circuit based on the experimental device parameters extracted indicate that the oscillator is capable of self-starting, as shown by the data in Figure 10.23.

This indicates that the transconductance of the SiC JFETs used in the circuit is sufficient to ensure that the Barkhausen criteria is met and the oscillations continue indefinitely. By modifying the Colpitts oscillator circuit to that shown by the schematic in Figure 10.24, it is possible to utilize a second active component to produce AM of the carrier waveform.

The amplitude of the oscillations is varied by changes in the amplitude of the feedback signal seen at the gate of JFET denoted by Q_1. This is achieved by utilizing JFET Q_2 as a variable resistor. A negative bias on the gate of Q_2 (in respect of the source of the JFET), which is designated as V_{SIG} in Figure 10.24, it is possible to control the resistance of the JFET channel and therefore the magnitude of the feedback signal to the gate of Q_1. This can be clearly observed in the data shown in Figure 10.24, where a 1 MHz sinusoidal signal

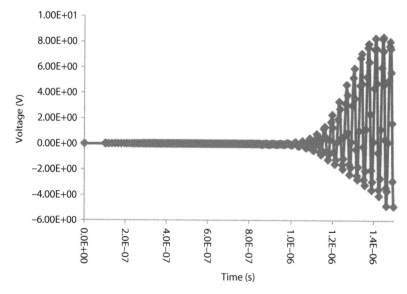

FIGURE 10.23
SPICE simulation of the self-starting of a Colpitts oscillator.

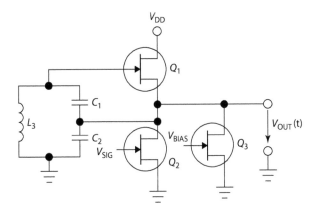

FIGURE 10.24
Circuit schematic used in a high-temperature modulation experiment.

was fed into V_{SIG}. This resulted in changes to the channel resistance of JFET Q_2 and this directly modulated the feedback seen at the capacitance tap. The effect of this change in the capacitance feedback can be observed by increasing the negative bias on V_{SIG}, which results in the amplitude of the carrier signal increasing. This can ultimately be used to create an amplitude-modulated signal at the JFET gate, which is also the signal transmitted through the inductor.

A hybrid module was then assembled using a 1000 µm thick aluminum nitride substrate onto which a seed layer of 10 nm chrome was deposited by physical vapor deposition followed by 250 nm of gold. The gold layer on the substrate was then electroplated to a thickness of approximately 8 µm, to reduce the track resistance. Capacitors fabricated with a 60 nm thick HfO_2 dielectric were selected for their low leakage characteristics (sub µA at 5 V) with values of 68 pF (C_1) and 82 pF (C_2), respectively. The inductor, denoted by L_3, was a gold spiral patterned directly onto the substrate. The frequency characteristics of the inductor were determined at 1 MHz using an Agilent 4284A LCR bridge, which demonstrated an inductance of 1.4 µH and a resistance of 4.8 Ω. The capacitors and JFETs were then attached to the circuit using silver epoxy, which was baked at 150°C for 1 h, prior to the electrical connections being made by gold wire bonding. The frequency spectrum of the oscillator was measured using an Anritsu MS2721B Spectrum Master and the amplitude of the RF signal was measured through an external aerial attached to a Tektronix TDS3045C oscilloscope.

With the gate to Q_3 held at −8 V, the application of a modulation signal to the gate of JFET Q_2 varied the amplitude of the output signal, as can be observed from the data in Figure 10.25. For an applied bias between −4 and −6 V (denoted by region B in Figure 10.26), increasing the magnitude of V_{SIG} increases the resistance of the JFET channel.

As previously described, this increases the magnitude of the output signal, with a behavior that is reasonably linear. For bias levels with a magnitude below 2.5 V, the resistance of the channel reduces the current flowing through the circuit, resulting in a reduction in the amplitude of the oscillations – as can be observed in the region denoted by A in Figure 10.26. The data indicate that it is possible to achieve AM by varying the voltage applied to Q_2 in two regions. However, in region B (the higher bias region) the frequency of the oscillations was observed to vary with the applied bias.

This is a nondesired result from a systems perspective as it will require a high level of sophistication in the receiver circuit and so the performance of the circuit operating in

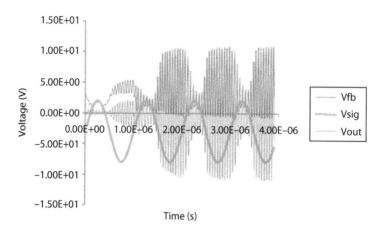

FIGURE 10.25
Simulated output characteristics of the circuit shown in Figure 10.24.

region A was investigated further. As shown by the data in Figure 10.27, the variation of the amplitude with the applied voltage is not linear. This originates in the physics behind the operation of the JFET, where the channel characteristics vary with the square of the applied voltage.

The frequency spectrum of the oscillator while operating at 280°C is shown by the data in Figure 10.28. The gate of Q_2 was held at a constant −3.5 V, resulting in the maximum oscillator amplitude. During the measurements, the amplitude of the peak frequency was 65 dBm above the background noise, with a full half width maximum of approximately 7.2 kHz.

The oscillation frequency was measured as a function of temperature to determine the temperature coefficient of the peak frequency. As can be seen from the data in Figure 10.29,

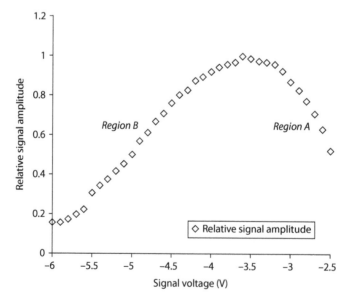

FIGURE 10.26
Relative amplitude of an output signal with varying *V* bias.

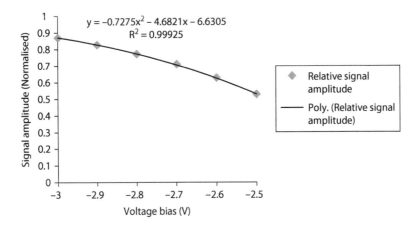

FIGURE 10.27
Signal amplitude vs voltage bias of region A.

the frequency decreases with increasing temperature, linked to the increased capacitance of the HfO_2 metal-insulator-metal (MIM) capacitors [28,29].

The data in Figure 10.30 show that in addition to the frequency shift with temperature, the amplitude of the oscillations also decreases. The data point for 353 K is influenced by a lack of temperature stability in the oven used for the testing, related to the proportional-integral-derivative (PID) parameters being unoptimized for low temperatures. The data for the decrease in signal amplitude with temperature were taken with Q_2 biased at a gate-source potential of −3.5 V to obtain maximum oscillation amplitude and hence maximize the transmitted power. For a high-temperature communications system based on AM to be effective, a feedback system that can maintain the signal amplitude with varying temperatures is essential. With the current maturity level of SiC technology this is not possible, effectively ruling out the possibility of high-fidelity data transmission using an AM communications link.

Extreme environments are not just classified in terms of high ambient temperature, high radiation dose rate and the existence of chemically corrosive species, they also often

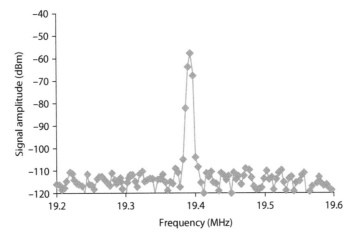

FIGURE 10.28
Frequency spectrum of AM Colpitts oscillator at 280°C.

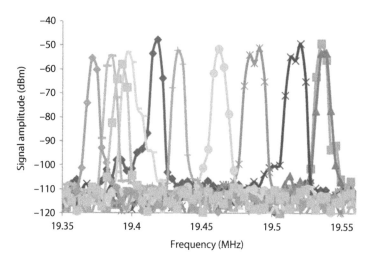

FIGURE 10.29
Change in frequency with increasing temperature.

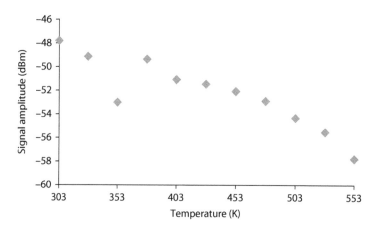

FIGURE 10.30
Maximum signal amplitude as a function of temperature.

include high intensity sources of RF interference or noise. FM communication schemes are inherently more resilient to these sources of noise than AM schemes [30].

Figure 10.31 shows a schematic representation of a simple frequency-modulated communication system that can be implemented in SiC technology.

By replacing the oscillator and mixer blocks in the schematic circuit diagram shown in Figure 10.21 with a VCO circuit, direct FM of the oscillations can be achieved [30]. In a simple sensor system, the sensor output is applied directly to the varactor diode, varying

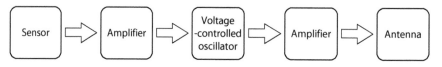

FIGURE 10.31
Block diagram of a simple FM transmitter.

the depletion capacitance and hence the frequency of a VCO. This frequency-modulated signal is then amplified to an antenna, resulting in an analog frequency-modulated communications system.

However, the system is also capable of transmitting data in digital form, as shown by the schematic in Figure 10.20b. The digital transmission of data by means of FM of the carrier wave is called frequency shift keying (FSK), in contrast to the amplitude-modulated transfer of digital data, which is referred to as amplitude shift keying (ASK). FSK can be achieved by choosing and transmitting two distinct frequencies to represent the "1" and "0" binary bits. Similarly, ASK can be achieved by selecting two distinct amplitudes at the same frequency to represent digital information.

The FM circuit shown in Figure 10.17 was demonstrated experimentally and assessed at high temperatures, with a view to estimating the possibility of FSK communications in extreme environments. A hybrid FM circuit, fabricated using the same techniques as the AM circuit previously described, was tested at temperatures up to 300°C. As shown by the data in Figure 10.32, the frequency of the oscillator varies linearly with the increasing magnitude of the reverse bias applied (between −0.75 and −1.5 V) to the Schottky diode (D_1). The shift in oscillator frequency for an applied bias of −0.75 and −1.5 V does not show a significant variation with temperature, which indicates that FM is a suitable technique for data transmission across a wide temperature range. The observed drift in the oscillator frequency with temperature is caused by the shift in the capacitances in the circuit, as described for the AM circuit.

The data in Figure 10.33 show the only reported frequency shift keyed behavior from an all SiC circuit that was operating at 300°C, captured using an Anritsu MS21721B Spectrum Master. The data clearly show two distinct peaks, one located at a frequency of 20 kHz with respect to that of the carrier waveform that represents a digital "1" and the carrier waveform itself, which represents a digital "0". The peaks are formed by the application of a square wave of 1 kHz to the diode in the VCO circuit, shown in Figure 10.19. Hence, it is possible to transmit frequency shift keyed data using an SiC oscillator over a range of temperatures that are beyond the capability of conventional, silicon-based electronic systems.

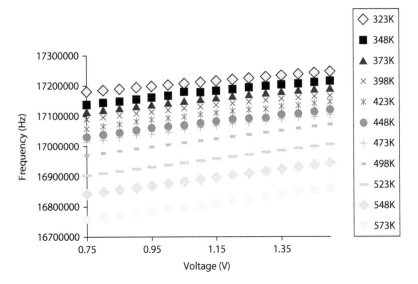

FIGURE 10.32
Linear regions of oscillator frequency change.

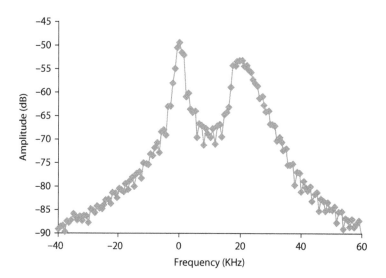

FIGURE 10.33
Spectrum showing frequency shift keyed modulation.

Of the different available modulation schemes, FSK systems are the most power efficient. This is a critical consideration for a wireless sensor node that may be powered by energy harvested from the ambient environment. The disadvantage of this modulation scheme is its low bandwidth efficiency, as it requires significant bandwidth for each sensor to operate in a unique frequency range [31]. However, there is a special case of frequency shift keyed modulation that addresses this issue and can also be viewed as an attractive alternative to the phase shift keyed modulation that commonly prevails in digital communication systems, such as mobile telephones, due to the higher data rates possible. Minimum-shift keying (MSK) has been shown to be a special case of continuous phase FSK, where the frequency deviation equals the bit rate. MSK can also be viewed as a form of offset quadrature-phase shift keying signaling in which the symbol pulse is a half-cycle sinusoid rather than the usual rectangular form. This method has the potential to combine in a single modulation format a number of attractive attributes, including constant envelope, compact spectrum, the error rate performance BPSK and simple demodulation and synchronization circuits. These features make MSK an excellent modulation technique for digital links in which bandwidth conservation and the use of efficient transmitters with nonlinear (amplitude-saturated) devices are important design criteria [31]. However, at this stage of the technology maturity, the requirements for this technique cannot be met by SiC devices and so this must remain an option for future developments.

10.5 Conclusions

This chapter has demonstrated the possibility of fabricating high-temperature electronic oscillators using SiC technology. The low maturity of SiC in comparison to silicon precludes the use of microprocessor-controlled oscillators; however, analog circuits based on discrete components have been shown to offer performance that is suitable for a wide range of applications. Predictions as to the frequency of the oscillations require accurate knowledge of the parasitic capacitances in the circuit, but more crucially the variation with temperature of these parasitics and the discrete components used in the circuit. Utilizing device characteristics that were determined over the temperature range of interest, a high-temperature, Colpitts-based VCO was demonstrated. The circuit was used to show that it is possible to control the frequency drift at high temperature along with AM and FM of two high-temperature oscillators working at temperatures up to 300°C. The physical design of the Colpitts oscillator specifically lends itself to miniaturization, featuring a capacitive feedback path, offering greater frequency stability and physically smaller components than the inductive feedback path found in the Hartley oscillator. The Colpitts oscillator also demonstrates an inherently more powerful self-starting-up ability than the Clapp oscillator, resulting in the allowed utilization of lower components with larger tolerances, which is typical in the low technological maturity SiC devices. LC oscillators provide a simple solution for producing high-frequency sine waves; these circuits contain a tuned LC tank and an active device arranged in an amplifier layout; and they are particularity useful in situations where the energy supply can be intermittent due to their self-starting ability.

References

1. N.G. Wright and A.B. Horsfall, *Journal of Physics D*, vol. 40 (2007) pp. 6345.
2. Littelfuse. Silicon carbide (SiC) diodes. www.littelfuse.com/products/power-semiconductors/silicon-carbide.aspx.
3. Wolfspeed. www.wolfspeed.com.
4. H.K. Chan, N.G. Wood, K.V. Vassilevski, N.G. Wright, A. Peters, and A.B. Horsfall, *Proceedings of IEEE Sensors Conference*, Busan, South Korea, November 1–4 (2015).
5. M.H. Weng, R. Mahapatra, N.G. Wright, and A.B. Horsfall, *IEEE Sensors Journal*, vol. 7 (2007) pp. 1395.
6. J.A. Cooper, M.R. Melloch, R. Singh, A. Agarwal, and J.W. Palmour, *IEEE Transactions on Electron Devices*, vol. 49 (2002) pp. 658.
7. C.M. Zetterling, in *Process Technology for Silicon Carbide Devices*, Inspec Publishing (2002), ISBN 0852969988.
8. M.H. Weng *et al.*, *Semiconductor Science and Technology*, vol. 32 (2017) 054003.
9. O. Mostaghimi, N.G. Wright, and A.B. Horsfall, *Proceedings of ECCE Conference*, Raleigh, North Carolina, September 15–20 (2012) pp. 3956.
10. D.R. Brennan, B. Miao, K.V. Vassilevski, N.G. Wright, and A.B. Horsfall, *Materials Science Forum*, vols. 653–956 (2010) pp. 953.
11. V. Blahm and T.P. Sylvan, *Solid State Design*, vol. 5 (1964) pp. 26.
12. A.A. Andronov, A.A. VItt, and S.E. Khaikin, in *Theory of Oscillators*, Dover Publications (2009), ISBN 0486655083.
13. T. Kimoto, K. Yamada, H. Niva, and J. Suda, *Energies*, vol. 9 (2017) pp. 918.
14. G. Sozzi, M. Puzzanghera, G. Chiorboli, and R. Nipoti, *IEEE Transactions on Electron Devices*, vol. 64 (2017) pp. 2572.
15. J. Jensen, *IRE Trans Circuit Theory*, vol. 4 (1957) pp. 276.
16. I.M. Gottlieb, in *Practical Oscillator Handbook*, Newnew (1997) ISBN0750631020.
17. L.S. Senhouse, *IEEE Transactions on Electron Devices*, vol. 16 (1969) pp. 161–165.
18. K.V. Vassilevski, *IJHSES*, vol. 15 (2005) pp. 899.
19. K.V. Vassilevski, *IEEE Electron Device Letters*, vol. 21 (2000) pp. 485–487.
20. B. Razavi, in *Design of Analog CMOS Integrated Circuits*, McGraw-Hill, Boston, MA (2016), ISBN 0072524932.

21. A.H. Morrish, in *The Physical Principles of Magnetism*, Wiley, New York, NY (2001), ISBN 9780780360297.
22. R. Trew, *Proceedings of the IEEE*, vol. 79 (1991) pp. 598–620.
23. D. Leenarts, J. van der Tang, and C. Vacher, in *Circuit Design for RF Transceivers*, Kluwer, Boston, MA (2001), ISBN 9780306479786.
24. K. Sanitram, in *Basic Electronics: Devices, Circuits and Fundamentals*, Dover Publications (1972), ISBN 0486210766.
25. S. Barker, B, Miao, D.R. Brennan, N.G. Wright, and A.B. Horsfall, *Proceedings of IEEE Sensors Conference* (2009), pp. 777.
26. C.M. Anderson, *IEEE Transactions on Electron Devices*, vol. 32 (2011) pp. 788–790.
27. A.P. Knights, A.G. O'Neill, and C.M Johnson, *Proceedings of EDMO Conference* (1999), pp. 301.
28. B. Miao, R. Mahapatra, N.G. Wright, and A.B. Horsfall, *Journal of Applied Physics*, vol. 104 (2008) 054510.
29. B. Miao, R. Mahapatra, R. Jenkins, J. Silvie, N.G. Wright, and A.B. Horsfall, *IEEE Transactions on Nuclear Science*, vol. 56 (2009) pp. 2916.
30. D.R. Brennan, K.V. Vassilevski, N.G. Wright, and A.B. Horsfall, *Materials Science Forum*, vol. 717 (2012) pp. 1269.
31. R. deBuda, *IEEE Transactions on Communications*, vol. 7 (1972) pp. 429–435.

11

The Use of Error Correction Codes within Molecular Communications Systems

Yi Lu, Matthew D. Higgins, and Mark S. Leeson

CONTENTS

Molecular communications (MC) is a promising area with applications in health-care [1–4], the environment [5,6] and manufacturing [7–9]. As such, it has received an increasing amount of attention in recent years. Considering any kind of communication system, the reliability of the data at the receiver (R_X) is one of the most important factors. Thus, this chapter focuses on the study of error correction codes (ECCs) within diffusion-based MC (DBMC) systems. Introducing ECCs is a fast and efficient way to deal with data reliability and in this chapter it is shown that such codes can be taken forward into the MC system. In addition, due to the energy limitation of nano-machines [10], which are used as the transmitter (T_X) and the R_X in the MC system, here, the energy consumption needed for the introduction of ECCs is also considered.

This chapter begins with a brief introduction to MC systems and the context of the problem. Then, four popular ECCs are introduced into MC systems with an associated analysis of the propagation and communication channel models. Finally, the performance of the coded system with regard to bit error rate (BER) and critical distance is presented.

11.1 Introduction

This section provides an overview of the MC system under consideration. Furthermore, related literature on the use of ECCs in the MC system is also presented.

11.1.1 Architecture of the MC System

To achieve the transmission process between the T_X and the R_X, a carrier signal needs to be generated and released from the T_X and, through propagation, then delivered to the R_X where the original information can be decoded. Considering this transmission process, a conventional communication system should include three main components: the T_X, the communication channel and the R_X. A basic diagram of the communication system is presented in Figure 11.1.

For MC, the molecules themselves are used as the information carrier to be transmitted between the T_Xs and the R_Xs, and these kinds of molecules are called information molecules. The three elements shown in Figure 11.1 can also be defined within the MC system. However, the T_X for the MC system is a bio-nanomachine, such as a modified living cell or a biological nano-robot, which is able to synthesize, store and release information molecules. The original information can be modulated and then released from the T_X.

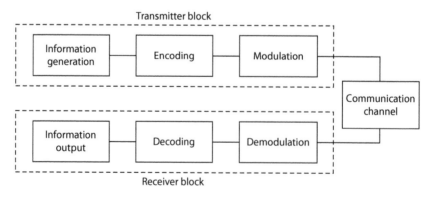

FIGURE 11.1
Block diagram of a basic conventional communication system.

In addition, encoding techniques may be included in the T_X to enhance the reliability of the transmission process.

There are three different types of MC based on different propagation schemes (details are provided in Section 11.1.2). Each propagation scheme can be considered as the information transport between the T_X and the R_X. The R_X in the MC system is another bio-nanomachine. It aims to capture the information molecules from the surrounding area, which are then demodulated and decoded if needed to recover the original information. Here, the R_X is considered an absorbing receiver, where the R_X's surface contains several receptors, the information molecules can be captured by these receptors and these then form chemical bonds to trigger the detection process in the R_X. These information molecules will be destroyed after dissociation.

11.1.2 MC Types

Due to different propagation schemes, the MC types can be divided into three categories [11]: walkway-based MC [12–14], advection-based MC [15] and DBMC [16–18].

In walkway-based MC, the information molecules propagate through active transports. For this type of MC, the pathway between the T_X and the R_X is pre-designed. An example is using molecular motors as the transport mechanism for transmitting information molecules between the T_X and the R_X. Another example of this MC type shows that the information molecules are carried by microtubules, propelled by molecular motors, and are absorbed on a flat surface between the T_X and the R_X. This type of MC is normally used for intra- or inter-cell communications.

For advection-based MC, the molecules propagate through diffusion in a fluidic medium. The use of gap junctions [19–21] and self-propelled microorganisms [22–24] (e.g. bacteria) as the transport mechanisms are two examples of advection-based MC. For the gap junction transport mechanism, the information molecule can be transmitted from the T_X to the R_X through cells that are in contact. The way that the information is inserted into bacteria and the propagation of the bacteria to complete the transmission process is called the bacterial transport mechanism. The bacteria are guided by attractant molecules that are released from the R_X.

In DBMC, the information molecules propagate through their spontaneous diffusion in the medium [11,25,26]. The focus here is on DBMC.

11.1.3 Literature Review of ECCs in the MC System

One of the early works that considered the use of ECCs in the MC system [27] considered the Hamming code as a simple block code for use in MC systems. The results indicated that introducing Hamming codes can improve the performance of the MC system. In [28], Ko et al. proposed a molecular coding distance function that considers the transition probability between codewords. Using this distance function, a suitable code for the MC system can be constructed. Due to the issue of the energy limitation of the nano-machine, the designs of simple codes for the MC system were given in [29]. Minimum energy codes were introduced into MC systems by Bai et al. [30] to reduce the energy consumption by minimizing the average code weights. The work from each study all concluded that the employment of coding techniques in the MC system can improve the BER performance compared to the uncoded system.

In this chapter, four selected ECCs from the block code family and the convolutional code family are introduced with the aim to improve the BER performance and reduce the energy consumption of the MC system. The details for BER and the critical distance of the MC system are considered in Section 11.6.

11.2 Design of the Point-to-Point DBMC System

The basic design of the three-dimensional (3D) point-to-point (PTP) DBMC system is shown in Figure 11.2. This system contains one T_X and one R_X. The center of the T_X (the release point) is at a distance d from the R_X, which has a radius of R.

The T_X can be considered as a functional bio-nanomachine, where the information molecules can be modulated, encoded and released. The propagation process is completed by the free diffusion of molecules. The R_X used in this system is a perfect absorbing receiver, where the information molecules are captured when they reach the capturing area and are removed from the environment. The whole transmission process can be divided into three phases: the emission phase, the diffusion phase and the reception phase.

Before emitting the carrier molecules, the information is modulated and encoded in the T_X, and then the molecules are released into the transmission medium. After emission from the T_X, the information molecules propagate in the environment via diffusion. During this process, they are considered to move randomly and independently. The receptors on the surface of the R_X will detect and absorb information molecules once they arrive in the capturing area. Here, the capturing surface is considered a sphere that has the same radius as the R_X. After that, the information molecules are demodulated and decoded within the R_X where the original information can be recovered.

Three communication scenarios are introduced depending on the different types of T_X and R_X: Scenario 1, a nano-machine communicates with a nano-machine (N2N communication); Scenario 2, a nano-machine communicates with a macro-machine (N2M communication); and Scenario 3, a macro-machine communicates with a nano-machine (M2N communication). The communication links for all the foregoing scenarios are based on the molecular diffusion process.

An intra-body health monitoring system shown in Figure 11.3 is one of the applications that involves all three scenarios. This system contains two sizes of machines, viz. nano-machines and macro-machines. The components of the former, nano-sensors or nano-robots, are all at the scale of a nano-meter. The latter, typically macro-robots, are

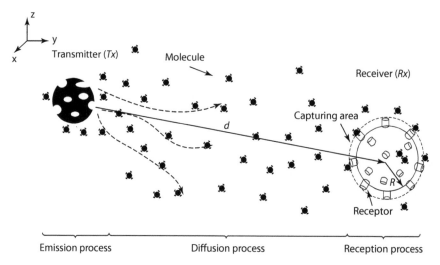

FIGURE 11.2
The 3D PTP DBMC system model.

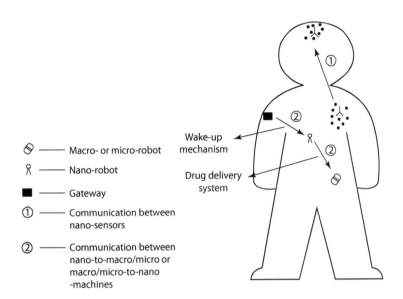

Macro- or micro-robot

Nano-robot

Gateway

① Communication between nano-sensors

② Communication between nano-to-macro/micro or macro/micro-to-nano -machines

Wake-up mechanism

Drug delivery system

FIGURE 11.3
The communication scenarios of the intra-body nano-network.

manufactured by using a collection of nanoscale components. In drug delivery systems, a set of nano-robots as beacons located around the body can transmit information to guide macro-scale drug delivery robots working around human blood vessels [31,32]. Conversely, some applications, such as macro-machines, which act as gateways need to communicate with nano-robots and transmit information between the outside and inside of the body. This kind of macro-machine is not designed to be mobile and is most likely found on (or just under) the skin. Through emitting the information, nano-robots can be polled to get ready for a specified operation.

Here, the three communication scenarios are considered using the same propagation and communication models, which are introduced in the next two sections.

11.3 The Propagation Model

Modeling molecular propagation is one of the key challenges in predicting MC system performance. To simplify the system analysis, the information molecules are considered to move randomly and independently. The propagation of these information molecules is governed by Brownian motion. As shown in Figure 11.2, when the information molecules reach the capturing area of the R_X, there is a probability that the information molecule escapes the absorption of the R_X; this probability is called the survival probability, which can be denoted as $P_{su}(d, t)$. This probability with time t in the 3D diffusion medium satisfies the backward diffusion equation [27]:

$$\frac{\partial P_{su}(d,t)}{\partial t} = D\nabla^2 P_{su}(d,t), \tag{11.1}$$

where ∇^2 is the Laplace operator, t is the transmission time and D is the diffusion coefficient. The initial condition and the boundary conditions of Equation 11.1 are

$$P_{su}(d,0) = 1, \ \forall \ |d| > R, \tag{11.2}$$

$$P_{su}(|d| = R, t) = 0 \text{ and } P_{su}(|d| \to \infty, t) = 1 \ \forall t. \tag{11.3}$$

Exploiting radial symmetry, the solution to Equation 11.1 is

$$P_{su}(d,t) = 1 - \frac{R}{d}\text{erfc}\left(\frac{(d-R)}{2\sqrt{Dt}}\right), \tag{11.4}$$

where erfc is the complementary error function.

For this MC system model, the capture probability rather than the escape probability is more important. Thus, the capture probability $P_{ca}(d, t)$ can be obtained by

$$P_{ca}(d,t) = 1 - P_{su}(d,t) = \frac{R}{d}\text{erfc}\left(\frac{(d-R)}{2\sqrt{Dt}}\right). \tag{11.5}$$

Figure 11.4 illustrates that for a certain distance, the capture probability increases with the increase in time, which means that as time increases, more and more molecules will arrive at the receiver; then, by increasing the transmission time, the capture probability will become stable. In addition, for a certain period of time, the capture probability decreases when the distance between the T_X and the R_X increases.

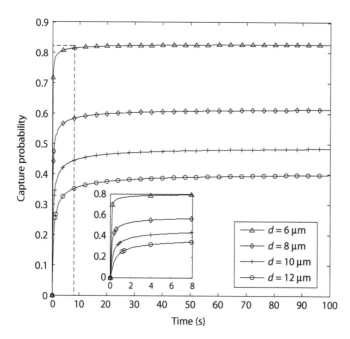

FIGURE 11.4
Capture probabilities vs. time for different distances with $R = 5\,\mu m$ and $D = 79.4\,\mu m^2\,s^{-1}$.

11.4 The Communication Channel Model

For a PTP DBMC system, the transmitted information is represented by a sequence of symbols that are distributed over sequential and consecutive time slots, which have equal length with one symbol in each slot; the duration of each time slot is denoted as t_s. Here, the on-off keying is used for modulating the information, where "1" represents a specific number of molecules released from T_X, and "0" represents no molecules released. Specifically, if the number of information molecules arriving at the R_X at a certain time slot exceeds a pre-designed threshold τ, the symbol is interpreted as a "1". Otherwise, it will be interpreted as a "0". However, during the transmission process, errors may be caused by intersymbol interference (ISI), which is caused by the remaining molecules from the previously transmitted symbols. The ISI effect is related to the properties of the medium used, the distance of the symbol propagation and the selection of the pre-designed threshold value. Considering a memory limited channel with ISI length I, the current transmitted symbol can be affected by the previous I symbols.

The communication channel used here is a binomial one, where each molecule arrives at the R_X or does not. It has been previously stated that the previous bits can have an influence on the current bit due to ISI. Considering that N_{tx} information molecules are released at the start of the current time slot, the number of molecules received during the current time slot, N_0, follows a binomial distribution given by [33]

$$N_0 \sim \mathcal{B}\left(N_{tx}, P_{ca,0}\right), \tag{11.6}$$

where $P_{ca,0} = P_{ca}(d, t_s)$.

If N_{tx} is large enough, a binomial distribution can be approximated by a normal distribution, thus,

$$N_0 \sim \mathcal{N}\left(N_{tx}P_{ca,0}, N_{tx}P_{ca,0}\left(1 - P_{ca,0}\right)\right). \tag{11.7}$$

The values of t_s for different distance, d, can be selected by the time at which 60% of the information molecules arrive at the R_X [33]. Considering the capture probability shown in Equation 11.5, if t goes to infinity $t \to \infty$, the analytical result R/d is obtained for the probability that an information molecule is received by R_X. Thus,

$$0.6\frac{R}{d} = \frac{R}{d}\text{erfc}\left(\frac{d-R}{2\sqrt{Dt_s}}\right), \tag{11.8}$$

then t_s can be derived as

$$t_s = \frac{(d-R)^2}{4D\left(\text{erfc}^{-1}(0.6)\right)^2}, \tag{11.9}$$

where erfc^{-1} is the inverse of the complementary error function.

As mentioned previously, the transmitted molecules at the start of the time slot cannot be guaranteed to reach the R_X within the current time slot, and information molecules may still exist in the environment and may arrive in future time slots. Thus, the number of information molecules that are released at the start of the ith time slot before the current one but arrive in the current time slot is denoted as N_i, and described by

$$N_i \sim \mathcal{B}\left(N_{tx}, P_{ca,i} - P_{ca,i-1}\right), \tag{11.10}$$

where $P_{ca,i} = P_{ca}\left(d, (i+1) t_s\right)$ for $i = \{1, 2,\ldots, I\}$.

The corresponding normal approximation can be obtained as

$$N_i \sim \mathcal{N}\left(N_{tx}\left(P_{ca,i} - P_{ca,i-1}\right), N_{tx}\left(P_{ca,i} - P_{ca,i-1}\right)\left(1 - P_{ca,i} + P_{ca,i-1}\right)\right) = \mathcal{N}\left(\varpi_i, \gamma_i\right), \tag{11.11}$$

where:
$$\varpi_i = N_{tx}(P_{ca,i} - P_{ca,i-1})$$
$$\gamma_i = N_{tx}(P_{ca,i} - P_{ca,i-1})(1 - P_{ca,i} + P_{ca,i-1})$$

Thus, the total number of information molecules received in the current time slot, N_T, comprises the number of received information molecules that were sent at the start of the current time slot and the number of received information molecules sent from all I previous time slots [34]:

$$N_T = a_c N_0 + \sum_{i=1}^{I} a_{c-i} N_i$$

$$\sim \mathcal{N}\left(a_c N_{tx} P_{ca,0} + \sum_{i=1}^{I} a_{c-i}\varpi_i, \ a_c N_{tx} P_{ca,0}\left(1 - P_{ca,0}\right) + \sum_{i=1}^{I} a_{c-i}\gamma_i\right), \tag{11.12}$$

where $\{a_{c-i}, i=0, 1, 2,\ldots, I\}$ represents the binary transmitted information sequence, which includes current and all previous I symbols.

11.5 BER Analysis for the MC System

BER is considered a key metric to assess the performance of systems that transmit information between the T_X and the R_X. Here, the ISI is considered as the main noise source.

11.5.1 BER Analysis

Considering a channel with an ISI length equal to I, error patterns can be obtained by the different permutations of the previous I symbols, so the number of error patterns is 2^I. The errors during the transmission process occur when there is a discrepancy between the transmitter and receiver signals. For a binary transmission, there are two cases, firstly, when a "0" is transmitted, but a "1" is received. Secondly, when a "1" is transmitted, but a "0" is received.

For the first case, the error probability for error pattern j, $P_{e01,j}$, $j = \{1, 2,\ldots, 2^I\}$ shows that the number of received information molecules exceeds τ, which means $N_{T,j} > \tau$, thus,

$$P_{e01,j} = p_{tx}^{\alpha_j} \left(1 - p_{tx}\right)^{I-\alpha_j} P\left(N_{T,j} - \tau > 0\right)$$

$$= p_{tx}^{\alpha_j} \left(1 - p_{tx}\right)^{I-\alpha_j} \Phi\left(\frac{\mu_{01,j} - \tau}{\sigma_{01,j}}\right), \tag{11.13}$$

where:

$$\mu_{01,j} = \sum_{i=1}^{I} a_{c-i,j} \varpi_i, \ \sigma_{01,j} = \sqrt{\sum_{i=1}^{I} a_{c-i,j} \gamma_i}. \tag{11.14}$$

For the second case, the error probability of the normal and Poisson approximations for the error pattern j, $P_{e10,j}$, where $N_{T,j} \leq \tau$, thus it can be obtained as

$$P_{e10,j} = p_{tx}^{\alpha_j} \left(1 - p_{tx}\right)^{I-\alpha_j} P\left(N_{T,j} - \tau \leq 0\right)$$

$$= p_{tx}^{\alpha_j} \left(1 - p_{tx}\right)^{I-\alpha_j} \Phi\left(-\frac{\mu_{10,j} - \tau}{\sigma_{10,j}}\right), \tag{11.15}$$

where:

$$\mu_{10,j} = N_{tx} P_{ca,0} + \sum_{i=1}^{I} a_{c-i,j} \varpi_i, \ \sigma_{10,j} = \sqrt{N_{tx} P_{ca,0}\left(1 - P_{ca,0}\right) + \sum_{i=1}^{I} a_{c-i,j} \gamma_i}. \tag{11.16}$$

Thus, the average BER of the uncoded MC system, P_e, can be derived as

$$P_e = p_{tx} P_{e10} + \left(1 - p_{tx}\right) P_{e01}$$

$$= p_{tx} \sum_{j=1}^{2^I} P_{e10,j} + \left(1 - p_{tx}\right) \sum_{j=1}^{2^I} P_{e01,j}, \tag{11.17}$$

where P_{e01} and P_{e10} are represented as

$$P_{e01} = \sum_{j=1}^{2^I} P_{e01,j}, \ P_{e10} = \sum_{j=1}^{2^I} P_{e10,j}. \tag{11.18}$$

11.5.2 Numerical Results

As mentioned in Section 11.4, the previous transmitted symbols will affect the current transmitted symbol, which causes the ISI. Using an infinite ISI length in the system analysis will substantially increase the complexity of the analysis process. Thus, a limited memory channel is considered here with an ISI length I. In order to obtain an accurate result, the value of I should be evaluated.

Figure 11.5 shows the BER with the number of information molecules per bit for different ISI lengths from 1 to 10 at $d = 15 \ \mu m$, $R = 5 \ \mu m$ and $D = 79.4 \ \mu m^2/s$. The results show that the longer the ISI length I, the higher the BER. It can also be noted that as the ISI length increases considerably, the effect it has upon the BER becomes less prominent, i.e. the BER

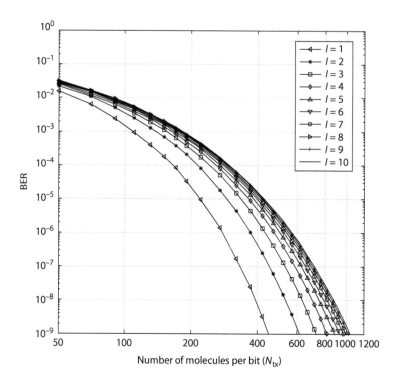

FIGURE 11.5
BER vs. number of molecules per bit for different ISI length $I = \{1, 2, ..., 10\}$ at $d = 15\,\mu m$, $p_{tx} = 0.5$.

value begins to converge. Therefore, choosing $I = 10$ for analysis of the channel is enough to produce an accurate result.

The value of the threshold used at the R_X is a pre-designed threshold. It can be obtained by searching the minimum BER for a specific N_{tx} in the range $\tau \in [1, N_{tx}]$.

The BER results shown in all the BER figures from here on are based on a set of parameters that are shown in Table 11.1.

Figure 11.6 demonstrates that the communication system has a better performance if large numbers of molecules are sent at the start of a one-time slot. In addition, for a chosen value of N_{tx}, the BER increases as the distance between the T_X and the R_X increases. Thus, to achieve a better performance of the communication system (lower BER), it is better to have a smaller propagation distance, d, and use a larger number of transmitted molecules per bit, N_{tx}.

TABLE 11.1

Parameter Values

Parameter	Definition	Value
R	Radius of the R_X	$5\,\mu m$
D	Transmission distance	$\{6, 8, 10, 12\}\,\mu m$
D	Diffusion coefficient	$79.4\,\mu m^2 s^{-1}$
N_{tx}	Number of molecules per bit	$50\sim10^3$
I	ISI length	10

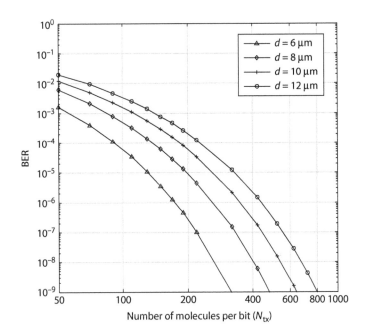

FIGURE 11.6
BER vs. number of molecules per bit.

11.6 ECCs in the PTP DBMC System

11.6.1 Overview

We now introduce four ECCs into the MC system. Three of them are from the block code family: Hamming codes, cyclic Reed–Muller (C-RM) codes and low-density parity-check (LDPC) codes. One of them is from the convolutional codes family, which is self-orthogonal convolutional codes (SOCCs). A brief review of each code is given in the following paragraphs.

Hamming codes, as one of the simplest linear codes devised for error correction, have been widely used in conventional communications systems [35]. Although these simple block codes can only correct a single error and are not powerful codes for the conventional communications system, the encoder and decoder can be implemented much more easily and efficiently in terms of the energy budget for the MC system. Here, the Hamming codes, as the most basic coding technique, are employed in the MC system and are also provided as a comparison.

RM codes [36,37] are a class of linear codes that can be constructed using Galois fields. Here, the RM codes are constructed in cyclic form (C-RM) to show that they are also a subset of Bose, Chaudhuri and Hocquenghem (BCH) codes. By using the shift-register encoder and the majority logic decoder, this kind of C-RM code can be easily encoded and decoded, respectively. From the perspective of energy, the main advantage of such C-RM codes is that the encoder is simpler than the original RM codes, which holds benefits for the above-mentioned applications [38–40].

The random and structured LDPC codes [41,42] are two kinds of LDPC codes. The Euclidean geometry LDPC (EG-LDPC) codes, which are considered for use in the MC system, are one of the structured LDPC codes. Several advantages over random LDPC codes are shown in EG-LDPC codes, such as the existence of several decoding algorithms (cyclic or quasi-cyclic), a simpler decoding scheme and the ability to extend (or shorten) the code in order to adapt to an application [43–45]. A comprehensive account of the implementation of a cyclic EG-LDPC code has been given in [46]. Thus, in this chapter, the focus will be on one specific construction, namely the cyclic EG-LDPC code.

SOCCs are a type of convolutional codes that have the property of being easy to implement, thus satisfying one of the key design requirements, simplicity [47,48]. Furthermore, this kind of convolutional code has been shown to have an equal, or superior, performance to block codes in low-cost and low-complexity applications. Examples can be found detailing their competitiveness in practical applications [49–52].

11.6.2 The Construction of Logic Gates in the Biological Field

In conventional communication systems, the coding techniques can be implemented using electric circuits that are composed of large numbers of transistors; through the interconnection of these transistors, the functions of Boolean logic can be realized. However, considering the T_X, R_X and the channel of the MC system, these electronic circuits cannot be realized due to their complexity level. Thus, the implementation of these logic gates needs to be re-investigated from the biological view.

In [53], the authors considered protein-based signaling networks within biological cells. It is shown that the fundamental motif in all signaling networks is based on the protein phosphorylation and dephosphorylation cycle, which is also called a cascade cycle via kinases and phosphatases, respectively. The cascade cycle can complete this transfer cycle very quickly when operating in ultrasensitive mode, which satisfies the requirement of the logic gates: a fast changing between two states. Thus, it is possible to construct various control and computational analog and digital circuits by combining these cascade cycles.

Figure 11.7 shows how a NAND gate could be formed from a cascade cycle. 'A' and 'B' are the inputs to the kinase step that can cause the cascade cycle to switch. A NAND gate is a universal gate, thus it is possible to build all future logical circuits. In [53], the author presented a basic memory unit, the binary counter and NOT gate by combining cascade cycles.

Furthermore, for the output of electronic circuits, a clear high- or low-level voltage should be clearly seen in order to guarantee the reliability of the transmission process. However, thermal noise and intrinsic distortion of the transistor exist in the circuits, which cause an

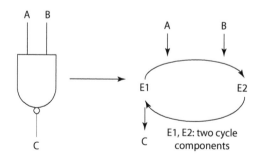

FIGURE 11.7
NAND gate using cascade cycle. (Adapted from Sauro, H.M. and Kholodenko, B.N., *Progress in Biophysics and Molecular Biology*, 86, 5–43, 2004.)

unstable level of output. In the biological field, the circuits formed of cascade cycles also suffer from noise, which is called the cascade cycle intrinsic fluctuation distortion [54]. One of the methods to reduce the effects from this kind of noise is to increase the signal intensity by increasing the number of substrate molecules, N_{sm}, at the input stage. This kind of molecule is the input signal of the cascade cycle. Here, such molecules can be looked upon as the code generation molecules that are required to encode and decode the transmission data. They are different from the information molecules, which are introduced in Section 11.4, as they are internal molecules and are only used for the encoding and decoding processes within the T_X and the R_X, and do not suffer from any effects caused by the diffusion process.

The impact of the number of substrate molecules on the output signal of the cascade cycle has been investigated in [54]. Each substrate molecule is either unmodified or modified at the output stage. The results shown in [54] indicate that for different numbers of substrate molecules used at the input stage, the performance of the output signal is different. When the N_{sm} is small ($N_{sm} = 30$), the high and low outputs overlap and blur due to the cascade cycle's intrinsic fluctuation distortion; when N_{sm} increases to 300, a clear transmission output pulse is given. A further increase of N_{sm} will cause a slow output response and information loss. Overall, the selected value of $N_{sm} = 300$ is sufficient to reduce the effects that come from the biochemical intrinsic distortion and obtain a clear output signal.

11.6.3 Energy Model

Other than BER, energy consumption is another key metric for a designed system. Introducing coding techniques can reduce the BER; however, in the meantime, an extra cost is also incurred due to the encoding and decoding process within the T_X and the R_X. This cost is proportional to the complexity of the design of the encoder and decoder.

Adenosine triphosphate (ATP) [55] is normally used to measure the energy transfer between cells in living organisms. Here, it is used to calculate the energy requirements of the proposed coding DBMC systems. The energy consumption during the encoding and decoding process can be composed of two parts: the energy cost of the coded circuits construction and the energy cost for generating the substrate molecules.

One cascade cycle can form a NAND gate, and one ATP is cost to activate one cascade cycle [55,56]. The energy cost from one ATP reaction, in Joules, is approximately equal to 8.3×10^{-20} J. Considering such small quantities, here the energy is worked in $k_B T$. Thus, the energy cost for one ATP reaction is $20 k_B T$, where k_B is the Boltzmann constant and it is assumed here that the system is operating at an absolute temperature, $T = 300$ K. As mentioned in Section 11.6.2, all further logic circuits can be devised from combinations of NAND gates based on the principles of Boolean algebra [57]. So, the total energy cost for the logic circuits can be computed.

Synthesizing the substrate molecules also results in energy consumption. The energy cost of synthesizing a molecule is approximately $2450 k_B T$ [57]. In this case, the total energy cost of synthesizing the substrate molecules is $2450 N_{sm} k_B T$.

11.6.4 ECCs in MC Systems

In block coding, the transmitted binary information sequence is segmented into information blocks with a fixed length, k. After the encoding process, each information block is transformed into an n bits codeword [50] by inserting $(n-k)$ parity check bits to improve the reliability of the information. Most known block codes belong to the class of linear codes and the cyclic code is one of the famous linear codes. This class of codes has a strong structural property and is usually used in practice. Compared with the general linear block codes, cyclic codes can be obtained by imposing an additional strong structural element on the code [58]. In this section, Hamming codes, C-RM codes and EG-LDPC codes as the cyclic codes will be described in turn.

11.6.4.1 Hamming Codes

Hamming codes, which are denoted in the form (n_H, k_H), where, $n_H = 2^m - 1$ is the coded length output for the number of parity check bits m, $m \geq 2$, and $k_H = 2^m - 1 - m$ is the number of date bits per block. The minimum distance, d_{minH}, of the Hamming code is 3, which means that only one error can be corrected in each block. The data rate of Hamming codes can be calculated as $r_H = k_H / n_H$.

The Hamming codes constructed as cyclic codes can be encoded by multiplying the information polynomial with the generator polynomial. Here, three Hamming codes are selected for use in MC systems: (7,4), (15,11) and (31,26) Hamming codes. The generator polynomials for these codes are given by $g_{m=3}(x) = x^3 + x + 1$, $g_{m=4}(x) = x^4 + x + 1$ and $g_{m=5}(x) = x^5 + x^2 + 1$, respectively. The Meggitt decoder is used here. Considering the Meggitt theorem, the syndrome polynomial for testing the error patterns of the foregoing Hamming codes are configured as $S_{m=3}(x) = x^2 + 1$, $S_{m=4}(x) = x^3 + 1$ and $S_{m=5}(x) = x^4 + x$, respectively. Figure 11.8 shows a general form of encoders and decoders for Hamming codes.

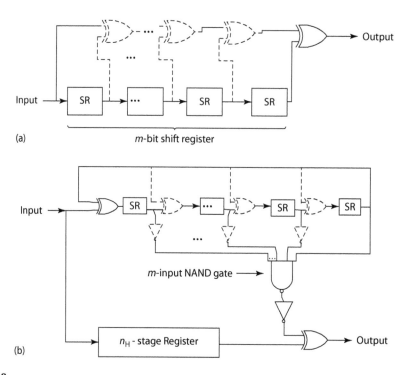

FIGURE 11.8
General nonsystematic encoder (a) and Meggit decoder (b) for Hamming code. (Adapted from Blahut, R.E., *Algebraic Codes for Data Transmission*, Cambridge University Press, 2003.)

11.6.4.2 C-RM Codes

The RM codes considered here are constructed as cyclic codes to achieve multiple error correction capabilities.

For any integer, $l \geq 2$ and $0 \leq z < l-1$, the zth-order C-RM codes can be represented as (z, l) C-RM, with a data rate $r_R = k_R/n_R$, a block length $n_R = 2^l - 1$ and the information length can be calculated as

$$k_R = 1 + \binom{l}{1} + \binom{l}{2} + \cdots + \binom{l}{r} = n_R - \sum_{i=1}^{l-z-1} \binom{l}{i}. \tag{11.19}$$

The minimum distance of a C-RM code is $d_{minR} = 2^{l-z} - 1$ and the error capability can be calculated as

$$E_{cR} = [d_{minR} - 1]/2. \tag{11.20}$$

Here, two C-RM codes, (1,4)C-RM and (2,5)C-RM codes are considered for MC systems. The generator and check polynomial for (1,4) and (2,5)C-RM are given by

$$g_{(1,4)\text{C-RM}}(x) = M_1 \cdot M_2 \cdot M_3 = x^{10} + x^8 + x^5 + x^4 + x^2 + x + 1. \tag{11.21}$$

$$h_{(1,4)\text{C-RM}}(x) = (x+1) \cdot (x^4 + x^3 + 1) = x^5 + x^3 + x + 1. \tag{11.22}$$

$$g_{(2,5)\text{C-RM}}(x) = x^{15} + x^{11} + x^{10} + x^9 + x^8 + x^7 + x^5 + x^3 + x^2 + x + 1. \tag{11.23}$$

$$h_{(2,5)\text{C-RM}}(x) = x^{16} + x^{12} + x^{11} + x^{10} + x^9 + x^4 + x + 1. \tag{11.24}$$

A general design of C-RM codes is shown in Figure 11.9, where the encoding process can be achieved using simple shift registers and the decoding process can be realized using a multiple-step majority logic method.

11.6.4.3 EG-LDPC Codes

In this section, a special case called cyclic two-dimensional EG-LDPC codes is considered [43,59]. To simplify the nomenclature, the following LDPC codes are assumed to be cyclic two-dimensional EG-LDPC codes.

LDPC codes can be represented as (n_L, k_L) with a block length $n_L = 2^{2s} - 1$ and an information length $k_L = 2^{2s} - 3^s$. The data rate of LDPC codes can be represented as $r_L = k_L/n_L$, and the error correction capability can be denoted as E_{cL}, where

$$E_{cL} = [d_{minL} - 1]/2, \tag{11.25}$$

where the minimum distance $d_{minL} = 2^s + 1$.

The (15,7), (63,37) and (255,175)LDPC codes are considered in this section. The generator polynomial for the above codes is given as

$$g_{(15,7)\text{LDPC}}(x) = x^8 + x^7 + x^6 + x^4 + 1. \tag{11.26}$$

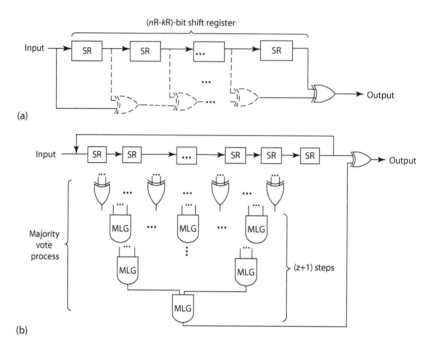

FIGURE 11.9

(a) General encoder and (b) general majority logic decoder for C-RM codes.

$$g_{(63,37)\text{LDPC}}\left(x\right) = x^{26} + x^{24} + x^{16} + x^{15} + x^{14} + x^{13} + x^{12}$$
$$+ x^{10} + x + 1. \tag{11.27}$$

$$g_{(255,175)\text{LDPC}}\left(x\right) = x^{80} + x^{78} + x^{76} + x^{74} + x^{71} + x^{69} + x^{68} + x^{67}$$
$$+ x^{66} + x^{64} + x^{63} + x^{61} + x^{59} + x^{58} + x^{55} + x^{54} + x^{51}$$
$$+ x^{49} + x^{47} + x^{45} + x^{42} + x^{40} + x^{39} + x^{38} + x^{37}$$
$$+ x^{36} + x^{27} + x^{26} + x^{25} + x^{23} + x^{22} + x^{21} + x^{19}$$
$$+ x^{18} + x^{17} + x^{16} + x^{15} + x^{14} + x^{13} + x^{11} + x^{10}$$
$$+ x^9 + x^7 + x^6 + x^3 + 1. \tag{11.28}$$

Figure 11.10 shows a general encoder and decoder design for LDPC codes. The encoding process can be achieved using simple feedback shift registers and the decoding process can be implemented using a one-step majority logic decoder.

11.6.4.4 SOCCs

One of the convolutional codes considered here is the SOCC. A SOCC is a kind of convolutional code that has the property of being easy to implement, thus satisfying one of the key design requirements of code simplicity [47,48].

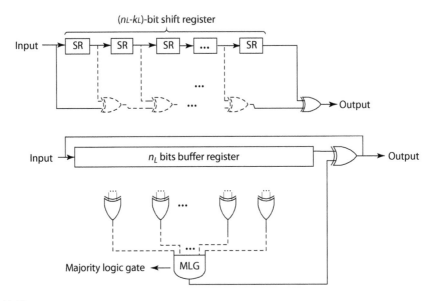

FIGURE 11.10
(a) General encoder design and (b) decoder design for LDPC code.

For an (n_S, k_S, b) SOCC, n_S is the code length, k_S is the information length and b is the number of input memory blocks. The code rate is $r_S = k_S/n_S$ [50]. The error correction capability, E_{cS}, is

$$E_{cS} = [J/2], \tag{11.29}$$

where J is the number of check sums that orthogonal on one error.

The effective constraint length, n_E, represents the total number of channel error bits checked by the orthogonal check sum equations, where

$$n_E = \tfrac{1}{2}(J^2 + J)k_S + 1. \tag{11.30}$$

In this work, only SOCCs with an information length $k_S = n_S - 1$ are considered. Here, four SOCCs are presented: (2,1,6) and (2,1,17)SOCCs, both with $r_S = 1/2$, and (3,2,2) and (3,2,13)SOCCs, both with $r_S = 2/3$.

The generator polynomials of (2,1,6) and (2,1,17)SOCCs are

$$g_{(2,1,6)\text{SOCC},1}{}^{(2)}(x) = x^6 + x^5 + x^2 + 1, \tag{11.31}$$

$$g_{(2,1,17)\text{SOCC},1}{}^{(2)}(x) = x^{17} + x^{16} + x^{13} + x^7 + x^2 + 1, \tag{11.32}$$

where $g_{\text{SOCC},i}{}^{(ns)}$ is the generator polynomials with $i = 1, 2, \ldots, k_S$.

For (3,2,2) and (3,2,13)SOCCs, the generator polynomial pairs are

$$g_{(3,2,2)\text{SOCC},1}{}^{(3)}(x) = x + 1, \tag{11.33}$$

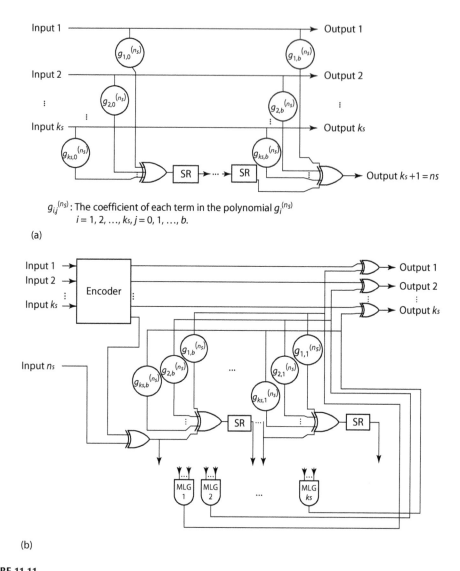

$g_{i,j}^{(n_s)}$: The coefficient of each term in the polynomial $g_i^{(n_s)}$
$i = 1, 2, ..., k_s, j = 0, 1, ..., b.$

(a)

(b)

FIGURE 11.11
The general (a) encoder and (b) decoder design for SOCCs.

$$g_{(3,2,2)SOCC,2}^{(3)}(x) = x^2 + 1. \tag{11.34}$$

The circuits of the encoder and feedback majority logic decoder for a general SOCC are shown in Figure 11.11.

11.6.5 BER and Coding Gain

Considering a linear block code with a decoder that can correct all errors less or equal to the error correction capacity, the decoded BER for the system with linear block codes can be expressed by the following approximation:

$$P_{\text{e-coded}} \approx \frac{1}{n} \sum_{j=E_c+1}^{n} j \binom{n}{j} \left(P_e^*\right)^j \left(1-P_e^*\right)^{n-j}, \tag{11.35}$$

where n is the block length, E_c is the error correction capacity and P_e^* is the one bit error probability in the uncoded case.

For a fair comparison with the uncoded operation, the number of molecules used for the calculation of P_e^* for the coded system should be evaluated with a reduction in the number of molecules used for an uncoded system, (11.17), by multiplying with the code rate.

On the other hand, for the convolutional code with a feedback majority logic decoder, the theoretical analysis of BER of the convolutional code can be upper bounded by [50]

$$P_{\text{e-coded}} \leq \frac{1}{k} \sum_{j=E_c+1}^{n_E} j \binom{n_E}{j} \left(P_e^*\right)^j \left(1-P_e^*\right)^{n_E-j}, \tag{11.36}$$

where k is the information length of the code. $P_{\text{e-coded}}$ can be approximated as [50]

$$P_{\text{e-coded}} \approx \frac{1}{k} \binom{n_E}{E_c+1} \left(P_e^*\right)^{E_c+1}. \tag{11.37}$$

The coding gain is also introduced as a way to measure the BER performance. For MC systems, the coding gain aims to measure the difference between the number of molecules for the uncoded and coded system required to reach the same BER level. It can be directly obtained as

$$G_{\text{coding}} = 10 \times \log\left(\frac{N_{\text{uncoded}}}{N_{\text{coded}}}\right), \tag{11.38}$$

where N_{uncoded} and N_{coded} are the number of information molecules for the uncoded and the coded system at a chosen BER level.

11.6.6 Energy Consumption Analysis

The energy consumption for different coding techniques is given in this section with respect to the energy model that was introduced in Section 11.6.3. Based on that model, Table 11.2 shows four basic logic gates and their corresponding ATPs' consumption.

TABLE 11.2

Logic Gates and Corresponding ATPs' Consumptions

Logic Gate	Cost in ATPs
NAND	1
NOT	1
XOR	4
Shift-register unit	4

11.6.6.1 Energy Consumption for Hamming Codes

With reference to Figure 11.8, for $m = \{3, 4, 5\}$ Hamming codes, two XOR gates and m shift-register units are needed for each encoder circuit, which implies that the energy consumption of the encoder circuits is

$$E_{\text{en-H}} = 20\,N_{\text{sm-en}}\left(4m+8\right)+2450\,N_{\text{sm-en}}, \tag{11.39}$$

and three XOR gates, $(m+n_{\text{H}})$ shift-register units, $(m-1)$ NOT gates and one multi-input NAND gate are needed for each decoder circuits, which implies the energy cost of the decoder is

$$E_{\text{de-H}} = 20N_{\text{sm-de}}\left(4\left(m+n_{\text{H}}\right)+\left(m-1+1\right)+12\right)+2450N_{\text{sm-de}}$$

$$= 20N_{\text{sm-de}}\left(5m+4n_{\text{H}}+12\right)+2450N_{\text{sm-de}}, \tag{11.40}$$

where $N_{\text{sm-en}}$ and $N_{\text{sm-de}}$ are the numbers of substrate molecules used for the encoder and decoder, respectively. As referenced in Section 11.6.2, $N_{\text{sm-en}} = N_{\text{sm-de}} = 300$.

11.6.6.2 Energy Consumption for C-RM Codes

Consider the hardware requirement in Figure 11.9. For C-RM codes, the encoder can be achieved using simple feedback shift registers and the decoding process can be completed using a multi-step majority logic method. For a majority logic gate (MLG), an output of one is produced when more than half of its inputs are equal to one, otherwise, the output is zero.

For any J-input MLG, the number of NAND gates can be calculated as

$$N_{\text{NAND-MLGs}} = \begin{cases} \displaystyle\sum_{i=J/2+1}^{J-1}\binom{J}{i}+1, & J \neq 2 \\ 2, & J = 2 \end{cases}. \tag{11.41}$$

For C-RM codes, $(n_{\text{R}}-k_{\text{R}})$ shift registers are used, and the number and the location of the two-input XOR gates in the circuits are dependent upon the generator polynomial of each code. The energy cost of encoding is

$$E_{\text{en-(1,4)RM}} = 20N_{\text{sm-en}}\left(4\left(n_{\text{R}}-k_{\text{R}}\right)+24\right)+2450N_{\text{sm-en}}, \tag{11.42}$$

$$E_{\text{en-(2,5)RM}} = 20N_{\text{sm-en}}\left(4\left(n_{\text{R}}-k_{\text{R}}\right)+40\right)+2450N_{\text{sm-en}}. \tag{11.43}$$

In general, the zth-order C-RM code can be decoded with a $(z+1)$-step majority logic decoder. For these decoding circuits, the total number, N_{ML}, of J-input MLGs used in the circuit can be analyzed as [50]

$$N_{\text{ML}} = 1 + \sum_{i=1}^{L-1} J^{i}, \tag{11.44}$$

where $J = d_{\text{minR}}-1$, and $L = z+1$ is the number of steps used in the majority logic decoder.

The multi-input XOR gates used in the majority vote process can be obtained by using the combination of multiple two-input XOR gates, and the number of inputs of the XOR gate is dependent on the check polynomial. In this work, the two-input MLGs are used in (1,3), (2,4) and (3,5) C-RM decoders' design and six-input MLGs are used in the (1,4) and (2,5) C-RM decoders' design. According to (11.41), the two-input MLG and the six-input MLG can be formed by 2 and 22 NAND gates.

In addition, n_R-stage buffer registers and an extra two-input XOR gate are also needed. Here, for C-RM codes, the energy cost of decoding is

$$E_{de\text{-}(1,4)RM} = 20N_{sm\text{-}de}\left(4n_R + 590\right) + 2450N_{sm\text{-}de}, \tag{11.45}$$

$$E_{de\text{-}(2,5)RM} = 20N_{sm\text{-}de}\left(4n_R + 6998\right) + 2450N_{sm\text{-}de}. \tag{11.46}$$

11.6.6.3 Energy Consumption for LDPC Codes

As shown in Figure 11.10, for (15,7), (63,37) and (255,175)LDPC codes, $(n_L - k_L)$ shift registers are used, and the number of two-input XOR gates in the circuits is dependent upon the generator polynomial of each code. The energy cost of encoding is

$$E_{en\text{-}(15,7)LDPC} = 20N_{sm\text{-}en}\left(4(n_L - k_L) + 16\right) + 2450N_{sm\text{-}en}, \tag{11.47}$$

$$E_{en\text{-}(63,37)LDPC} = 20N_{sm\text{-}en}\left(4(n_L - k_L) + 40\right) + 2450N_{sm\text{-}en}, \tag{11.48}$$

$$E_{en\text{-}(255,175)LDPC} = 20N_{sm\text{-}en}\left(4(n_L - k_L) + 180\right) + 2450N_{sm\text{-}en}. \tag{11.49}$$

In addition, for different LDPC codes, the decoding circuits can be modified with ρ-input XOR gates, γ-input MLGs and n_L buffer registers. The multi-input XOR gate can be obtained by using the combination of multiple two-input XOR gates. Here for (15,7), (63,37) and (255,175)LDPC codes, the energy cost of decoding is

$$E_{de\text{-}(15,7)LDPC} = 20N_{sm\text{-}de}\left(4n_L + 57\right) + 2450N_{sm\text{-}de}, \tag{11.50}$$

$$E_{de\text{-}(63,37)LDPC} = 20N_{sm\text{-}de}\left(4n_L + 321\right) + 2450N_{sm\text{-}de}, \tag{11.51}$$

$$E_{de\text{-}(255,175)LDPC} = 20N_{sm\text{-}de}\left(4n_L + 27297\right) + 2450N_{sm\text{-}de}. \tag{11.52}$$

11.6.6.4 Energy Consumption for SOCCs

Referring to the description of SOCCs in Section 11.6.4, for each encoder, the number of shift-register units for the encoder is b and the number of XOR gates is dependent upon on the generator polynomials. The energy cost of the encoding is thus

$$E_{en\text{-}(2,1,6)SOCC} = 20N_{sm\text{-}en}\left(4b + 12\right) + 2450N_{sm\text{-}en}, \tag{11.53}$$

$$E_{en\text{-}(2,1,17)SOCC} = 20N_{sm\text{-}en}\left(4b + 20\right) + 2450N_{sm\text{-}en}, \tag{11.54}$$

$$E_{\text{en-}(3,2,2)\text{SOCC}} = 20N_{\text{sm-en}}\left(4b+12\right)+2450N_{\text{sm-en}}, \tag{11.55}$$

$$E_{\text{en-}(3,2,13)\text{SOCC}} = 20N_{\text{sm-en}}\left(4b+28\right)+2450N_{\text{sm-en}}. \tag{11.56}$$

The decoder is composed of two parts, one part is the same as the encoder and the other part contains b register units, k_S MLGs; the MLGs used here are two-input MLGs, where each one can be looked on as an AND gate. The number of XOR gates is dependent on the polynomial generator and the information length. So, the energy cost of the decoding is

$$E_{\text{de-}(2,1,6)\text{SOCC}} = 20N_{\text{sm-de}}\left(8b+37\right)+2450N_{\text{sm-de}}, \tag{11.57}$$

$$E_{\text{de-}(2,1,17)\text{SOCC}} = 20N_{\text{sm-de}}\left(8b+70\right)+2450N_{\text{sm-de}}, \tag{11.58}$$

$$E_{\text{de-}(3,2,2)\text{SOCC}} = 20N_{\text{sm-de}}\left(8b+36\right)+2450N_{\text{sm-de}}, \tag{11.59}$$

$$E_{\text{de-}(3,2,13)\text{SOCC}} = 20N_{\text{sm-de}}\left(8b+74\right)+2450N_{\text{sm-de}}. \tag{11.60}$$

11.6.7 Critical Distance

In order to analyze when the coding becomes beneficial, the critical distance [60] is utilized as a measure of the real transmission distance at which the coding gain matches the extra energy requirements introduced by the ECCs.

The total energy cost for an uncoded, E_{uncoded}, and a coded, E_{coded}, system can be calculated as

$$E_{\text{uncoded}} = 2450N_{\text{uncoded}}, \tag{11.61}$$

$$E_{\text{coded}} = 2450N_{\text{coded}} + E_{\text{en}} + E_{\text{de}}, \tag{11.62}$$

where N_{uncoded} and N_{coded} are the numbers of molecules used for the uncoded and coded system at a chosen BER level. E_{en} and E_{de} are the energy consumption for the encoding and decoding process.

To reach the same BER level, the energy saving for a coded system compared with an uncoded system can be defined as

$$\begin{aligned}\Delta E &= E_{\text{uncoded}} - E_{\text{coded}} \\ &= 2450\left(N_{\text{uncoded}} - N_{\text{coded}}\right) - E_{\text{en}} - E_{\text{de}}.\end{aligned} \tag{11.63}$$

It is clear that when $\Delta E \geq 0$, the use of ECC is beneficial to the MC system. When $\Delta E = 0$, Equation 11.63 reduces to

$$N_{\text{uncoded}} - N_{\text{coded}} = \left(E_{\text{en}} + E_{\text{de}}\right)/2450. \tag{11.64}$$

Therefore, the relationship between N_{uncoded} and N_{coded} can be obtained by substituting the energy consumption values for different coding schemes that were introduced in Section 11.6.6.

11.7 Numerical Results

The performance of a coded PTP DBMC system is evaluated from two aspects: the BER and the critical distance for N2N, N2M and M2N communication scenarios. The BER results are obtained based on the set of parameters in Table 11.3.

11.7.1 BER for Coded DBMC Systems

Figure 11.12 shows the BER results for the coded PTP DBMC system. Figure 11.12a gives the BER comparison between the system with block codes and the uncoded system, and Figure 11.12b shows the BER comparison between the system with SOCCs and the uncoded system. The coding gains are shown to be 1.35, 1.68, 1.68, 1.68, 2.30, 1.68, 2.43, 2.57, 1.46, 1.68, 1.47, and 1.90 dB for (7,4)Hamming, (15,11)Hamming, (31,26)Hamming, (1,4) C-RM, (2,5)C-RM, (15,7)LDPC, (63,37)LDPC, (255,175)LDPC, (2,1,6)SOCC, (2,1,17)SOCC, (3,2,2)SOCC and (3,2,13)SOCC, respectively, at the BER level of 10^{-3}. And for a BER level of 10^{-9}, the coding gains for the foregoing codes are 1.59, 2.43, 2.78, 2.30, 4.29, 2.76, 5.33, 6.95, 2.89, 4.04, 2.09 and 3.80 dB, respectively. It can be seen that (15,11), (31,26)Hamming codes, (1,4)C-RM code, (15,7)LDPC code and (2,1,17)SOCC have the same coding gain at 10^{-3}; this is because the number of molecules used for achieving that BER level is the same. The results also clearly show that the (255,175)LDPC code gives the largest coding gain compared to other codes.

11.7.2 Critical Distance

For N2N communication, the extra energy requirements introduced by the encoder and decoder need to be taken into account. For N2M communication, the T_Xs are considered much simpler than the R_Xs, so when calculating the energy, only the encoder consumption needs to be taken into account by setting $E_{de} = 0$ in Equation 11.64. Moreover, for M2N communication, the R_X needs to be much simpler than the T_X, so only the decoder consumption needs to be included, so E_{en} is set to zero in Equation 11.64.

As Equation 11.64 shows, the critical distance is affected by two factors. Firstly, the BER performance for different coding schemes at different distances is shown in Figure 11.12. Secondly, the encoder and decoder circuitry for each of the codes is different, with varying levels of complexity such as those in Figures 11.8 through Figure 11.11.

The critical distance for each code can be treated as a baseline level. When the designed parameters fall to the left of this, it is considered worthwhile to apply a particular code in the system designed, otherwise not. This means that the code is worth applying only when the application has a transmission distance equal or larger than the critical distance.

TABLE 11.3

Parameter Setting for BER

Parameter	Definition	Value
R	Radius of R_X	5 μm
D	Transmission distance	6 μm
D	Diffusion coefficient	79.4 μm²s⁻¹
I	ISI length	10
N_{tx}	Number of molecules per bit	10~400

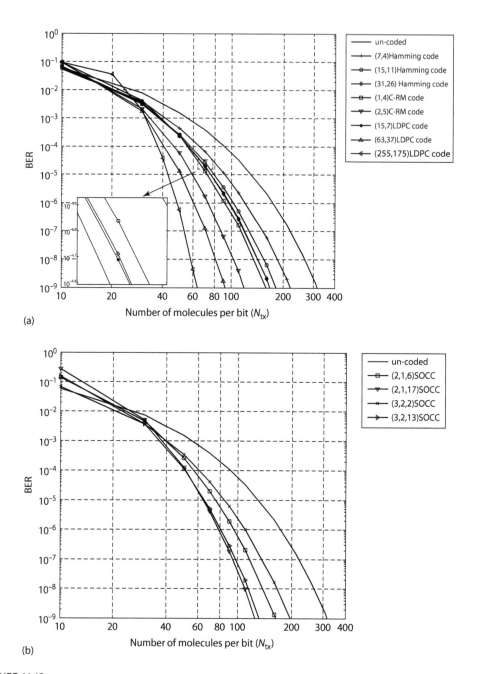

(a)

(b)

FIGURE 11.12
(a) BER comparison for a coded system with block codes and an uncoded system. (b) BER comparison for a coded system with SOCCs and an uncoded system.

On the other hand, for each communication scenario, there exists a lowest critical distance level for BERs from 10^{-9} to 10^{-3} and the corresponding code is considered as a best-fit ECC.

This best-fit ECC is defined as the code that has a wider application range.

Figures 11.13 through 11.15 show the critical distances of the system with block codes under three different communication scenarios. The results presented in Figure 11.13

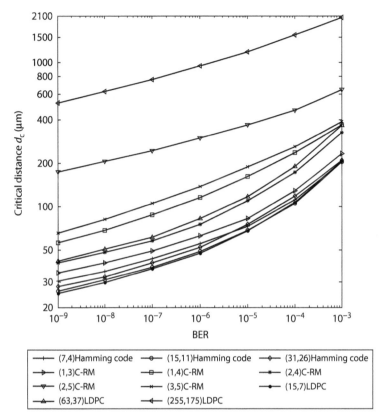

FIGURE 11.13
Critical distance with BER for block codes when considering N2N communication.

consider the N2N communication scenario. In this scenario, the (15,7)LDPC gives the lowest critical distance, and the (255,175)LDPC code from the same code family presents the longest critical distance based on the high complexity of the circuit design. Figures 11.14 and 11.15 illustrate the critical distance for the system when considering the N2M and M2N communication scenarios, respectively. In these cases, the (31,25)Hamming and (15,7) LDPC codes are best with the lowest critical distances, respectively.

Using the concept of the "best-fit ECC" previously introduced, the best-fit block codes for N2N, N2M and M2N communication scenarios are (15,7)LDPC, (31,26)Hamming code and (15,7)LDPC, respectively. These codes are also compared with SOCCs in different communication scenarios, and the results are illustrated in Figures 11.16 through 11.18.

Figure 11.16 considers the critical distance comparison for the N2N communication scenario. In this case, the energy consumption for both the encoder and decoder circuits needs to be considered. The lowest critical distance is given by the use of the (3,2,2)SOCC, which means under this communication scenario that the (3,2,2)SOCC has a wider application range.

Figures 11.17 and 11.18 show the other scenarios: N2M and M2N communications. The lowest critical distance belongs to the (31,26)Hamming code and the (3,2,2)SOCC for the N2M communication scenario, and (3,2,2)SOCC for the M2N communication scenario.

In addition, all the results indicate that the level of critical distance for the N2M communication scenario is lower than the levels of critical distance for the N2N and M2N

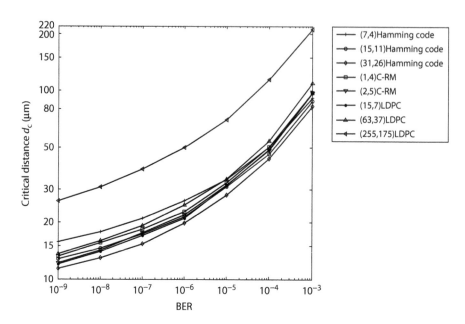

FIGURE 11.14
Critical distance with BER for block codes when considering N2M communication.

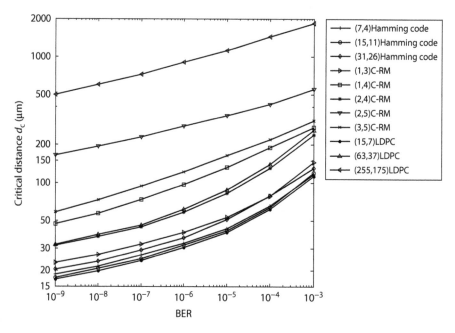

FIGURE 11.15
Critical distance with BER for block codes when considering M2N communication.

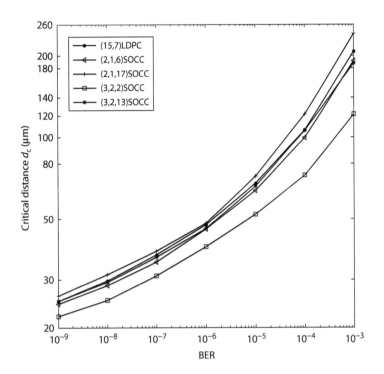

FIGURE 11.16
Critical distance comparisons between (15,7)LDPC and selected SOCCs when considering N2N communication.

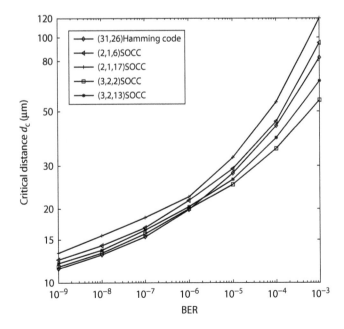

FIGURE 11.17
Critical distance comparisons between (31,26)Hamming code and selected SOCCs when considering N2M communication.

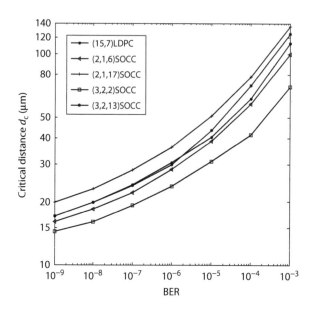

FIGURE 11.18
Critical distance comparisons between (15,7)LDPC and selected SOCCs when considering M2N communication.

TABLE 11.4

Best-Fit ECC for Different DBMC Scenarios

Communication Scenario	Lower BER Operating Region	Higher BER Operating Region
N2N		(3,2,2)SOCC	
N2M	(31,26)Hamming code	(3,2,2)SOCC
M2N		(3,2,2)SOCC	

communication scenarios. This means that the encoder design for all selected ECCs is simpler than the decoder design. Through the comparison, the (3,2,2)SOCC is selected as the best-fit ECC for all three communication scenarios except the N2M communication system with a BER level from 10^{-9} to 10^{-6}.

Table 11.4 gives the best-fit ECCs for different communication scenarios. The lower BER operating region for the N2M communication scenario is from 10^{-9} to 10^{-6}, and the higher BER operating region is from 10^{-6} to 10^{-3}. As shown in Table 11.4, the (3,2,2)SOCC is the superior code as it is the best-fit ECC for N2N, M2N communications and also the N2M communication system with a higher operating BER level.

11.8 Summary

This chapter presented an overview of PTP DBMC systems and introduced coding techniques. After a brief introduction to MC and the relevant state of the art of the ECCs in an MC system, a 3D PTP DBMC system was established. Following this, propagation and communication models were presented. Then, the selected block codes, Hamming, C-RM and LDPC codes, and the selected convolutional code, SOCC, were introduced into the

PTP DBMC system and the performance was compared among them regarding both coding gain and critical distance. The results indicate that the coding techniques do improve the performance of the PTP DBMC systems.

The critical distance for three different communication scenarios was also analyzed by considering the energy consumption caused by the introduction of ECCs. Critical distance was defined as the real transmission distance at which the use of coding techniques becomes beneficial. It was indicated that an increase in the operating BER results in a longer critical distance. For the N2M and M2N communication scenarios, the critical distance decreases compared with the N2N communication scenario. The best-fit ECC for a proposed system can be determined by analyzing these performance metrics.

Overall, this chapter presented the fundamental aspect of the use of coding techniques within the DBMC system. It is believed that this study will help the system designer to understand the basic DBMC system and will be beneficial to the further investigation of new coding techniques within this emerging communication paradigm.

References

1. R. Freitasjr, What is nanomedicine? Nanomedicine: Nanotechnology, *Nanomedicine: Nanotechnology, Biology and Medicine*, vol. 1, pp. 2–9, 2005.
2. Y. Moritani, S. Hiyama, and T. Suda, Molecular communication for health care applications, in *4th Annual IEEE International Conference on Pervasive Computing and Communications Workshops*, Pisa, Italy, 2006, pp. 549–553.
3. T. M. Allen and P. R. Cullis, Drug delivery systems: Entering the mainstream, *Science*, vol. 303, pp. 1818–1822, 2004.
4. B. Atakan, O. B. Akan, and S. Balasubramaniam, Body area nanonetworks with molecular communications in nanomedicine, *IEEE Communications Magazine*, vol. 50, pp. 28–34, 2012.
5. J. Han, J. Fu, and R. B. Schoch, Molecular sieving using nanofilters: Past, present and future, *Lab on a Chip*, vol. 8, pp. 23–33, 2008.
6. T. Nakano, A. W. Eckford, and T. Haraguchi, *Molecular Communication*, Cambridge University Press, 2013.
7. D. Tessier, I. Radu, and M. Filteau, Antimicrobial fabrics coated with nano-sized silver salt crystals, in *Proceedings of the NSTI Nanotechnology Conference*, 2005, pp. 762–764.
8. J. W. Aylott, Optical nanosensors: An enabling technology for intracellular measurements, *Analyst*, vol. 128, pp. 309–312, 2003.
9. M. Kocaoglu, B. Gulbahar, and O. B. Akan, Stochastic resonance in graphene bilayer optical nanoreceivers, *IEEE Transactions on Nanotechnology*, vol. 13, pp. 1107–1117, 2014.
10. J. M. Jornet and I. F. Akyildiz, Joint energy harvesting and communication analysis for perpetual wireless nanosensor networks in the terahertz band, *IEEE Transactions on Nanotechnology*, vol. 11, pp. 570–580, 2012.
11. I. F. Akyildiz, F. Brunetti, and C. Blázquez, Nanonetworks: A new communication paradigm, *Computer Networks*, vol. 52, pp. 2260–2279, 2008.
12. C. Bustamante, Y. R. Chemla, N. R. Forde, and D. Izhaky, Mechanical processes in biochemistry, *Annual Review of Biochemistry*, vol. 73, pp. 705–748, 2004.
13. M. Moore, A. Enomoto, T. Nakano, R. Egashira, T. Suda, A. Kayasuga, et al., A design of a molecular communication system for nanomachines using molecular motors, in *Proceedings of the IEEE Conference on Pervasive Computing and Communications*, Pisa, Italy, 2006, pp. 554–559.
14. V. Serreli, C.-F. Lee, E. R. Kay, and D. A. Leigh, A molecular information ratchet, *Nature*, vol. 445, pp. 523–527, 2007.

15. T. Nakano, T. Suda, M. Moore, R. Egashira, A. Enomoto, K. Arima, Molecular communication for nanomachines using intercellular calcium signaling, in *5th IEEE Conference on Nanotechnology, 2005*, vol. 2, pp. 478–481, 2005.

16. M. Ş. Kuran, H. B. Yilmaz, T. Tugcu, and B. Özerman, Energy model for communication via diffusion in nanonetworks, *Nano Communication Networks*, vol. 1, pp. 86–95, 2010.

17. B. Atakan and O. B. Akan, Deterministic capacity of information flow in molecular nanonetworks, *Nano Communication Networks*, vol. 1, pp. 31–42, 2010.

18. M. Pierobon and I. F. Akyildiz, Diffusion-based noise analysis for molecular communication in nanonetworks, *IEEE Transactions on Signal Processing*, vol. 59, pp. 2532–2547, 2011.

19. T. Nakano, T. Suda, M. Moore, R. Egashira, A. Enomoto, and K. Arima, Molecular communication for nanomachines using intercellular calcium signaling, in *5th IEEE Conference on Nanotechnology, 2005*, pp. 478–481.

20. T. Nakano, T. Suda, T. Koujin, T. Haraguchi, and Y. Hiraoka, Molecular communication through gap junction channels, in *Transactions on Computational Systems Biology X*, Springer, 2008, pp. 81–99.

21. T. Nakano, Y.-H. Hsu, W. C. Tang, T. Suda, D. Lin, T. Koujin, *et al.*, Microplatform for intercellular communication, in *3rd IEEE International Conference on Nano/Micro Engineered and Molecular Systems (NEMS 2008)*, 2008, pp. 476–479.

22. L. C. Cobo and I. F. Akyildiz, Bacteria-based communication in nanonetworks, *Nano Communication Networks*, vol. 1, pp. 244–256, 2010.

23. M. Gregori and I. F. Akyildiz, A new nanonetwork architecture using flagellated bacteria and catalytic nanomotors, *IEEE Journal on Selected Areas in Communications*, vol. 28, pp. 612–619, 2010.

24. M. Gregori, I. Llatser, A. Cabellos-Aparicio, and E. Alarcón, Physical channel characterization for medium-range nanonetworks using flagellated bacteria, *Computer Networks*, vol. 55, pp. 779–791, 2011.

25. M. J. Berridge, The AM and FM of calcium signalling, *Nature*, vol. 386, pp. 759–760, 1997.

26. T. Nakano, M. J. Moore, F. Wei, A. V. Vasilakos, and J. Shuai, Molecular communication and networking: Opportunities and challenges, *IEEE Transactions on NanoBioscience*, vol. 11, pp. 135–148, 2012.

27. M. S. Leeson and M. D. Higgins, Forward error correction for molecular communications, *Nano Communication Networks*, vol. 3, pp. 161–167, 2012.

28. P.-Y. Ko, Y.-C. Lee, P.-C. Yeh, C.-H. Lee, and K.-C. Chen, A new paradigm for channel coding in diffusion-based molecular communications: Molecular coding distance function, in *IEEE Global Communications Conference (GLOBECOM)*, 2012, pp. 3748–3753.

29. P.-J. Shih, C.-H. Lee, P.-C. Yeh and K.-C. Chen, Channel codes for reliability enhancement in molecular communication, *IEEE Journal on Selected Areas in Communications*, vol. 31, pp. 857–867, 2013.

30. C. Bai, M. S. Leeson, and M. D. Higgins, Minimum energy channel codes for molecular communications, *Electronics Letters*, vol. 50, pp. 1669–1671, 2014.

31. A. G. Thombre, J. R. Cardinal, A. R. DeNoto, S. M. Herbig, and K. L. Smith, Asymmetric membrane capsules for osmotic drug delivery: I. Development of a manufacturing process, *Journal of Controlled Release*, vol. 57, pp. 55–64, 1999.

32. S. Nain and N. N. Sharma, Propulsion of an artificial nanoswimmer: A comprehensive review, *Frontiers in Life Science*, pp. 1–16, 2014.

33. M. Ş. Kuran, H. B. Yilmaz, T. Tugcu, and B. Özerman, Energy model for communication via diffusion in nanonetworks, *Nano Communication Networks*, vol. 1, pp. 86–95, 2010.

34. Y. Lu, M. D. Higgins, and M. S. Leeson, Comparison of channel coding schemes for molecular communications systems, *IEEE Transactions on Communications*, vol. 63, pp. 3991–4001, 2015.

35. R. W. Hamming, Error detecting and error correcting codes, *Bell System Technical Journal*, vol. 29, pp. 147–160, 1950.

36. D. E. Muller, Application of Boolean algebra to switching circuit design and to error detection, *Transactions of the IRE Professional Group on Electronic Computers*, vol. EC-3, pp. 6–12, 1954.

37. I. Reed, A class of multiple-error-correcting codes and the decoding scheme, *Transactions of the IRE Professional Group on Information Theory*, vol. 4, pp. 38–49, 1954.
38. S. Boztas and I. E. Shparlinski, Applied algebra, algebraic algorithms and error-correcting codes, in *Proceedings of the 14th International Symposium (AAECC-14)*, Melbourne, Australia, November 26–30, Springer, 2001.
39. W. W. Peterson and E. J. Weldon, *Error-Correcting Codes*, MIT Press, 1972.
40. T. Kasami, L. Shu, and W. Peterson, New generalizations of the Reed–Muller codes—I: Primitive codes, *IEEE Transactions on Information Theory*, vol. 14, pp. 189–199, 1968.
41. R. G. Gallager, Low-density parity-check codes, *IRE Transactions on Information Theory*, vol. 8, pp. 21–28, 1962.
42. D. J. MacKay and R. M. Neal, Near Shannon limit performance of low density parity check codes, *Electronics Letters*, vol. 33, pp. 457–458, 1997.
43. K. Yu, L. Shu, and M. P. C. Fossorier, Low-density parity-check codes based on finite geometries: A rediscovery and new results, *IEEE Transactions on Information Theory*, vol. 47, pp. 2711–2736, 2001.
44. T. K. Moon, *Error Correction Coding: Mathematical Methods and Algorithms*, Wiley, 2005.
45. J. C. Moreira and P. G. Farrell, *Essentials of Error-Control Coding*, Wiley, 2006.
46. P. Reviriego, J. A. Maestro, and M. F. Flanagan, Error detection in majority logic decoding of Euclidean geometry low density parity check (EG-LDPC) codes, *IEEE Transactions on Very Large Scale Integration (VLSI) Systems*, vol. 21, pp. 156–159, 2013.
47. K. Ganesan, P. Grover, and J. Rabaey, The power cost of over-designing codes, in *IEEE Workshop on Signal Processing Systems (SiPS)*, 2011, pp. 128–133.
48. P. Grover and A. Sahai, Green codes: Energy-efficient short-range communication, in *IEEE International Symposium on Information Theory (ISIT)*, 2008, pp. 1178–1182.
49. S. Bougeard, J. F. Helard, and J. Citerne, A new algorithm for decoding concatenated CSOCs: Application to very high bit rate transmissions, in *1999 IEEE International Conference on Personal Wireless Communication*, 1999, pp. 399–403.
50. S. Lin and D. J. Costello, *Error Control Coding: Fundamentals and Applications*, Prentice-Hall, 1983.
51. M. Kavehrad, Implementation of a self-orthogonal convolutional code used in satellite communications, *IEE Journal on Electronic Circuits and Systems*, vol. 3, pp. 134–138, 1979.
52. R. Townsend and E. Weldon, Self-orthogonal quasi-cyclic codes, *IEEE Transactions on Information Theory*, vol. 13, pp. 183–195, 1967.
53. H. M. Sauro and B. N. Kholodenko, Quantitative analysis of signaling networks, *Progress in Biophysics and Molecular Biology*, vol. 86, pp. 5–43, 2004.
54. J. Levine, H. Y. Kueh, and L. Mirny, Intrinsic fluctuations, robustness, and tunability in signaling cycles, *Biophysical Journal*, vol. 92, pp. 4473–4481, 2007.
55. D. L. Nelson, A. Lehninger, M. M. Cox, M. Osgood, and K. Ocorr, *Lehninger Principles of Biochemistry / The Absolute, Ultimate Guide to Lehninger Principles of Biochemistry*, Macmillan Higher Education, 2008.
56. E. Shacter, P. B. Chock, and E. R. Stadtman, Energy consumption in a cyclic phosphorylation/dephosphorylation cascade, *Journal of Biological Chemistry*, vol. 259, pp. 12260–12264, 1984.
57. M. Ş. Kuran, H. B. Yilmaz, T. Tugcu, and B. Özerman, Energy model for communication via diffusion in nanonetworks, *Nano Communication Networks*, vol. 1, pp. 86–95, 2010.
58. R. E. Blahut, *Algebraic Codes for Data Transmission*, Cambridge University Press, 2003.
59. W. Ryan and S. Lin, *Channel Codes: Classical and Modern*, Cambridge University Press, 2009.
60. S. L. Howard, C. Schlegel, and K. Iniewski, Error control coding in low-power wireless sensor networks: When is ECC energy-efficient?, *EURASIP Journal on Wireless Communications and Networking*, vol. 2006, pp. 29–29, 2006.

12

Miniaturized Battery-Free Wireless Bio-Integrated Systems

Philipp Gutruf

CONTENTS

12.1 Introduction: Background and Driving Forces

Contemporary electronic component and devices have recently stagnated in performance increase due to physical limitations in miniaturization and a development al goals solely based on Moore's law are not the only consideration when benchmarking gains of new hardware. A focus on integration has become apparent and the commercial success of new devices relies on form factor and compatibility with biological systems, mostly to improve human–device interactions. This development can be observed by the booming wearables market and health-monitoring devices in various form factors. These devices repurpose and simplify ordinary circuits and technology currently available on the market in wearable housings affixed with flexible bands to limbs. This technology has not changed in the last 20 years and progress is long overdue. Considerable efforts are being made toward the development of a new class of electronics that is intrinsically flexible and stretchable [1]. The research challenge is to overcome the intrinsically rigid nature of contemporary technology. However, the integration of conventional electronic components, passive or active, with biological systems is limited due to the inherent incompatibility in mechanical properties. Most biological systems such as the human skin and the surface of organs are intrinsically soft [2], whereas popular electronic components and their housings are hard. For the creation of these next-generation devices, this fundamental mechanical mismatch has to be overcome. Examples of such soft systems can be found in the literature and cover basic electrical functionality and sensing capabilities [3]. Examples include capacitive epidermal electronics for electrically safe, long-term electrophysiological measurements [4] and human–machine interfaces [5], approaches that leverage the ability of these systems to be laminated directly onto the skin. Other notable applications are the direct lamination of these electronics onto organs, which results in high-quality recordings due to intimate mechanical contact [6,7]. However, for this technology to break through into mainstream electronics for consumers and healthcare, a suitable integration strategy has to be found. Due to the inherent stretchability and soft nature of these circuits, a long-term, chronic wired connection is not possible due to the interfacial stress of the connection and the small force required to break these ultrathin devices, which can be quickly reached by tugging at a wired interface without strain relief. Furthermore, such devices will find their application mainly in intimate monitoring of biological systems either on the skin or on the surface of organs where a wired connection could be a significant infection risk and warrant secondary surgery for device removal. This necessitates a ubiquitous strategy for wireless integration with the currently available infrastructure.

Here, the search for a suitable power supply and communication strategy is critical. The main criteria for the selection of such a power supply and communication strategy are form factor, mechanical compliance and speed.

Table 12.1 compares common wireless standards. We can see that low-power communication protocols are favorable to increase the lifetime of battery-operated devices and for passive device increased access power for peripherals; however, most popular choices such as Bluetooth still require an electrochemical power supply that would not be suitable in form factor for epidermal electronics unless highly complex strain isolation approaches are used to decouple the movement of the target organ from the large stiff energy storage. Thus, the need for energy harvesting that arises for this class of systems. Current technologies that can harvest sufficient power to enable computing and sampling of information are rare. One of these current technology standards that allows for energy harvesting as well as digital communication is near field communication (NFC). Here, magnetic resonant coupling of a primary and secondary antenna allows for considerable transfer of power (up to 20 mW), and data can also be pushed from sender to receiver via amplitude shift keying (ASK) and frequency shift keying (FSK). This makes the technology highly attractive for fully bio-integrated devices. An additional advantage of this technology standard is widely availability due to common integration in smartphones, tablets, pay stations and access control systems.

12.2 NFC: Viable Standard for Highly Integrated Bioelectronics

In direct comparison, NFC is an outstanding technology because it provides digital communication and energy harvesting often in a highly integrated one-chip solution. This avoids large aggregates of passive components and the energy management associated with electrochemical energy storage that increases the device size and limits operational time.

TABLE 12.1

Wireless Connectivity for Bio-Integrated Devices

	NFC	**Bluetooth**	**Wifi**
Form factor	1 chip solution + antenna	1 chip solution + antenna	1 chip solution + antenna
Bio-integration	Very good Epidermal form factor	Medium Epidermal form factor with advanced technique	Unknown
Communication distance	Near field (0–1 m)	Far field (<3 m)	Far field (<3 m)
Speed	Low	Medium	High
Energy consumption	+10 mW	−20 mW	−30 mW
Energy harvesting	Yes	No	No

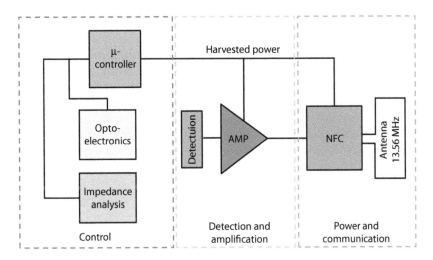

FIGURE 12.1
Electrical working principle scheme of NFC-enabled bio-integrated electronic devices.

In Figure 12.1, a template of a typical NFC-based system for bio-integrated devices is presented. Here, the NFC system-on-a-chip (SOC) is harvesting energy that is passed on to the periphery. A microcontroller is often found in these systems that orchestrates the timing for optoelectronics and impedance analysis for sensing applications such as the fingernail-mounted oximetry presented later in this chapter. Detection systems such as operational amplifiers schemes can also be powered with the harvested energy. The acquired signal is then digitalized by an analog-to-digital converter (ADC) and sent over air to a receiver. The common harvested power available in these systems can exceed 15 mW on a regular basis and provides suitable voltages in the range of 2.5–5 V to power most common optoelectronic, analog and digital electronic components. Receivers can be specialized high-powered systems for long-range applications as well as small, handheld systems commonly present in all households, such as NFC-enabled smartphones. This makes these systems very versatile due to the available infrastructure. The communication range of these systems can vary from a few centimeters up to 3 m for high-powered systems.

12.3 Mechanical Design: Transforming Flexible Electronics to Stretchable Devices

The commercial availability of small form factor–integrated NFC circuits, such as the bare die SOC, allows for the construction of highly stretchable devices that can be intimately bonded to biological systems. Here, the concept is to replace large, traditionally rigid key components of an electrical circuit with intrinsically stretchable elements. For an NFC circuit, this is the antenna, which is spatially the largest part of the device, the highly integrated silicon-based semiconductor that supports data communication, energy harvesting and sensing function can be treated as a rigid island that can be tolerated by soft systems as the strain is accommodated by the surrounding tissue if skin mounted.

An example of such a system is displayed in Figure 12.2. An IC was backpolished resulting in a small form factor 1.5×3 mm thin (125 µm) highly integrated system with the capability to yield temperature measurements and analog-to-digital conversion.

The development of a stretchable antenna that is capable of enduring 30% strain is critical to render an epidermal device that conforms intimately to the skin. The device geometry is critical for mechanical compliance; here, a mutual mechanical plane design can be chosen to distribute strain in the sandwich of the polyimide support with the copper core and polyimide encapsulation that also acts as a passivation layer for the copper bridge that completes the antenna. Such a stack can be achieved using a sophisticated microfabrication technique that requires the careful management of adhesion to a sacrificial or carrier substrate and the final release of the device and transfer onto an ultrathin adhesive or low modulus silicone material that bonds the device intimately to the skin. The finalized structure can be seen in an exploded layer schematic in Figure 12.2a.

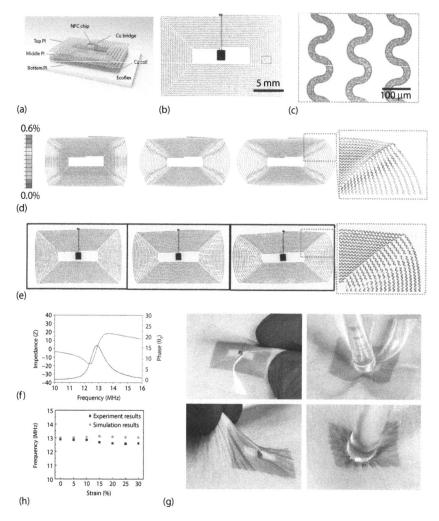

FIGURE 12.2

Stretchable NFC-enabled epidermal electronics. (a) Layered makeup of device, (b) photographic image of the stretchable antenna with closeup and (c) of the serpentine structure facilitating out-of-plane deformation that yields stretchability. (d) FEM simulation of the device with consecutive application of strain up to 30%. (e) Corresponding bench top experiments showing good correlation. (f) Electromagnetic properties of the device. (h) Antenna performance under strain. (g) Application examples with and without flexible backing, demonstrating intimate contact with the skin. (Adapted from Kim J., et al., *Small*, 11, 906–912, 2015 [10].)

The device uses serpentines to facilitate stretchability, a well-documented and explored method to transform flexible substrates into stretchable systems [8]. A photographic image of the device including a magnified view of the serpentines can be seen in Figure 12.2b and c, respectively. The device design can be aided by finite element modeling (FEM), which can be a very useful tool to aid the serpentine geometry to achieve the designed stretchability, which in the case of skin should be around 30% to accommodate the natural motion of the skin. A close match between the computer model and a uniaxial strain testing the application of 10%, 20% and 30% strain states can be observed in Figure 12.2d and e, respectively. Here, we can see that the intrinsic strain occurring in the model does not exceed 0.6%, which is well below the yield strain of copper, the material used for the antenna, demonstrating a successful design and sufficient headroom for on-body applications where strains rarely exceed 30% on the human body. The good correlation between deformations observed in bench top experiments with FEM simulation establishes a valuable tool for future epidermal designs. This method enables a route for computer-aided design that allows for the rapid prototyping of such systems. Mechanical compliance with biological systems is not the only figure of merit for these systems. In order to satisfy the energy requirements of especially optoelectronic systems, the electromagnetic properties of such an antenna are equally important because they define the working distance, the harvesting power and the space occupied by the system. Here, the quality factor (Q factor) is critical. A better Q factor allows for more power to be transferred into the system. At the same time, the bandwidth for the data transmission must also be guaranteed. In the case of the ISO/IEC 15693 standard, it was originally designed for vicinity cards and allows for communication distances of up to 1–1.5 m and is therefore ideal for wearable devices. The base frequency of the resonant magnetic coupling scheme is 13.56 MHz and features an ASK and FSK modulation with a bandwidth of 423.75 kHz. This means that the bandwidth of the antenna has to support this modulation. Therefore, the ideal antenna is perfectly tuned to 13.56 MHz and has a good Q factor supporting a bandwidth of over 423.75 kHz. The successful realization of such electrical properties coupled with an ultrathin layout and system-level stretchability can be seen in Figure 2.2f. Here, the Q factor is good and the bandwidth is around 2 MHz, which allows for good data transmission and high efficiency of power transfer, and was acquired using a passive acquisition method. It can also be seen that the resonant frequency of the system does not shift with strain on the host substrate (Figure 12.2h), a major advantage over systems operating in the far-field regime.

The mechanical compliance of these devices is displayed in Figure 2.2g. Here, the device that is laminated on ultrathin elastomeric substrates is compared to a similar device bonded to a flexible polyethylene terephthalate (PET) substrate, commonly found in radio-frequency identification (RFID)-tagged consumer electronics. It can be seen that the flexible backing is not sufficient for conformal contact. Here, the deformation experienced in the regular use of the device will frustrate the skin interface and cause delamination at the fringes of the flexible backing. The stretchable system, however, offers system-level stretchability and conformal mechanics that work complementary with the skin, showing that there is no mechanical impact of the device on the skin. Therefore, when worn over an extended time period, the device is imperceptible to the subject.

12.4 Miniaturization: Benefits for Bio-Integration

The skin as an interface to access various health status information is very popular because it acts as a window to the underlying vascular structure and as a transmitter of electrical signals giving clinically relevant information about pulse shape, heart rate, heart rate variability and oxygenation. Also, liquids such as sweat and interstitial fluid can be extracted to gather critical information. However, the skin as a mechanical platform is challenging due to its intrinsic movement and its constant renewal of the upper layer, the epidermis. This frustrates interfacial bonding and limits the lifetime of any device attached to it. Therefore, the adhesive plays a critical role in bonding the device to the skin; however, its ultimate lifetime is restricted to a maximum of 1–2 weeks. Adhesives that accommodate device adhesion over an extended period on the skin without causing irritation are currently the subject of investigation in science and industry. To enable a more permanent mounting location that maintains intimate contact to the body, the fingernail or other rigid areas of the body have been identified as a suitable alternative to the skin-mounted devices. Here, a lifetime of up to 4 months is possible, limited solely by the growth rate of the fingernail, which may vary between individual subjects.

Examples of such devices can be seen in Figure 12.3a and b. Here, the NFC electronics are highly miniaturized to minimize the overall footprint of the device to enable application to the fingernail. This is achieved with a multilayer device makeup that increases the inductance of the coil, resulting in a longer communication distance and higher power

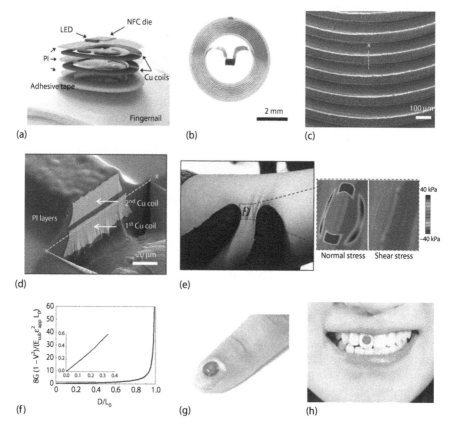

FIGURE 12.3
Highly miniaturized NFC systems. (a) Layered makeup of the device with components and adhesive. (b) Photographic image of the device. (c) SEM picture of the antenna coil with FIB cutaway revealing layered structure (d). (e) Device laminated onto the skin with corresponding normal stress and shear stress FEM simulation. (f) Calculated energy release rates with device size. (g) Application on the fingernail and tooth (h). (Adapted from Kim J., et al., *Advanced Functional Materials*, 25, 4761–4767, 2015 [11].)

reserves for energy harvesting. In this use case, the stretchability of the system is not required; however, flexibility is critical to enable imperceptible devices.

Critically with circuit geometry optimized for such uses, microfabrication strategies allowing for an overall thin device geometry are translatable to commercial-scale manufacturing. The flexibility of the circuitry is ensured with a mutual mechanical plane design, a strategy where in-plane symmetry is used to keep layer stress to a minimum analogous to the stretchable design described earlier in the chapter. Figure 12.3c shows a secondary electron microscope picture with the topographical nature of the coil enabled by a focused ion beam cutaway in Figure 12.3d revealing the layer structure.

A notable benefit of such system miniaturization is the potential use of a flexible device in low strain areas of the skin. The system miniaturization yields favorable mechanics due to the scaling of the underlying physical processes of device delamination and shear stress. Figure 12.3e shows such a device mounted on the forearm, the corresponding FEM simulations showing a low level of shear stress because of the small form factor; the normal stress is also below the sensation threshold, which results in a good user' experience when wearing the device. The small form factor also translates to low energy release rates, which means in practice a device that sticks well to the skin and is not prone to delamination. The scaling of this property can be seen in Figure 12.3f, which demonstrates favorable mechanics at small device diameters.

Critically, the high level of miniaturization allows for mounting locations such as the fingernail (Figure 12.3g) and the tooth (Figure 12.3h), which is a stable mechanical platform and offers rich opportunities for biosensing with potential device lifetimes of multiple years in the case of the tooth.

12.5 Applications: Contactless Vital Information Sensing

As discussed previously, the fingernail is a stable platform for chronic operation, and this presents an opportunity for use in sensing applications. There is a significant need for the continuous recording of vital information such as heart rate, heart rate variability and oxygenation during a chronic illness. Pulse oximetry offers the possibility to capture this vital information. A pulse oximeter operates on the working principle of signal attenuation of two narrowband light sources that interact with hemoglobin in the target region such as the fingers or earlobes, regions chosen due to their easy accessibility for external hardware. Hemoglobin, the main carrier of oxygen in the blood, alters its optical properties based on its oxygenation state. This can be leveraged for sensing by measuring the absorption at wavelengths that are adversely affected by this change. By then measuring the pulsatile component of the signal, the oxygen saturation of the subject can be extracted.

Current sensing solutions are bulky and intrinsically prone to motion artifacts due to heavy sensing strategies such as fingertip clamps that are, due to inertia, inherently unstable because the mass of the wireless hardware and the associated electrochemical energy storage cause relative motion of the clamp relative to the finger. The underlying vascular structure that is nonhomogeneous then results in variation in the optical path length, and a probing volume that shifts the optical signals levels resulting in motion artifacts. Here, ultrathin, flexible, highly miniaturized, lightweight and wireless technologies offer a distinct advantage over contemporary solutions.

Such a solution is presented in Figure 12.4a. Here, we can see the highly integrated layered makeup that consists of the antenna coil, which is, as previously discussed, a dual coil for maximum energy harvesting capability and footprint. The components needed for the pulse oximeter are also carefully chosen to meet the strict requirements for energy consumption, footprint and height requirements.

The electrical working principle is shown in Figure 12.4b, where the NFC scheme introduced at the beginning of the chapter is utilized. In this specific embodiment, a low-power, small outline microcontroller is used to control the red and infrared light-emitting diode (LED) light sources in a time-sequenced manner and a customized analog frontend with a transimpedance amplifier scheme coupled to a broadband semiconductor photodiode is used to capture the back-reflected light emerging from the nail bed. The signal is then digitalized in the bare die NFC chip that also provides power by means of energy harvesting from the ambient electrical field provided by the readout hardware. This makeup allows for a completely battery-free operation. Figure 12.4c shows a photograph of the device,

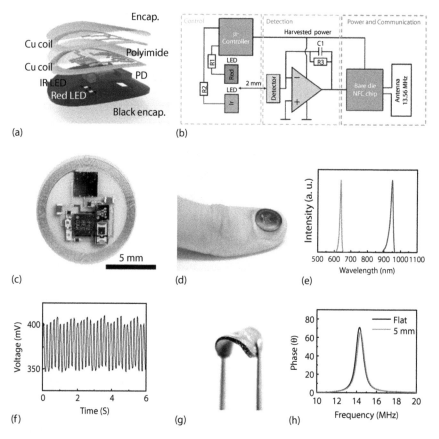

FIGURE 12.4

Highly miniaturized, battery-free, and flexible pulse oximeter system. (a) Layered makeup of the device. (b) Electrical operating scheme. (c) Photographic image of the device before encapsulation. (d) Photographic image of encapsulated device laminated to the fingernail during operation. (e) Spectral analysis of the LED light sources used for this device. (f) Raw signal obtained wirelessly from the device showing the pulsatile component of both excitation wavelengths. (g) Device in bent state demonstrating flexibility. (h) Electromagnetic device performance in flat and bent state. (Adapted from Kim J., et al., *Advanced Functional Materials*, 27, 2017 [12].)

illustrating the compact layout built around the optimized sensor with a light source distance of 2 mm.

To avoid a light short circuit, which manifests itself in a total reflection at the device fingernail or skin interface resulting in no interaction with the target tissue, which is a problem in reflectance-based oximeters, the device is encapsulated in black elastomeric material that also has the added benefit of conformal contact through the low modulus nature of the material. Optically, the light sources are now effectively isolated and suppress direct current light short circuits at the interface.

A characterization of the red and infrared LED can be seen in Figure 12.4e, showing the spectrally narrow bandwidth nature of the light source, which is essential for a good signal-to-noise ratio. Furthermore, a carefully chosen wavelength can yield a higher difference in absorption for the oxygenated and deoxygenated hemoglobin species. The resulting raw signal that is wirelessly acquired is shown in Figure 12.4f, where we can see the distinct pulsatile component of both infrared and red light reflection, which can be used to calculate the pulse oxygenation of the subject.

Figure 12.4g and h shows the bending characteristics of the device. The phase information shows that the device does not suffer from performance loss, judged by evaluating the quality factor of the antenna in its flat and bent state, which shows nearly no decay in performance. This is an important attribute because fingernail curvature can vary greatly from subject to subject, with bending radii of just 5 mm observed in some individuals.

The operational device performance evaluated in Figure 12.5a shows the extracted data from an experiment where the subject was deprived of oxygen intake by holding his or her breath. We can see that the pulsatile amplitude ratio of red and infrared reflectance shifts as the subject's oxygenation decreases. The corresponding extraction of SpO_2 can be seen in Figure 12.5b, with a comparison to a commercially available oximeter shown in Figure 12.5c. We can see that a much higher temporal resolution is achieved with the wireless device. This is possible due to the intrinsically superior signal quality facilitated by the intimate contact of the sensor system with the body yielding better signal-to-noise ratios. The commercial system requires a longer sampling time and averaging to compensate for motion artifacts and higher gain levels due to lower signal amplitudes that result in higher baseline noise.

Another advantage is shown in Figure 12.5d. The SpO_2 extracted from the device mounted on the fingernail is shaken vigorously, and the magnitude of the movement can be seen by simultaneously recorded accelerometer traces. We can observe that the signal is stable throughout the measurement, which is not feasible with commercial systems since they lack mechanical stability due to either high mass or cables associated with the device, which cause motion artifacts.

Application examples are shown in Figure 12.5e and f, where the device is mounted on the fingernail and the data is read out by a computer mouse with NFC functionality. This is an example of a use case that would facilitate continuous readout during daily activity. This capability is critical to address the clinical need of prolonged monitoring of postoperative or chronically ill patients. This information can be relayed via the internet to healthcare providers allowing for much less intrusive care, which enables more flexible surgery recovery scenarios as well as remote diagnosis and treatment. A subtler mounting location is demonstrated in Figure 12.5f, where the device is placed inside the earlobe, allowing for signal extraction when an NFC-enabled smartphone is brought into close proximity, a use case where intermittent oxygenation and heart rate monitoring is sufficient.

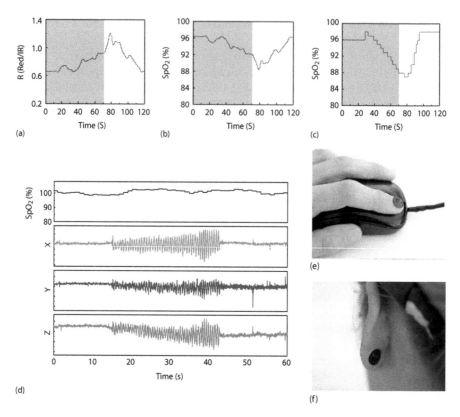

FIGURE 12.5

In situ device characterization. (a) Pulsatile component amplitude ratio over the course of an experiment involving restricted oxygen intake. (b) Corresponding calculation of the pulse oxygen saturation. (c) Control recorded with a commercial pulse oximeter (Apple I health). (d) Oxygenation readout with wireless battery-free device during elevated activity shown by accelerometer traces. (e) Application scenario on an NFC-enabled mouse. (f) Application scenario of the device mounted behind the earlobe. (Adapted from Adapted from Kim J., et al., *Advanced Functional Materials*, 27, 2017.)

12.6 Discussion: Battery-Free Miniaturized Electronics – Applications Beyond the Skin

The utility of the schemes presented in this chapter has a much wider scope than the applications discussed in this chapter. The magnetic resonant power transfer used here is little affected by surrounding moisture, meaning that it is an excellent candidate for the powering and data communication of implants, which will open the door for smart sensor nodes that can assess the health state instantaneously with currently unmet precision in a minimally invasive fashion. The utility of this approach has already been demonstrated in the context of devices for neuroscience [9]. The optogenetic stimulation of targeted neuronal populations has been accomplished. Applications beyond the brain can be envisioned, namely cardiac and peripheral device nodes that evaluate the health status simultaneously and deliver a complete insight into the physiological state of a patient with chronic interfaces that facilitate recordings over multiple years. This would be a leap forward in terms of medical device technology. The data generated from such sensor systems over multiple subjects can build powerful datasets that enable disease diagnostics including early warning systems for high-risk cases with unmatched precision and the removal of human error in data interpretation.

The technology also provides fundamental capabilities in access control and hygiene management where fingernail-mounted devices can give advanced insights into human behavior. Moreover, the technology can be used in industrial applications in hard-to-reach or vacuum-sealed systems where wired connections are prohibitive due to process line content. The advances in this field over recent years present a leap forward in device capability and suggest the acceleration of such sensors in the near future.

References

1. Kim J, Ghaffari R, and Kim D-H: The quest for miniaturized soft bioelectronic devices. *Nature Biomedical Engineering* 2017, 1: s41551-41017-40049.
2. Kim D-H, Lu N, Ma R, Kim Y-S, Kim R-H, Wang S, Wu J, Won SM, Tao H, et al.: Epidermal electronics. *Science* 2011, 333: 838–843.
3. Yeo W-H, Kim Y-S, Lee J, Ameen A, Shi L, Li M, Wang S, Ma R, Jin SH, et al.: Multifunctional epidermal electronics printed directly onto the skin. *Advanced Materials* 2013, 25: 2773–2778.
4. Jeong J-W, Kim MK, Cheng H, Yeo W-H, Huang X, Liu Y, Zhang Y, Huang Y, and Rogers JA: Capacitive epidermal electronics for electrically safe, long-term electrophysiological measurements. *Advanced Healthcare Materials* 2014, 3: 642–648.
5. Jeong JW, Yeo WH, Akhtar A, Norton JJ, Kwack YJ, Li S, Jung SY, Su Y, Lee W, et al.: Materials and optimized designs for human-machine interfaces via epidermal electronics. *Advanced Materials* 2013, 25: 6839–6846.
6. Kim D-H, Ghaffari R, Lu N, Wang S, Lee SP, Keum H, D'Angelo R, Klinker L, Su Y, et al.: Electronic sensor and actuator webs for large-area complex geometry cardiac mapping and therapy. *Proceedings of the National Academy of Sciences* 2012, 109: 19910–19915.
7. Xu L, Gutbrod SR, Bonifas AP, Su Y, Sulkin MS, Lu N, Chung H-J, Jang K-I, Liu Z, et al.: 3D multifunctional integumentary membranes for spatiotemporal cardiac measurements and stimulation across the entire epicardium. *Nature Communications* 2014, 5: 3329.
8. Gutruf P, Walia S, Nur Ali M, Sriram S, and Bhaskaran M: Strain response of stretchable micro-electrodes: Controlling sensitivity with serpentine designs and encapsulation. *Applied Physics Letters* 2014, 104: 021908.
9. Shin G, Gomez AM, Al-Hasani R, Jeong YR, Kim J, Xie Z, Banks A, Lee SM, Han SY, et al.: Flexible near-field wireless optoelectronics as subdermal implants for broad applications in optogenetics. *Neuron* 2017, 93: 509–521. e503.
10. Kim J, Banks A, Cheng H, Xie Z, Xu S, Jang KI, Lee JW, Liu Z, Gutruf P, et al.: Epidermal electronics with advanced capabilities in near-field communication. *Small* 2015, 11: 906–912.
11. Kim J, Banks A, Xie Z, Heo SY, Gutruf P, Lee JW, Xu S, Jang KI, Liu F, et al.: Miniaturized flexible electronic systems with wireless power and near-field communication capabilities. *Advanced Functional Materials* 2015, 25: 4761–4767.
12. Kim J, Gutruf P, Chiarelli AM, Heo SY, Cho K, Xie Z, Banks A, Han S, Jang KI, et al.: Miniaturized battery-free wireless systems for wearable pulse oximetry. *Advanced Functional Materials* 2017, 27: 1604373.

13

A Low-Power Vision- and IMU-Based System for the Intraoperative Prosthesis Pose Estimation of Total Hip Replacement Surgeries

Shaojie Su, Hong Chen, Hanjun Jiang, and Zhihua Wang

CONTENTS

13.1 Introduction

13.1.1 Background of Total Hip Replacement

As the average age of the population increases, the number of people suffering from severe pain in their damaged hip joints caused by a variety of human diseases and activities has increased steadily. The last resort to regain full mobility and alleviate the pain is total hip replacement (THR) surgery. According to the Agency for Healthcare Research and Quality (AHRQ), more than 285,000 THRs are performed in the United States annually, and the number is forecasted to double in the next two decades [1]. However, THR surgery still has a failure rate of about 10%, leading to serious complications such as dislocation, prosthetic impingement, intraoperative fracture, infection and leg length discrepancy along with the long-term complications of wearing and loosening [2]. Among them, dislocation and impingement, of which the incidence is reported to range from 0.5% to 11% [3–5], are two common complications.

During THR surgery, the placement of hip prostheses is the key step that will affect the postoperative hip range of motion (ROM), as shown in Figure 13.1. As Widmer pointed out in [6], the malposition of prosthesis placement is a primary cause of dislocation and prosthetic impingement. Researchers consider the hip ROM as an important indicator of joint stability and conclude that THR with a prosthetic hip ROM larger than that for everyday activities can minimize the risk of dislocation and impingement [3,4]. As a result, an intraoperative-assisted system in THR for visualizing the prosthesis placement and estimating the hip ROM for surgeons is strongly needed.

13.1.2 Related Research Work

Many efforts have been made to help the placement of prostheses in the safe zone. Image-guided surgical navigation systems such as computed tomography (CT)-based and CT-free intraoperative navigation systems have brought significant improvements in motion tracking [7]. However, the CT-based navigation system has the disadvantages of high cost, increased operation time and exposure to radiation [8], while the CT-free navigation is even more expensive and requires the affixing of markers to bony anatomical landmarks, leading to more complexity in data collection, calibration and processing, and extra harm to the patients as well [9]. In the 1960s, people began to design instrumented prostheses

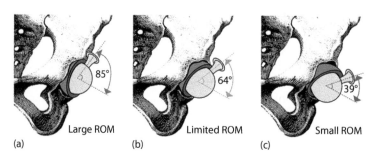

| (a) | (b) | (c) |

FIGURE 13.1
(a) One can raise the thigh up high if the acetabular cup is implanted in the right orientation, indicating a large hip ROM. (b) Impingement may occur when one raises the thigh up as the abduction of the acetabular cup is a bit larger than the safe zone, indicating a limited hip ROM. (c) Impingement or dislocation occurs during one's everyday activities as the abduction of the acetabular cup is too large, indicating a small hip ROM.

that could collect data in vivo and transmit the data wirelessly outside the body. Graichen et al. [10] have proposed a hip endoprosthesis instrumented with sensors for joint contact forces measurement. Damm et al. [11] have applied six semiconductor strain gauges inside the hollow neck of a new instrumented hip joint prosthesis to measure the deformation of the neck. Bergmann et al. [12] have developed an instrumented hip implant with a thermistor inside to measure the implant temperature. However, these instrumented prostheses are designed for postoperative monitoring and provide no intraoperative help to the surgeons.

Another strong need in THR surgeries is the estimation of hip ROM. One traditional solution is using the standard goniometer during THR surgery [13,14]. However, the use of such a device in current clinical practice is extremely rare and has been taken over by visual estimation [14]. As sensor technology has developed, researchers have tried to deploy miniature inertial measurement units (IMUs) on body segments adjacent to a joint as another solution to calculate the joint angles and ROM [15–18]. The accuracy of IMU drift correction methods are evaluated in [15] by comparing estimates of pelvis, thigh and shank orientation from IMUs and navigation systems during maximal hip flexion and abduction, walking, squatting and standing on one leg. The accuracy of a commercially available IMU system for estimating rotations across the hip, knee and ankle during level walking, stair ascent and stair descent is revealed in [16] with its angle estimation errors ranging from 1.38° to 6.69°. Nevertheless, the relative motion between the soft tissue and the underlying bony anatomy can be a source of error for the wearable IMU solution [18].

13.1.3 Introduction of the Proposed System

In order to help surgeons with the implantation of hip prostheses and the estimation of hip ROM in THR surgeries, a monocular vision- and IMU-based prosthesis pose estimation system has been designed. As shown in Figure 13.2, the system consists of three parts: an acetabular cup with a series of customized patterns printed on its inner surface, a hollow femoral head trial with circuits rigidly mounted inside and a computer with a hip prosthesis pose estimation software [19]. The femoral head trial is the same size as a conventional femoral head prosthesis, and its upper hemisphere is made of transparent materials so that the camera mounted in the center can take images of the reference patterns. During THR

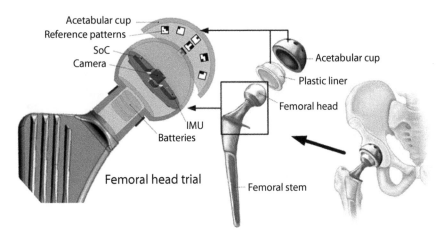

FIGURE 13.2
Structure of the new femoral head trial.

surgery, the sensor data will be obtained in vivo and transmitted wirelessly to a computer where the data will be processed and displayed on a screen. In addition, some low-power technologies are adopted in both the hardware and software implementation to ensure the battery lifetime of the system can fully cover the whole THR surgery. More details about the system will be discussed in the rest of the chapter. Section 13.2 introduces the system architecture. Section 13.3 gives details of the vision- and IMU-based prosthetic pose estimation methods. In Section 13.4, the low-power technologies adopted in the system are described. The implementation and experimental results are presented in Section 13.5. Finally, a conclusion is drawn in Section 13.6.

13.2 System Design

13.2.1 System Architecture

The system architecture is shown in Figure 13.3. The acetabular cup and the femoral head trial are designed for data collection, while the computer is in charge of the data processing and results display. Inside the hollow femoral head trial, a camera, an IMU, a data packer and a customized system-on-a-chip (SoC) are included. The monocular camera has a resolution of 240 × 240 and works with an adaptive frame rate, while the IMU works at a sampling rate of 100 Hz. The data packer synchronizes the data from the two sensors and packs them for transmission. The SoC is composed of three functional blocks: a power management unit, a microcontroller and an radio frequency (RF) transceiver. It manages the sensors and transmits the data packets wirelessly to the computer. The femoral head trial is powered by a battery, which is mounted in the neck of the femoral head trial.

On the computer, the raw sensor data are first extracted from the received data and then used to estimate the pose of the implanted hip prosthesis. The result is displayed to the surgeons in 3D in real time. During this process, an alarm will be given if any potential dislocation or impingement is detected. With the help of the system, the surgeons can

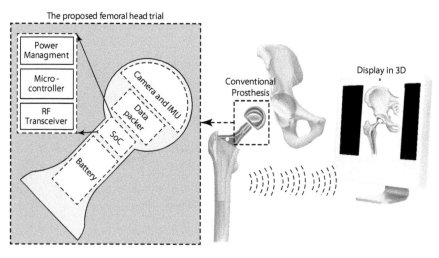

FIGURE 13.3
System architecture.

easily know the poses of both the acetabular cup and the femoral head trial, and make a judgment on the prosthesis placement better than just by their experiences. The femoral head is named as a trial because it will not be left inside the human body. After the prostheses are put within the safe zone, the surgeons will fix the implanted acetabular cup and replace the femoral head trial with a conventional femoral head prosthesis.

13.2.2 Design of the Reference Patterns

Two kinds of reference patterns are designed and printed on the internal surface of the acetabular cup liner [19], as shown in Figure 13.4a. Pattern I is a circular ring that is used as a landmark to adjust the femoral head trial to a preset pose at the beginning of the surgery, while pattern II is designed for relative pose estimation. As shown in Figure 13.4b, pattern II has the following three key design features. Firstly, each pattern II is a square surrounded by a black border. Secondly, there are nine small black or white blocks inside the border, identifying the pattern as black representing "0" and white representing "1", respectively. Thirdly, each pattern II must not be a rotational symmetric figure. A pattern II becomes distinguishable if it has the three features. All pattern II are of the same known size and are printed on the known positions inside the acetabular cup. The four vertexes of each pattern II are used as feature points in the vision-based pose estimation algorithms. Figure 13.3c gives examples of both distinguishable and indistinguishable pattern II.

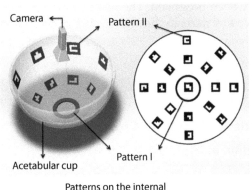

Patterns on the internal
surface of the acetabulum

(a)

An example of pattern II
whose ID is "011111111".

(b)

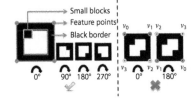

Distinguishable (left) and indistinguishable
(right) pattern II.

(c)

FIGURE 13.4

Reference patterns design on acetabulum cup. (a) Patterns on the internal surface of the acetabulum. (b) An example of pattern II whose ID is "011111111". (c) Distinguishable (left) and indistinguishable (right) pattern II.

13.2.3 Design of the SoC

The functional block diagram of the application-specific SoC with necessary peripheral components can be referred to [20]. As shown in Figure 13.5, the SoC has three main functional blocks. The first one is a power management unit, consisting of three programmable low-dropout (LDO) linear regulators supplying other functional blocks, and one programmable boost charge pump [21] for light-emitting diode (LED) driving. All the regulated voltages are generated by linear regulators except the boosted voltage to drive the LEDs as the flash light, as the switch regulator using inductors is not a good choice due to the number of external components and the consideration of electromagnetic interference (EMI). The second block is a programmable ultra-high frequency (UHF) band transceiver with 3 Mbps minimum-shift keying (MSK) transmitting and 64 kbps on-off keying (OOK) receiving. The third block is an ultra-low-power digital core, which takes care of the system control, the data processing and the communication protocol. The power management unit and the RF transceiver are controlled by the digital core as well. The digital core architecture is based on the Wishbone bus. Three master modules (including the microcontroller unit [MCU], the inter-integrated circuit [I²C] controller for test purposes and the watchdog timer) and six slave modules (the medium access control [MAC] controller, the image processor, etc.) are connected to the bus. Since wireless data transmission uses most of the energy, it is necessary to apply low-power technologies to this RF transceiver. The transceiver will be discussed in Section 13.4.1 on low-power technologies.

In this section, the structure and architecture of the system have been introduced and details of the customized SoC have been presented. The pose estimation methods and the low-power technologies are the two key parts in the system design and will be detailed in the following sections.

13.3 Pose Estimation Methods for Hip Joint Prostheses

13.3.1 Modeling and Representation of Hip Joint Motions

In a 3D reference frame, a rigid body is known to have six degrees-of-freedom (DOF), including translations along three axes and rotations about three axes. A hip joint prosthesis is a ball and socket joint, connecting the pelvis with the femur. An artificial hip joint consists of a femoral head prosthesis and an acetabular cup, both of which are supposed to be rigid. Since the round femoral head ball fits well into the fixed acetabular cup, the femoral head is only capable of rotating and only three rotation angles about the three axes are adequate to determine its orientation during surgery. However, dislocation may occur when the surgeons drag or rotate the patient's thigh to test the hip ROM during surgery. In this case, more information other than the three rotation angles is needed to describe whether the femoral head ball is in the center of the acetabular cup socket.

To describe the motions of the femoral head prosthesis, an acetabular cup coordinate system {A} is built, of which the origin is at the center of the acetabular cup and the z-axis is the normal vector of the cup margin plane, as shown in Figure 13.6a. Since the dislocation is taken into consideration, the joint is modeled as a 4-DOF joint, including the rotations about the x-, y- and z-axes, and the translation d along the z-axis. Once d is larger than a

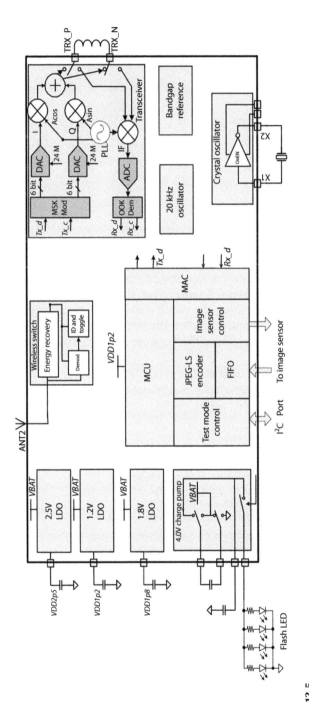

FIGURE 13.5
Block diagram of the SoC (inside the thick line) and the peripheral components.

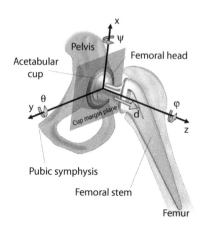

FIGURE 13.6
The 4-DOF model of a hip joint prostheses.

certain value, it means the dislocation occurs and the system will give an alarm to the surgeons until the femoral head trial is put back into the acetabular cup. The translations along the x- and y-axes can be left out of consideration. To help the surgeons judge whether the hip prostheses have been placed into the safe zone, the system should estimate the real-time orientation of the femoral head prosthesis with respect to the acetabular cup and show it to the surgeons. Besides, the depth d should be calculated for the detection of potential dislocations. In kinematics, such a problem is categorized as the forward kinematics problem, which is to determine the position and orientation of the end-effector in terms of the joint variables [22].

Four coordinate systems are defined (shown in Figure 13.7 and listed in Table 13.1) to describe the pose of the hip prosthesis mathematically. The camera inside the femoral

TABLE 13.1

Definitions of Four Coordinate Systems

Name and Notation	Definition
Femoral head coordinate system {F}	Origin: the center of the femoral head ball x_f: orthogonal to z_f, pointing in the medial direction y_f: orthogonal to x_f and z_f z_f: connecting the origin to the femoral neck
Acetabular cup coordinate system {A}	Origin: the center of hemisphere inside the cup x_a: orthogonal to z_a, defined by a specific pattern y_a: orthogonal to x_a and z_a z_a: the normal vector of cup margin plane
Pelvic coordinate system {P}	Origin: the mid-point of left and right anterior superior iliac spine x_p: orthogonal to y_p and z_p y_p: orthogonal to the plane defined with left and right anterior superior iliac spine and the mid-point left and right posterior superior iliac spine z_p: connecting left anterior superior iliac spine to right anterior superior iliac spine
World coordinate system {W}	Origin: a defined point at the objective location x_w: pointing to the east y_w: pointing to the north z_w: the direction of gravitational force

Note: {F} and {A} are used to estimate the relative pose between prostheses, {P} is used by surgeons to judge orientations of prostheses and {W} is used to present absolute poses of prostheses.

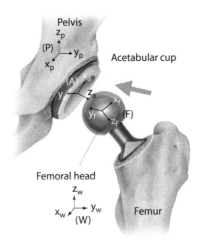

FIGURE 13.7
Definition of the four related coordinate systems.

head estimates the orientation of the femoral head trial with respect to the acetabular cup coordinate system {A} by taking pictures of the inner surface of the acetabular cup, while the IMU estimates the orientation of {F} with respect to {W} (representing the corresponding coordinate system, the same below). Therefore, proper methods are needed to represent the orientation and position of a rigid body with respect to a space-fixed coordinate system.

The femoral head coordinate system {F} is a body-fixed frame attaching to the femoral head trial. Suppose the system always starts with {F} totally overlapping the space-fixed frame {A} (the same below). A point fixed to the femoral head can be represented as $p^T = (x_f, y_f, z_f, 1)$ in both {F} and {A} with homogeneous coordinates applied. {F} has 6-DOF in total, including the 3-DOF translations and the 3-DOF rotations. A translation of the femoral head can be generally represented by three parameters:

$$\text{Trans}(a,b,c) = ai_a + bj_a + ck_a, \tag{13.1}$$

where i_a, j_a and k_a are the orthonormal basis of {A}. Since the femoral head is a rigid body, the coordinates of the point in {F} will remain p. After the translation of the femoral head trial, the new coordinates of the point in {A}, denoted as p', can be obtained as

$$p' = \begin{bmatrix} x_f + a \\ y_f + b \\ z_f + c \\ 1 \end{bmatrix} = \begin{bmatrix} 1 & 0 & 0 & a \\ 0 & 1 & 0 & b \\ 0 & 0 & 1 & c \\ 0 & 0 & 0 & 1 \end{bmatrix} p = H_T p, \tag{13.2}$$

where H_T is a 4×4 matrix and can be used to represent the translation of the femoral head trial with respect to {A}.

Similar to the translation matrix H_T, a rotation can also be represented by a 4×4 matrix, H_R. The following three matrices represent three elemental rotations, which are rotations about the x-, y- and z-axes by angles of Ψ, θ and φ, respectively.

$$H_{Rx}(\psi) = Rot(x, \psi) = \begin{bmatrix} 1 & 0 & 0 & 0 \\ 0 & \cos\psi & -\sin\psi & 0 \\ 0 & \sin\psi & \cos\psi & 0 \\ 0 & 0 & 0 & 1 \end{bmatrix}, \tag{13.3}$$

$$H_{Ry}(\theta) = Rot(y, \theta) = \begin{bmatrix} \cos\theta & 0 & \sin\theta & 0 \\ 0 & 1 & 0 & 0 \\ -\sin\theta & 0 & \cos\theta & 0 \\ 0 & 0 & 0 & 1 \end{bmatrix}, \tag{13.4}$$

$$H_{Rz}(\varphi) = Rot(z, \varphi) = \begin{bmatrix} \cos\varphi & -\sin\varphi & 0 & 0 \\ \sin\varphi & \cos\varphi & 0 & 0 \\ 0 & 0 & 1 & 0 \\ 0 & 0 & 0 & 1 \end{bmatrix}. \tag{13.5}$$

Generally, a specific rotation of {F} can be divided into the foregoing three elemental rotations. The three elemental rotations may be extrinsic (rotations about the axes of the space-fixed coordinate system) or intrinsic (rotations about the axes of the body-fixed coordinate system). For example, let $H_R = H_{Rz}(\varphi)H_{Ry}(\theta)H_{Rx}(\Psi)$ be a composition of extrinsic rotations. In this case, H_R represents rotations that first rotate the femoral head Ψ about x_a-axis, then θ about y_a-axis and finally φ about z_a-axis of the space-fixed frame {A}. But if H_R is a composition of intrinsic rotations, then it can also be regarded as rotations of first φ about z_f-axis, then θ about y_f-axis and finally Ψ about x_f-axis of the body-fixed frame {F}. To avoid the confusion of extrinsic and intrinsic rotations, mathematicians have devised several orientation representation methods, among which Euler angles, Tait–Bryan angles and quaternions are the most famous [23].

In 1776, Leonhard Euler introduced the Euler angles to describe the orientation of a rigid body with respect to a space-fixed coordinate system. The Euler angles, typically denoted as (φ, θ, Ψ), are three elemental rotation angles whose first and third rotation axes are the same. The Euler angles following this definition are called proper or classic Euler angles. For instance, the proper Euler angles (φ, θ, Ψ) in Equation 13.6 describe a sequence of rotations that first φ about z-axis, then θ about y-axis and finally Ψ about z-axis if defined by extrinsic rotations.

$$H_R = H_{Rz}(\psi) H_{Ry}(\theta) H_{Rz}(\varphi). \tag{13.6}$$

In the 19th century, Peter Guthrie Tait and George H. Bryan devised a second type of formalism, called Tait–Bryan angles. Usually denoted as (α, β, γ), Tait–Bryan angles represent a composition of three rotations about three different axes, respectively. For instance, defined by intrinsic rotations, the Tait–Bryan angles (α, β, γ) in Equation 13.7 describe a sequence of rotations about z-, y- and x-axes for γ, β and α, respectively.

$$H_R = H_{Rz}(\gamma) H_{Ry}(\beta) H_{Rx}(\alpha). \tag{13.7}$$

The Tait–Bryan angles defined in Equation 13.7 are also called yaw, pitch and roll. They are very useful for aerospace applications. For an aircraft, a pose change can be obtained by three rotations about its principal axes if done in the proper order. A yaw will obtain the bearing angle, a pitch will yield the elevation angle and a roll gives the bank angle.

Both the proper Euler angles and Tait–Bryan angles are easy to understand, but they suffer from a problem called gimbal lock. To avoid this, Irish mathematician William Rowan Hamilton proposed the quaternions in 1843. The quaternion representation obtains a rotation matrix with a single rotation about one axis and does not have a singularity problem [24]. A quaternion q is defined as a four-dimensional vector, consisting of one real number and three imaginary numbers, as shown in Equations 13.8 and 13.9:

$$q = \begin{pmatrix} q_0 & q_1 & q_2 & q_3 \end{pmatrix}^T = q_0 + (q_1 i + q_2 j + q_3 k) = q_0 + q_v, \tag{13.8}$$

$$i^2 = j^2 = k^2 = ijk = -1, \tag{13.9}$$

where q_0, q_1, q_2 and q_3 are real numbers and i, j and k are imaginary units.

The quaternion q is called a unit quaternion if it satisfies Equation 13.10:

$$\begin{cases} q_0^2 + (q_1^2 + q_2^2 + q_3^2) = 1 \\ q = q_0 + q_v = \cos(\theta) + u\sin(\theta) \end{cases}, \tag{13.10}$$

where u is a 3D unit vector ($u = u_x i + u_y j + u_z k$). The quaternion components of q are

$$\begin{cases} q_0 = \cos\theta \\ q_1 = u_x \sin\theta \\ q_2 = u_y \sin\theta \\ q_3 = u_z \sin\theta \end{cases}. \tag{13.11}$$

The point p fixed to the femoral head trial can be regarded as a vector setting out from the origin of {F}, and can also be represented as a quaternion $p = (0, x_f, y_f, z_f)$ in {F}. If the femoral head trial is rotated by an angle of 2θ about the vector u, the new coordinates of the point in {A} after the rotation, denoted as p', can be calculated as

$$p' = q \cdot p \cdot q^*, \tag{13.12}$$

where q is a unit quaternion and $q^* = q_0 - q_1 i - q_2 j - q_3 k$ is the conjugate of q.

The dot is a specific operation of quaternions called Hamilton product [23], defined as

$$a \cdot b = (a_0 + a_1 i + a_2 j + a_3 k)(b_0 + b_1 i + b_2 j + b_3 k) = \begin{pmatrix} a_0 b_0 - a_1 b_1 - a_2 b_2 - a_3 b_3 \\ a_0 b_1 + a_1 b_0 + a_2 b_3 - a_3 b_2 \\ a_0 b_2 - a_1 b_3 + a_2 b_0 + a_3 b_1 \\ a_0 b_3 + a_1 b_2 - a_2 b_1 + a_3 b_0 \end{pmatrix}. \tag{13.13}$$

Such a rotation can also be achieved by a rotation matrix H_{Rq}:

$$p' = H_{Rq}p = \begin{bmatrix} 1-2(q_2{}^2+q_3{}^2) & 2(q_1q_2-q_3q_0) & 2(q_1q_3+q_2q_0) & 0 \\ 2(q_1q_2+q_3q_0) & 1-2(q_1{}^2+q_3{}^2) & 2(q_2q_3-q_1q_0) & 0 \\ 2(q_1q_3-q_2q_0) & 2(q_2q_3+q_1q_0) & 1-2(q_1{}^2+q_2{}^2) & 0 \\ 0 & 0 & 0 & 1 \end{bmatrix} p, \quad (13.14)$$

where the relationship between the quaternion and the rotation matrix is shown. As a result, no matter which orientation representation method is applied, a rotation of a rigid body can always be represented by a 4×4 matrix with homogeneous coordinates applied.

13.3.2 Problem Definition

Suppose the system starts with {F} totally overlapping {A}. One of the vertexes of the reference patterns printed on the inner surface of the acetabular cup is expressed as $p_a = (x_a, y_a, z_a, 1)^T$ in {A} and $p_f = (x_f, y_f, z_f, 1)^T$ in {F}; p_f varies with the femoral head trial being moved, while p_a remains unchanged. According to the discussion in the previous section, a 4×4 matrix H can be found to represent the transformation between p_f and p_a, and H can be divided into several translations H_T and rotations H_R as

$$p_f = Hp_a = H_{T1}H_{R1}H_{T2}H_{R2}\ldots H_{Tn}H_{Rn}p_f. \quad (13.15)$$

Because p_f and p_a are actually the same point expressed in different coordinate systems, the matrix H just represents the transformation from the {F} to the {A}, which contains information of relative pose between the femoral head prosthesis and the acetabular cup. Consequently, the prosthesis pose estimation problem of the system is finally defined as the problem of calculating the transformation matrix H in Equation 13.15.

The camera and the IMU in the system provide us with two ways to calculate H. One is to use IMU to directly measure the motion parameters (accelerations and angular velocities) to calculate H. The other is to find the coordinates of the corresponding feature points in the images and solve H with the theory of computer vision. These two solutions will be explained next in detail.

13.3.3 IMU-Based Pose Estimation Method

An IMU is composed of an accelerometer and a gyroscope and is widely used in navigation systems and robots. Mounted inside the femoral head trial, the IMU can be used to estimate the real-time orientation of the femoral head with respect to the world coordinate system {W}. In this part, the accelerometer-based orientation estimation method will be introduced, then the gyroscope-based one and finally the famous data fusion method called extended Kalman Filter (EKF) for the IMU data.

The Tait–Bryan angles (α, β, γ) are selected to explain the principle of the IMU-based orientation estimation method as they are easier to understand. Suppose the system starts with {F} overlapping the world frame {W}. The orientation of the femoral head trial is obtained by the following three intrinsic rotations in sequence of $z_f - y_f - x_f$: first rotates about z_f-axis by an angle γ, called yaw or heading angle ($\gamma \in [-\pi, \pi]$); then about the y_f-axis by an angle β, called pitch or elevation angle $\left(\beta \in \left[-\dfrac{\pi}{2}, \dfrac{\pi}{2}\right]\right)$; finally, about x_f-axis by a roll or bank angle α ($\alpha \in [-\pi, \pi]$). The rotation matrix is

$$
H = \begin{bmatrix} \cos\gamma & -\sin\gamma & 0 & 0 \\ \sin\gamma & \cos\gamma & 0 & 0 \\ 0 & 0 & 1 & 0 \\ 0 & 0 & 0 & 1 \end{bmatrix} \begin{bmatrix} \cos\beta & 0 & \sin\beta & 0 \\ 0 & 1 & 0 & 0 \\ -\sin\beta & 0 & \cos\beta & 0 \\ 0 & 0 & 0 & 1 \end{bmatrix} \begin{bmatrix} 1 & 0 & 0 & 0 \\ 0 & \cos\alpha & -\sin\alpha & 0 \\ 0 & \sin\alpha & \cos\alpha & 0 \\ 0 & 0 & 0 & 1 \end{bmatrix}
$$

$$
= \begin{bmatrix} \cos\beta\cos\gamma & \sin\alpha\sin\beta\cos\gamma-\cos\alpha\sin\gamma & \cos\alpha\sin\beta\cos\gamma+\sin\alpha\sin\gamma & 0 \\ \cos\beta\sin\gamma & \sin\alpha\sin\beta\sin\gamma+\cos\alpha\cos\gamma & \cos\alpha\sin\beta\sin\gamma-\sin\alpha\cos\gamma & 0 \\ -\sin\beta & \sin\alpha\cos\beta & \cos\alpha\cos\beta & 0 \\ 0 & 0 & 0 & 1 \end{bmatrix}. \tag{13.16}
$$

As the accelerometer is mounted inside the femoral head prosthesis, it measures the projection of the gravitational acceleration in the three axes of {F} if the femoral head trial stays still or moves slowly. Denoted as g_f, the gravitational acceleration vector in {F} is therefore represented as follows:

$$
g_f = \begin{bmatrix} g_x \\ g_y \\ g_z \\ 1 \end{bmatrix} = H^{-1} g_w = H^{-1} \begin{bmatrix} 0 \\ 0 \\ g \\ 1 \end{bmatrix} = \begin{bmatrix} -g\sin\beta \\ g\sin\alpha\cos\beta \\ g\cos\alpha\cos\beta \\ 1 \end{bmatrix}. \tag{13.17}
$$

By solving the equations in Equation 13.17, the roll α and the pitch β can be obtained in Equation 13.18:

$$
\begin{bmatrix} \alpha \\ \beta \end{bmatrix} = \begin{bmatrix} \tan^{-1}\left(\dfrac{g_y}{g_z}\right) \\ \tan^{-1}\left(\dfrac{-g_x}{\sqrt{g_y^2+g_z^2}}\right) \end{bmatrix}. \tag{13.18}
$$

However, the yaw γ cannot be determined without the help of other sensors. In addition, if the femoral head trial is in a variable speed motion, the IMU will have a measured acceleration besides the gravitational acceleration, leading to the invalidity of the method.

On the other hand, the gyroscope-based method is good at dealing with the variable motion. The gyroscope measures the angular velocity about the three axes of {F}, denoted as $\omega_f = [\omega_{fx}\ \omega_{fy}\ \omega_{fz}]^T$. The relationship between the Tait–Bryan angles (α, β, γ) and ω_f is shown in Equation 13.19. The Tait–Bryan angles (α, β, γ) can be obtained by solving the first-order nonlinear differential equations in 13.20.

$$
\begin{bmatrix} \omega_{fx} \\ \omega_{fy} \\ \omega_{fz} \end{bmatrix} = \begin{bmatrix} \cos\gamma & -\sin\gamma & 0 \\ \sin\gamma & \cos\gamma & 0 \\ 0 & 0 & 1 \end{bmatrix} \begin{bmatrix} \cos\beta & 0 & \sin\beta \\ 0 & 1 & 0 \\ -\sin\beta & 0 & \cos\beta \end{bmatrix} \begin{bmatrix} \dot\alpha \\ 0 \\ 0 \end{bmatrix}
$$

$$
+ \begin{bmatrix} \cos\gamma & -\sin\gamma & 0 \\ \sin\gamma & \cos\gamma & 0 \\ 0 & 0 & 1 \end{bmatrix} \begin{bmatrix} 0 \\ \dot\beta \\ 0 \end{bmatrix} + \begin{bmatrix} 0 \\ 0 \\ \dot\gamma \end{bmatrix} \tag{13.19}
$$

$$
= \begin{bmatrix} \cos\beta\cos\gamma & -\sin\gamma & 0 \\ \cos\beta\sin\gamma & \cos\gamma & 0 \\ \cos\beta & 0 & 1 \end{bmatrix} \begin{bmatrix} \dot\alpha \\ \dot\beta \\ \dot\gamma \end{bmatrix}.
$$

$$
\begin{bmatrix} \dot\alpha \\ \dot\beta \\ \dot\gamma \end{bmatrix} = \begin{bmatrix} \dfrac{\cos\gamma}{\cos\beta} & \dfrac{\sin\gamma}{\cos\beta} & 0 \\ -\sin\gamma & \cos\gamma & 0 \\ -\cos\gamma & -\sin\gamma & 1 \end{bmatrix} \begin{bmatrix} \omega_{fx} \\ \omega_{fy} \\ \omega_{fz} \end{bmatrix} \tag{13.20}
$$

Since the gyroscope has drifts, the gyroscope-based method may suffer from a large error if the femoral head trial stays still for a long time. Besides, if the pitch $\beta = \pm(\pi/2)$, the first two terms in the first row in Equation 13.20 will become infinite, leading to a singularity.

Although both the accelerometer and the gyroscope can be used for orientation estimation alone, each has its own limitations. From the foregoing discussion, a complementary property can be found between the accelerometer and the gyroscope. To make use of the complementary property and avoid the problem of singularities, a quaternion-based data fusion algorithm called extended Kalman filter (EKF) [25] is adopted in the system. The block diagram of the IMU data fusion method is shown in Figure 13.8.

In order to use EKF, a state vector X_k and its state transition need to be defined. Let $^{F(t)}_{A}q$ be the quaternion that represents the orientation of {F} with respect to {A} at time t. The difference quotient of $^{F(t)}_{A}q$ can be computed as

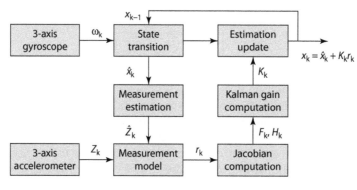

FIGURE 13.8
Block diagram of orientation estimation principles for an IMU.

$$
{}^{F(t)}_{A}\dot{q} = \lim_{\Delta t \to 0} \frac{1}{\Delta t}\left[{}^{F(t+\Delta t)}_{A}q - {}^{F(t)}_{A}q\right] = \lim_{\Delta t \to 0} \frac{1}{\Delta t}\left[{}^{F(t+\Delta t)}_{F(t)}q \cdot {}^{F(t)}_{A}q - {}^{F(0)}_{A}q \cdot {}^{F(t)}_{A}q\right],
\tag{13.21}
$$

where ${}^{F(t+\Delta t)}_{F(t)}q$ represents the rotation in the period Δt. In the limit of $\Delta t \longrightarrow 0$, the angle of rotation $\delta\theta$ is very small so that the first-order Taylor expansion can be used to approximate the sin and cos functions as

$$
\lim_{\Delta t \to 0} {}^{F(t+\Delta t)}_{F(t)}q = \begin{pmatrix} \cos\dfrac{\delta\theta}{2} \\ u\sin\dfrac{\delta\theta}{2} \end{pmatrix} \approx \begin{pmatrix} 1 \\ u\dfrac{\delta\theta}{2} \end{pmatrix} = \begin{pmatrix} 1 \\ \dfrac{1}{2}\delta\theta \end{pmatrix}.
\tag{13.22}
$$

The vector $\delta\theta$ has the direction of the axis of rotation and the magnitude of the angle of rotation. Dividing this vector by Δt, in the limit $\Delta t \longrightarrow 0$, will yield the rotational velocity as

$$
\boldsymbol{\omega} = \begin{bmatrix} \omega_x & \omega_y & \omega_z \end{bmatrix}^T = \lim_{\Delta t \to 0} \frac{\delta\theta}{\Delta t}.
\tag{13.23}
$$

By combining Equations 13.21 through 13.23, the quaternion derivative ${}^{F(t)}_{A}\dot{q}$ can be derived as

$$
{}^{F(t)}_{A}\dot{q} \approx \left[\begin{pmatrix} 1 \\ \dfrac{1}{2}\delta\theta \end{pmatrix} - \begin{pmatrix} 1 \\ 0 \end{pmatrix}\right] \cdot {}^{F(t)}_{A}q = \frac{1}{2}\begin{pmatrix} 1 \\ \boldsymbol{\omega} \end{pmatrix} \cdot {}^{F(t)}_{A}q = \frac{1}{2}\Omega(\boldsymbol{\omega})\,{}^{F(t)}_{A}q
\tag{13.24}
$$

with

$$
\Omega(\boldsymbol{\omega}) = \begin{bmatrix} 0 & -\omega_x & -\omega_y & -\omega_z \\ \omega_x & 0 & \omega_z & -\omega_y \\ \omega_y & -\omega_z & 0 & \omega_x \\ \omega_z & \omega_y & -\omega_x & 0 \end{bmatrix} = \begin{bmatrix} 0 & -\boldsymbol{\omega}^T \\ \boldsymbol{\omega} & -[\boldsymbol{\omega}_\times] \end{bmatrix}.
\tag{13.25}
$$

The discrete-time model corresponding to Equation 13.24 is

$$
\begin{cases} q_k = \left[I + \dfrac{T_s}{2}\Omega(\boldsymbol{\omega}_k)\right]q_{k-1}, \; k = 1, 2, \ldots \\ q_0 = 0, \end{cases}
\tag{13.26}
$$

where T_s is the sample interval. If the drifts of the gyroscope are taken into consideration, the state vector can be defined as the rotation quaternion of the femoral head trial q_k and bias vector of the gyroscope ${}^g b_k$. The state vector X_k is expressed as

$$
X_{k+1} = \begin{bmatrix} q_{k+1} \\ {}^g b_{k+1} \end{bmatrix} = f\left(x_k, \boldsymbol{\omega}_{k+1}\right) + \mathbf{w}_{k+1} = \begin{bmatrix} I + \dfrac{T_s}{2}\Omega(\hat{\boldsymbol{\omega}}_{k+1}) & 0 \\ 0 & I \end{bmatrix}\begin{bmatrix} q_k \\ {}^g b_k \end{bmatrix} + \mathbf{w}_{k+1},
\tag{13.27}
$$

where w_{k+1} is a noise vector and $\hat{\boldsymbol{\omega}}_{k+1} = \boldsymbol{\omega}_{k+1} - {}^g b_k$ is the calibrated angular velocity of {F} measured by the gyroscope.

The measurement vector Z_{k+1} is constructed by the output of the accelerometer as

$$Z_{k+1} = a_{k+1} = C_w^f [q_{k+1}] g + v_{k+1}, \tag{13.28}$$

where g is the gravitational acceleration with respect to {W} and v_{k+1} is a noise vector. The state vector is estimated iteratively. The specific processing steps are

1. Consider the last filtered state estimate x_k.
2. Calculate the Jacobian matrix F_k for the system dynamics around X_k to linearize the system dynamics, as shown in Equation 13.29:

$$F_k = \frac{\partial f_k}{\partial X_k} \tag{13.29}$$

3. Apply the prediction step of the Kalman filter to the linearized system dynamics obtained in step 2:

$$\hat{x}_{k+1} = \begin{bmatrix} I + \dfrac{T_s}{2} \Omega(\hat{\omega}_{k+1}) & 0 \\ 0 & I \end{bmatrix} x_k \tag{13.30}$$

$$P_{k+1|k} = F_k P_k F_k^T + Q_k \tag{13.31}$$

In Equation 13.31, Q_k is the process noise covariance matrix, P_k is the covariance of x_k and $P_{k+1|k}$ is the estimate of the covariance of $\hat{x}_{k+1}$

4. Linearize the observation vector $\hat{z}_{k+1} = h(\hat{x}_{k+1})$:

$$H_{k+1} = \frac{\partial h_{k+1}}{\partial \hat{x}_{k+1}} \tag{13.32}$$

5. Apply the filtering cycle of the Kalman filter to the linearized observation dynamics:

$$x_{k+1} = \hat{x}_{k+1} + K_{k+1} \left[z_{k+1} - h(\hat{x}_{k+1}) \right] \tag{13.33}$$

$$P_{k+1} = \left[I - K_{k+1} H_{k+1} \right] P_{k+1|k} \tag{13.34}$$

In Equations 13.33 and 13.34, K_{k+1} is Kalman gain for iteration $(k+1)$. After a series of iterations, the optimized state vector x_k will converge to a stable value, which is the estimation of the femoral head orientation.

13.3.4 Vision-Based Pose Estimation Method

The vision-based pose estimation method is another method widely used in navigation and automatic control systems. As the camera is rigidly attached to the femoral head trial, the femoral head pose estimation problem can be changed into the camera ego-motion estimation problem.

A camera model, which is used to describe the projection of a point from the 3D coordinate system to the 2D image plane, is the basis of the vision-based pose estimation method. Researchers have proposed several camera models, among which the pinhole model is the most popular. Consider a point $P_w = (x_w, y_w, z_w, 1)^T$ in the world coordinate system {W}. The pinhole camera model in Equation 13.35 shows how P_w is projected to the image:

$$p = \begin{pmatrix} u \\ v \\ 1 \end{pmatrix} = DK_0H_w^cP_w = \begin{pmatrix} \dfrac{f}{dx} & \dfrac{f\cot\theta}{dx} & u_0 & 0 \\ 0 & \dfrac{f}{\sin\theta dy} & v_0 & 0 \\ 0 & 0 & 1 & 0 \end{pmatrix} \begin{pmatrix} R & T \\ 0 & 1 \end{pmatrix} P_w \qquad (13.35)$$

In Equation 13.35,

$$D = \begin{pmatrix} \dfrac{1}{dx} & -\dfrac{\cot\theta}{dx} & 0 \\ 0 & \dfrac{1}{\sin\theta dy} & 0 \\ 0 & 0 & 1 \end{pmatrix}, \quad K_0 = \begin{pmatrix} f & 0 & x_0 & 0 \\ 0 & f & y_0 & 0 \\ 0 & 0 & 1 & 0 \end{pmatrix}, \quad and \; H_w^c = \begin{pmatrix} R & T \\ 0 & 1 \end{pmatrix}.$$

As shown in Figure 13.9, the pinhole model is divided into three steps. First, the matrix H_w^c, which contains the camera's rotation matrix R and translation vector T with respect to {W}, transforms the coordinates of P_w from {W} into the camera coordinate system {C}. Then the point is projected from {C} to the 2D image plane by K_0, in which f is the camera focal length and (x_0, y_0) is the optic center on the camera image plane. Finally, the 2D coordinates of the point in the image plane {I} is discretized by D, in which dx and dy represent the size of each pixel in a charge-coupled device/complementary metal-oxide-semiconductor (CCD/CMOS) image sensor with an intersection angle θ. The matrix $K = DK_0$ is

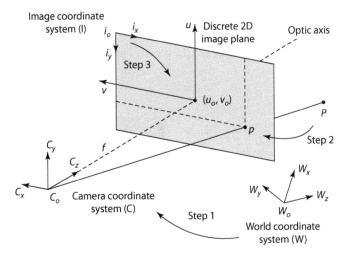

FIGURE 13.9
The general steps of the pinhole model.

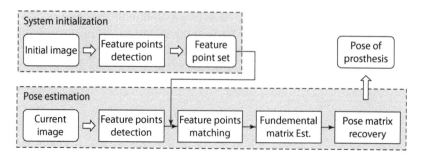

FIGURE 13.10
System initialization and pose estimation pipeline.

called the intrinsic parameter matrix, in which all the parameters are available by the camera calibration, while the matrix H_w^c is known as the extrinsic parameter matrix, containing the ego-motion information of the camera with respect to {W}. If a calibrated camera is given together with a set of 3D points whose coordinates in {W} are known information, the pose of the camera can be acquired by solving Equation 13.35. Researchers have named it a perspective-n-point (PnP) problem.

As introduced in Section 13.2.2, an array of reference patterns is designed on the inner surface of the acetabular cup. Since the pattern IDs and positions of the patterns are prior information, the four vertexes of each pattern II can be used as feature points for camera ego-motion estimation. However, if the patterns are covered by blood during surgery, pattern recognition will become difficult and the outcome of the PnP algorithms may be wrong. To avoid the interference of blood, another method based on epipolar geometry [26] is adopted in the system instead of a PnP algorithm. This method consists of two major steps with its pipeline shown in Figure 13.10.

13.3.4.1 Initialization

Two tasks are performed in initialization: the first is to determine whether the femoral head is inside the acetabular cup; the second is to adjust the femur to the initial pose and start tracking [27]. When the femoral head is put inside the acetabular cup, pattern I may appear in the visual range of the camera. If the camera's optic axis is not perpendicular to the circular ring's plane, the projection of it will become an elliptical ring (shown in Figure 13.11a). A pink circle will appear (shown in Figure 13.11b) when the femoral head

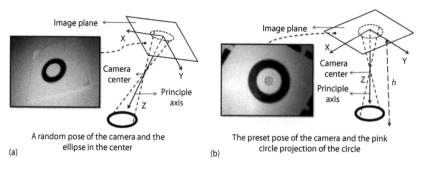

FIGURE 13.11
(a) A random pose of the camera and the ellipse in the center and (b) the preset pose of the camera and the pink circle projection of the circle.

is adjusted to the preset position, where the camera's axis is perpendicular to the circular ring's plane and the distance between the camera image plane and the center of the ring is exactly h (a preset value). The pink circle means that the femoral head is at the preset position inside the acetabular cup and the motion of the femur can be tracked from that initial position.

13.3.4.2 Feature Detection

After initialization, feature detection is done in three steps: contour detection, pattern recognition and vertexes calculation. Each image acquired will be converted from a red-green-blue (RGB) format into a gray scale one and processed by the polygonal approximation algorithm to search all the contours in the image. The quadrilaterals are chosen as candidates from the contours for further pattern detection. To recognize the patterns, samples are taken at the nine points within each possible pattern contour. The pattern is identified by the colors of the nine sample points, with white representing "1" and black representing "0". The 9-bit binary sequence is then converted to a decimal number, which is the ID of the reference pattern. Once a pattern is recognized, the 2D coordinates of its four vertexes will be calculated, which are denoted as (u_0, v_0), (u_1, v_1), (u_2, v_2) and (u_3, v_3). They will be used as feature points. The feature detection process is shown in Figure 13.12.

As the position of each reference pattern is known information, the 3D coordinates of their vertexes in {A} can also be extracted from the database, denoted as (x_0, y_0, z_0), (x_1, y_1, z_1), (x_2, y_2, z_2) and (x_3, y_3, z_3). The correspondences between the 3D feature points and their 2D projections are established according to the ID of each reference pattern:

$$c_i : (u_i, v_i) \leftrightarrow (x_i, y_i, z_i) \quad i = 0, 1, 2, 3 \dots \tag{13.36}$$

If there are n reference patterns recognized in the image, $4n$ correspondences will be established since each square has four squares. As at least four correspondences are needed to solve the equations in Equation 13.35, one recognized pattern is enough to estimate the camera pose with a PnP algorithm [27]. However, the PnP algorithm strictly relies on the accuracy of the 3D coordinates of the feature points. To get rid of the high dependency on the accuracy of manufacture, a different pose estimation method based on multiple view geometry is applied in this system [28].

13.3.4.3 Feature Matching

Suppose a calibrated camera takes pictures of the same object from two different positions. The camera first takes a picture at Position 1, then shifts to Position 2 by a rotation R and a

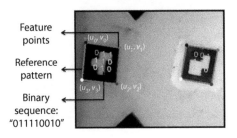

FIGURE 13.12
Find reference patterns and identify them.

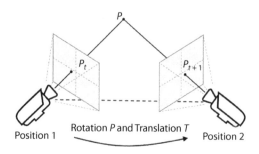

FIGURE 13.13
Principle of two view geometry.

translation T and takes a second picture at Position 2, as shown in Figure 13.13. The point P is fixed in the 3D world coordinate system {W}. Based on Equation 13.35, the projections of P in the two images, denoted as p_t and p_{t+1}, can be represented in Equations 13.37 and 13.38, respectively:

$$p_t = KH_1P, \tag{13.37}$$

$$p_{t+1} = KH_2P, \tag{13.38}$$

where H_1 and H_2 are the camera transformation matrices in Position 1 and 2, respectively, and K is the camera intrinsic parameter matrix.

The relationship between the correspondences in the two images is acquired by combining Equations 13.37 and 13.38:

$$p_{t+1} = KH_2P = K\begin{pmatrix} R & T \\ 0 & 1 \end{pmatrix}H_1P = K\begin{pmatrix} R & T \\ 0 & 1 \end{pmatrix}K^{-1}p_t. \tag{13.39}$$

If more than four correspondences between the two images can be found, the motion of the camera R and T from Positions 1 to 2 can be solved by Equation 13.39. Figure 13.14 shows the feature match between two adjacent image frames.

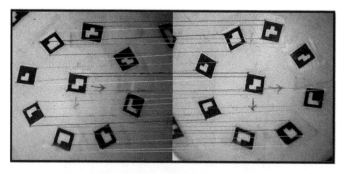

FIGURE 13.14
Matching feature points between two image frames.

13.3.4.4 Estimation of the Fundamental Matrix

After the features are matched, the 3×3 fundamental matrix F is defined as

$$x'^T Fx = 0, \tag{13.40}$$

where $x = (u,v,1)^T \leftrightarrow x' = (u',v',1)^T$ are corresponding points. From a set of n point matches, a set of linear equations can be obtained:

$$Af = \begin{bmatrix} u_1'u_1 & u_1'v_1 & u_1' & v_1'u_1 & v_1'v_1 & v_1' & u_1 & v_1 & 1 \\ \vdots & \vdots & \vdots & \vdots & \vdots & \vdots & \vdots & \vdots & \vdots \\ u_n'u_n & u_n'v_n & u_n' & v_n'u_n & v_n'v_n & v_n' & u_n & v_n & 1 \end{bmatrix} f = 0, \tag{13.41}$$

$$f^T = \begin{bmatrix} f_{11} & f_{12} & f_{13} & f_{21} & f_{22} & f_{23} & f_{31} & f_{32} & f_{33} \end{bmatrix}, \tag{13.42}$$

where A is an $n \times 9$ matrix. Given at least eight pairs of correspondences, which means at least two reference patterns should be found in the image, F can be calculated.

13.3.4.5 Recovery of Transformation Matrix

An essential matrix is a 3×3 matrix that captures the geometric relationship (a rotation R and a translation t) between two locations of one moving camera [29], and can be calculated from the fundamental matrix F acquired at the previous step and the camera calibration matrices K and K' as

$$E = [t]_\times R = K'^T FK, \tag{13.43}$$

where $[t]_\times$ is a 3×3 antisymmetric matrix defined by vector t. The transformation matrix $H = [R \mid t]$ can then be acquired by solving Equation 13.43, and the current camera pose can be calculated by

$$P = HP_0, \tag{13.44}$$

where P and P_0 are the current and past camera pose matrix.

13.3.5 Data Fusion of the Camera and the IMU

Both the camera and the IMU can be used alone to achieve pose estimation, but each of them has its own limitations. The IMU has large measurement uncertainty at slow motion due to drifts, and can only do pose estimation with respect to the world coordinate system {W} while surgeons are more interested in the relative pose between the femoral head and the acetabular cup. On the other hand, the camera is easily interfered with by blood. Furthermore, if the femoral head trial is moved at high speed, there will be motion blur on the images. Although the motion blur can be relieved by adjusting the camera to a higher frame rate mode, it will also increase the data throughput, leading to a higher power consumption and a higher complexity in the RF transceiver design.

Similar to the relationship between the accelerometer and the gyroscope, the camera and the IMU also share a complementary property in pose estimation. Better results could be obtained by fusing their data, as shown in Figure 13.15. Several algorithms can be used as the optimization filter. In this system, EKF is adopted for the optimization.

As mentioned in Section 13.3.3, a state vector and a measurement vector need to be defined in the EKF and it will iteratively estimate the state vector. The state vector is composed of the rotation quaternion of the femoral head trial q_{k+1}, the bias vector of the gyroscope ${}^g b_{k+1}$ and the accelerometer ${}^a b_{k+1}$. The state vector can be predicted by the IMU data, so X_{k+1} could be expressed as Equation 13.45.

$$X_{k+1} = \begin{bmatrix} q_{k+1} \\ {}^g b_{k+1} \\ {}^a b_{k+1} \end{bmatrix} = f_{k+1}(X_k, \omega_{k+1}) + w_k \tag{13.45}$$

where w_k is a noise vector and f_{k+1} is the state transition based on the previous state vector x_k and the calibrated angular velocity ω_{k+1} measured by the gyroscope.

For each image acquired from the camera, a set of features can be detected and the projections of the features are used for pose estimation. If there are N features detected in the current frame, the measurement vector can be defined as

$$Z_k = \begin{bmatrix} z_1^T & \cdots & z_N^T \end{bmatrix}^T, \tag{13.46}$$

$$z_i = \begin{bmatrix} u_i \\ v_i \end{bmatrix} + \eta_i = h_i(x, {}^A P_i) + \eta_i, \quad i = 1, 2, \ldots, N, \tag{13.47}$$

where z_i is the projection of the ith feature point and ${}^A P_i$ is its corresponding 3D coordinates in the acetabular cup frame {A}; η_i is the noise of the feature detection.

The iteration steps are similar to those in Section 13.3.3. The fusion of the camera and the IMU not only provides more accurate pose estimation of the femoral head prosthesis, but it also helps decrease the power consumption, which will be discussed in next section.

13.4 Low-Power Technology

Power consumption is a key factor for electronic devices, especially in medical applications. In this pose estimation system, a 500 mAh Li-ion battery is used to supply the whole system with the constraint of the femoral neck space. In order to ensure that the system

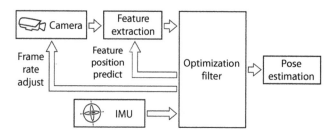

FIGURE 13.15
Fusion of the camera and the IMU data.

works for more than 20 minutes with this battery, low-power technologies are applied in both the hardware and the software design.

13.4.1 Low-Power RF Transceiver

Since most of the energy is consumed by the wireless data transmission, the power consumption of the RF transceiver is the key issue in the SoC design. The overall structure of the transceiver used in this design is shown in Figure 13.16. As the system needs to continuously transmit data to and sparsely receive configuration command from the PC, the system requires a high speed transmitter and a relatively low speed receiver. The MSK modulation has been chosen for the transmitter, since MSK is a constant-envelope modulation that can help to alleviate the design requirement of the data recorder on the other side. A high data rate of 3 Mbps is chosen such that the transmitter will work in burst-mode and the average power consumption of the transmitter will be low [30]. The receiver takes a 64 kbps OOK modulation for circuit simplification. In this system, the effort to lower the circuit power is equivalent to lowering the circuit current. For this reason, the current reusing techniques [31] have been greatly adopted in the transceiver circuit, mainly in the frequency synthesizer and the transmitter.

Figure 13.17 shows the structure of the frequency synthesizer. The voltage-controlled oscillator (VCO) is designed to run at ~800 MHz to generate quadrature-phase local oscillation signals. In this synthesizer, the quadratic frequency divider shares the same direct current (DC) path [32] with the VCO, as shown in Figure 13.3. A classical phase-locked loop (PLL) locks the VCO frequency. The PLL's divider is programmable, giving the SoC the capability of frequency hopping. The VCO has a coarse tuning circuit that helps to calibrate its center frequency automatically. Both the VCO tuning circuit and the phase frequency detector (PFD) take the reference frequency from an on-chip 24 MHz crystal oscillator.

Figure 13.18 shows the structure of the transmitter. The MSK modulator receives the data stream from the MCU, transforms it into the zero-IF baseband waveforms and then sends the 6-bit digital baseband data to the digital-to-analog converters (DACs). There are

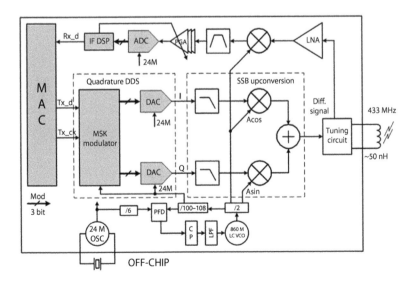

FIGURE 13.16
Transceiver architecture.

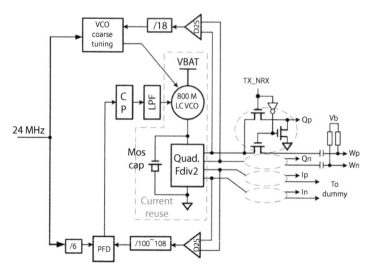

FIGURE 13.17
Quadrature LO oscillator.

two key points in this transmitter structure. First, the DCs of the two DACs are reused by the quadrature mixer (M1 and M2 are used to set the DC path). Secondly, there is no traditional power amplifier in this transmitter, and a coil that serves as the RF energy-emitting component is directly connected to the mixer. This specific structure has been proven effective in the special power-constrained application environment.

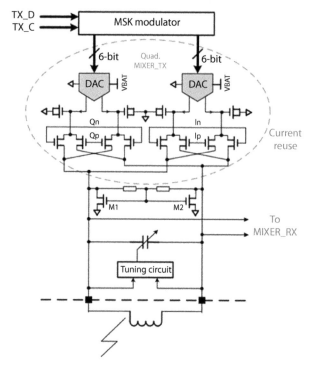

FIGURE 13.18
Transmitter circuit with current reusing.

13.4.2 Adaptive Sensor Control

Motion blur may occur in the images acquired from the camera if the femoral head is rotated at high speed. To avoid this, a simple solution is to set the camera to work at a higher frame rate. However, a higher frame rate means higher power consumption. In this system, an adaptive sensor control algorithm is applied to achieve the tradeoff between the camera frame rate and the power consumption.

When the system is initialized, the camera starts to work at a preset frame rate f_0. Suppose that ω_f and a_f are the sample of the IMU, representing the angular velocity and the acceleration of the femoral head, respectively. The camera frame rate is determined by the norm of these two vectors as

$$
f = \begin{cases}
f_0, & \text{if } \omega_f \leq \varepsilon_\omega \text{ and } \omega_f \leq \varepsilon_\omega \\[2mm]
\text{floor}\left(\dfrac{\omega_f}{\varepsilon_\omega} f_0\right), & \text{if } \omega_f > \varepsilon_\omega \\[2mm]
\text{floor}\left(\dfrac{a_f}{\varepsilon_a} f_0\right), & \text{if } \omega_f \leq \varepsilon_\omega \text{ and } \omega_f > \varepsilon_\omega
\end{cases}
\tag{13.48}
$$

where ε_ω and ε_a are two predefined thresholds and floor (x) is the round down function. With this adaptive sensor control solution, the camera will adjust itself to a low frame rate mode if the femoral head is moving at slow motion. This method is a low-power technology applied to the software.

13.5 Implementation and Experimental Results

13.5.1 Implementation Results of the SoC

The SoC has been implemented in the 0.18 μm CMOS technology. The digital core has 30k equivalent gates and 94 kB static random-access memory (SRAM) for image buffering. The die photo of the SoC is shown in Figure 13.19. It occupies a die area of 13.3 mm²

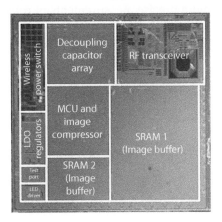

FIGURE 13.19
Die photo of the implemented SoC.

TABLE 13.2

Performance of the SoC

Parameters	Values
Supply voltage	2.5~3.3 V
External components	7
Type of RF link	Bidirectional
TX	
Bit rate	3 Mbps
Power consumption	3.9 mW
RX	
Bit rate	64 kbps
Power consumption	12 mW
MCU power consumption	240 µW
Technology	0.18 µm CMOS
Die area	13.3 mm²

(3.7×3.6 mm). The SoC can work at a battery voltage down to 2.5 V. The power consumption of the SoC has been broken down. The MCU consumes about 200 µA current from a 1.2 V power supply (from the on-chip regulator) when clocked at 24 MHz. The image compressor consumes about 900 µA current from the 1.2 V supply. The MSK transmitter (including all the functional blocks for TX) consumes a total power of 3.9 mW from the 2.5, 1.8 and 1.2 V supplies, while the OOK receiver consumes 12 mW. The performance of this SoC is summarized in Table 13.2.

13.5.2 Experimental Platform for the System

Figure 13.20 shows the experimental platform for the system, which is designed to test the accuracy of pose estimation and is manufactured with ±0.1 mm machining accuracy. The platform consists of a fixed acetabular cup in the center with patterns on its internal surface, a femoral head prosthesis with circuits inside, a rotatable holder and scales printed on the platform. This platform can provide the "femur" with 3-DOFs, including translation along z-axis in {F} and rotations around x- and z-axes in {A}. Vernier scales are designed on the platform to achieve more precise measurements of rotation angles and translation

FIGURE 13.20
The experimental platform for system accuracy analysis.

TABLE 13.3

Performance of Methods Using Different Sensors

Method	Relative Error		Power (MW)
	Rotation (%)	Translation (%)	
Monocular vision	7.0	8.0	380
Monocular vision and IMU	4.8	3.2	268

distance. With this platform, the observed values from the scales and the estimation values on the PC are compared and analyzed to verify the pose and depth estimation methods.

Experiments have been conducted to compare the performance of the monocular vision–based pose estimation method and the vision- and IMU-based one. During the experiments, the holder connected with the femoral head trial was first rotated to a preset position to calibrate the system. Next, the holder and the platform were rotated to several specific positions at different speeds. In each position, the system was kept still for a short interval (about 10 seconds), and the values read from the platform scales were recorded. Finally, the estimation errors were calculated by comparing the estimated values of the two methods with the recorded values read from the platform scales. The experimental results are listed in Table 13.3.

The experimental results show that the vision- and IMU-based method has a lower relative error and a lower power consumption than the monocular vision–based method. A sample that records the relative pose estimation results of the two methods is shown in Figure 13.21. As shown in Figure 13.21a, the monocular vision–based method has wavy error curves in rotation estimations, which means that the method suffers from more random errors. The random errors may result from the manufacture accuracy of the reference patterns and the distortion of the macro lens. Benefiting from the data fusion, the vision- and IMU-based method shows better accuracy in estimating rotation about x-axis and translation along z-axis, as shown in Figure 13.21b. As the accelerometer makes no contribution to the estimation of rotation around z-axis, the rotation estimation error of z-axis is larger than that of x-axis. With the adaptive sensor control adopted, the vision- and IMU-based method achieves a 29.5% decrease in power consumption.

13.5.3 Hip Joint Demonstration System

As a preliminary for clinical tests, a demonstration setup for hip ROM estimation has been designed, as shown in Figure 13.22. The demo setup consists of a human hip joint model and a computer. The monocular vision- and IMU-based system is contained inside the hip prosthesis, mounted on the hip joint model [33].

Experiments are conducted in three steps. First, the femur model is placed at a preset position to initialize the system. Second, the six typical femur postures (i.e. extension, flexion, adduction, abduction, internal rotation and external rotation) are simulated with the femur model for hip ROM estimation. Finally, the femur model is rotated to some certain poses, where dislocation is about to happen, to test the alarm signals. The hip poses are displayed in 3D mode on the screen in real time. The system is powered by batteries mounted in the neck of the femoral head prosthesis, and can work for more than 30 minutes, which meets the requirements of THR surgery.

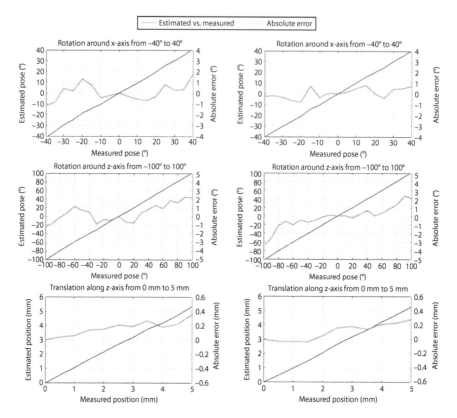

FIGURE 13.21
Comparison between the two proposed methods on estimations of rotations around x- and z-axes, and translations along z-axis. (a) Results of the monocular vision–based pose estimation method. (b) Results of the vision- and IMU-based pose estimation method.

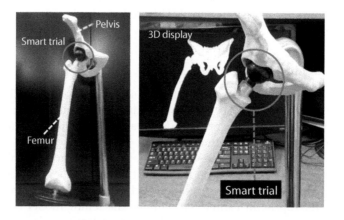

FIGURE 13.22
A demonstration setup of a human hip joint model with the femoral head trial (named smart trial) and a customized acetabular cup on it. The pose of the hip joint is displayed on the computer in real time.

13.6 Conclusion

In this chapter, a monocular vision- and IMU-based system is presented for the intra-operative estimation of the relative pose between a femoral head prosthesis and an acetabular cup. Multiple sensors, a data packer and an SoC with an RF transceiver are mounted inside the femoral head prosthesis. The data are obtained from the sensors and transmitted wirelessly to a computer outside the human body for pose estimation and display. As the sensors are rigidly fixed to the femoral head prosthesis, the prosthesis pose estimation problem is classified as a sensor ego-motion estimation problem and two pose estimation methods are proposed. One is purely based on monocular vision and the other one combines monocular vision with IMU. Experimental results show that the best approach for pose estimation is the method based on the monocular vision and IMU, which has a relative error of 4.8% in the 3-DOF rotation and 3.2% in the 1-DOF translation. In order to reduce the power consumption, the current reusing techniques are applied in the circuit design, and an adaptive sensor control solution is proposed.

References

1. S. Kurtz K. Ong, E. Lau, F. Mowat and M. Halpern, Projections of primary and revision hip and knee arthroplasty in the United States from 2005 to 2030, *The Journal of Bone and Joint Surgery (American)*, vol. 89, no. 4, p. 780, 2007.
2. J. Nutt, K. Papanikolaou and C. Kellett, (ii) Complications of total hip arthroplasty, *Orthopaedics and Trauma*, vol. 27, no. 5, pp. 272–276, 2013.
3. A. Bunn, C. Colwell and D. D'Lima, Bony impingement limits design-related increases in hip range of motion, *Clinical Orthopaedics and Related Research®*, vol. 470, no. 2, pp. 418–427, 2011.
4. M. Ghaffari, R. Nickmanesh, N. Tamannaee and F. Farahmand, The impingement-dislocation risk of total hip replacement: Effects of cup orientation and patient maneuvers, in *Proceeding of 34th Annual International Conference IEEE EMBS*, San Diego, CA, 2012, pp. 6801–6804.
5. O. Kessler, S. Patil, W. Stefan, E. Mayr, C. Colwell and D. D'Lima, Bony impingement affects range of motion after total hip arthroplasty: A subject-specific approach, *Journal of Orthopaedic Research*, vol. 26, no. 10, pp. 1419–1419, 2008.
6. K. H. Widmer, Is there really a 'safe zone' for the placement of total hip components? *Ceramics in Orthopaedics*, vol. 75, no. 5, pp. 249–252, 2006.
7. A. Martin and S. A. Von, CT-based and CT-free navigation in total knee arthroplasty: A prospective comparative study with respects to clinical and radiological results, *Zeitschrift fur Orthopadie und ihre Grenzgebiete*, vol. 143, no. 3, pp. 323–328, 2004.
8. H. Kiefer and A. Othman, OrthoPilot total hip arthroplasty workflow and surgery, *Orthopedics*, vol. 28 no. 28, pp. 1221–1226, 2005.
9. R. McGinnis, S. M. Cain, S. Tao, D. Whiteside, G. C. Goulet, E. C. Gardner, A. Bedi and N. C. Perkins, Accuracy of femur angles estimated by IMUs during clinical procedures used to diagnose femoroacetabular impingement, *IEEE Transactions on Biomedical Engineering*, vol. 62, no. 6, pp. 1503–1513, 2015.
10. F. Graichen, G. Bergmann and A. Rohlmann, Hip endoprosthesis for in vivo measurement of joint force and temperature, *Journal of Biomechanics*, vol. 32, no. 10, pp. 1113–1117, 1999.

11. P. Damm, F. Graichen, A. Rohlmann, A. Bender and G. Bergmann, Total hip joint prosthesis for in vivo measurement of forces and moments, *Medical Engineering and Physics*, vol. 32, no. 1, pp. 95–100, 2010.

12. G. Bergmann, F. Graichen, J. Dymke, A. Rohlmann, G. N. Duda and P. Damm, High-tech hip implant for wireless temperature measurements in vivo, *PLoS One*, vol. 7, no. 8, p. e43489, 2012.

13. R. D. Lea and J. J. Gerhardt, Range-of-motion measurements, *The Journal of Bone and Joint Surgery (American)*, vol. 77, no. 12, pp. 784–798, 1995.

14. I. Holm, B. Bolstad, T. Lütken, A. Ervik, M. Røkkum and H. Steen, Reliability of goniometric measurements and visual estimates of hip ROM in patients with osteoarthrosis, *Physiotherapy Research International*, vol. 5, no. 4, pp. 241–248, 2000.

15. F. Öhberg, R. Lundström and H. Grip, Comparative analysis of different adaptive filters for tracking lower segments of a human body using inertial motion sensors, *Measurement Science and Technology*, vol. 24, no. 8, pp. 50–61, 2013.

16. J. Zhang, A. Novak, B. Brouwer and Q. Li, Concurrent validation of Xsens MVN measurement of lower limb joint angular kinematics, *Physiological Measurement*, vol. 34, no. 8, pp. N63–N69, 2013.

17. J. Favre, R. Aissaoui, B. Jolles, J. de Guise and K. Aminian, Functional calibration procedure for 3D knee joint angle description using inertial sensors, *Journal of Biomechanics*, vol. 42, no. 14, pp. 2330–2335, 2009.

18. T. Seel, J. Raisch and T. Schauer, IMU-based joint angle measurement for gait analysis, *Sensors*, vol. 14, no. 4, pp. 6891–6909, 2014.

19. S. Su, J. Gao, H. Chen and Z. Wang, Design of a computer-aided visual system for total hip replacement surgery, *IEEE International Symposium on Circuits and Systems*, pp. 786–789, 2015.

20. H. Jiang, F. Li, X. Chen, Y. Ning, X. Zhang, B. Zhang, et al., A SoC with 3.9 mW 3Mbps UHF transmitter and 240 µW MCU for capsule endoscope with bidirectional communication, *IEEE International Solid State Circuits Conference*, Beijing, 2010, pp. 1–4.

21. X. Chen, X. Zhang, L. Zhang, X. Li, N. Qi, H. Jiang and Z. Wang, A wireless capsule endoscope system with low-power controlling and processing ASIC, *IEEE Transactions on Biomedical Circuits and Systems*, vol. 3, no. 1, pp. 11–22, 2009.

22. R. Paul, *Robot Manipulators*, Cambridge, MA: MIT Press, 1981.

23. J. Diebel, Representing attitude: Euler angles, unit quaternions, and rotation vectors, *Matrix*, vol. 58, 2006, pp. 14–18.

24. S. H. Won, N. Parnian, F. Golnaraghi and W. Melek, A quaternion-based tilt angle correction method for a hand-held device using an inertial measurement unit, *Industrial Electronics IECON 2008, Conference of IEEE*, Orlando, FL, pp. 2971–2975.

25. J. L. Marins, X. Yun, E. R. Bachmann, R. B. Mcghee and M. J. Zyda, An extended Kalman filter for quaternion-based orientation estimation using MARG sensors, in *2001 IEEE/RSJ International Conference on Intelligent Robots and Systems*, vol. 4, pp. 2003–2011.

26. P. J. Olver and A. Tannenbaum, *Mathematical Methods in Computer Vision*. Springer, New York, 2003.

27. J. Gao, S. Su, H. Chen, H. Jiang, C. Zhang, Z. Wang, et al., Estimation of the relative pose of the femoral and acetabular components in a visual aided system for total hip replacement surgeries, *New Circuits and Systems Conference*, vol. 59, pp. 81–84, 2014.

28. S. Su, Y. Zhou, Z. Wang and H. Chen, Monocular vision- and IMU-based system for prosthesis pose estimation during total hip replacement surgery, *IEEE Transactions on Biomedical Circuits and Systems*, vol. 11, no. 3, pp. 661–670, 2017.

29. Z. Zhang, Essential matrix, in *Computer Vision: A Reference Guide*, K. Ikeuchi, Ed. Springer, New York, 2014, pp. 258–259.

30. Z. Wang, H. Chen, M. Liu and H. Jiang, A wirelessly ultra-low-power system for equilibrium measurements in total hip replacement surgery, *New Circuits and Systems Conference 2012, IEEE*, Montreal, pp. 141–144.

31. K. G. Park, C. Y. Jeong, J. W. Park, J. W. Lee, J. G. Jo and C. Yoo, Current reusing VCO and divide-by-two frequency divider for quadrature LO generation, *IEEE Microwave and Wireless Components Letters*, vol. 18, no. 6, pp. 413–415, 2008.

32. K. Park, C. Jeong, J. Park, J. Lee, J. Jo, and C. Yoo, Current reusing VCO and divide-by-two frequency divider for quadrature LO generation, *IEEE Microwave and Wireless Components Letters*, vol. 18, no. 6, pp.413–415, 2008.

33. S. Su, J. Gao, Z. Weng, H. Chen and Z. Wang, Live demonstration: A smart trial for hip range of motion estimation in total hip replacement surgery, *Biomedical Circuits and Systems Conference 2015 IEEE*, Atlanta, GA, pp. 1–5.

Index

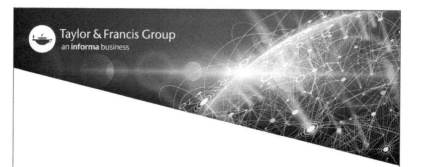

Milton Keynes UK
Ingram Content Group UK Ltd.
UKHW052020071024
449327UK00027B/2357